Cold Spring Harbor Organization Timeline

1890

The Brooklyn Institute
of Arts and Sciences
THE BIOLOGICAL LABORATORY

1904

Bashford Dean
1890

Carnegie Institution of Washington
STATION FOR
EXPERIMENTAL EVOLUTION

Herbert Conn
1891–1897

Charles B. Davenport
1904–1921

Established by
Mrs. E.H. Harriman
EUGENICS RECORD OFFICE

1910

Charles B. Davenport
1898–1924

Charles B. Davenport
1910–1918

1918

Carnegie Institution of Washington

DEPARTMENT OF
EXPERIMENTAL EVOLUTION

EUGENICS RECORD OFFICE

Charles B. Davenport
1904–1921

Charles B. Davenport
1918–1921

1921

DEPARTMENT OF GENETICS

Section of Experimental Evolution

Eugenics Record Office

Charles B. Davenport
1921–1934

1924

Long Island Biological Association
BIOLOGICAL LABORATORY

Albert F. Blakeslee
1934–1941

Reginald G. Harris
1924–1936

1940

DEPARTMENT OF GENETICS

Eric Ponder
1936–1941

Milislav Demerec
1941–1960

Milislav Demerec
1941–1960

Berwind Kaufman
1960–1862

Arthur Chovnick
1960–1962

1962

COLD SPRING HARBOR
LABORATORY
(OF QUANTITATIVE BIOLOGY)

GENETICS RESEARCH UNIT

Alfred D. Hershey
1962–1973

H. Edwin Umbarger
1962–1963

John Cairns
1963–1968

James D. Watson
1968–1994

Bruce Stillman
1994–Present

Illuminating
Life

Selected Papers from Cold Spring Harbor
(1903–1969)

Illuminating Life

Selected Papers from Cold Spring Harbor
(1903–1969)

Commentaries by

Jan Witkowski
Banbury Center
Cold Spring Harbor Laboratory

COLD SPRING HARBOR LABORATORY PRESS
Cold Spring Harbor, New York

ILLUMINATING LIFE:
Selected Papers from Cold Spring Harbor 1903–1969

Project Manager: Duncan Watson
Project Coordinator: Inez Sialiano
Production Editor: Patricia Barker
Desktop Editor: Susan Schaefer
Interior Book Designer: Denise Weiss
Cover Designer: Tony Urgo

Front Cover: Jun-ichi Tomizawa and Dale Kaiser deep in thought at the 1968 Symposium on Replication of DNA in Micro-Organisms. Photograph by Roger Hendrix.

Library of Congress Cataloging-in-Publication Data

Witkowski, Jan.
    Illuminating life: selected papers from Cold Spring Harbor Laboratory 1903-1969/
commentaries by Jan Witkowski.
        p.cm.
    Includes bibliographical references and indexes.
    ISBN 0-87969-566-8 (alk. paper)
    1. Molecular biology. 2. Molecular genetics. I. Cold Spring Harbor Laboratory. II.
Title.

QH506 .M6725 1999
572.8—dc21

99-045419

10 9 8 7 6 5 4 3 2 1

All Cold Spring Harbor Laboratory Press publications may be ordered directly from Cold Spring Harbor Laboratory Press, 10 Skyline Drive, Plainview, New York 11803-2500. Phone: 1-800-843-4388 in Continental U.S. and Canada. All other locations: (516) 349-1930. FAX: (516) 349-1946. E-mail: cshpress@cshl.org. For a complete catalog of Cold Spring Harbor Laboratory Press publications, visit our World Wide Web Site http://www.cshl.org

Contents

Foreword

Even today Cold Spring Harbor strikes the first-time visitor as an unlikely site for a world-famous institution. No major university is found less than an hour's drive away, and from across the inner harbor, on whose western shore it lies, the Cold Spring Harbor Laboratory still looks like a minor marine biology laboratory. Yet, it has long been a seminal site for the development of experimental biology; in particular, of genetics whose effective rebirth in 1900, following the rediscovery of Mendel's Laws, came just a decade after biology teaching first commenced here under the auspices of the Brooklyn Institute of Arts and Sciences. That Cold Spring Harbor's reputation for teaching as well as research now stands so high despite the modest ways in which it so long did science owes much to the extraordinary set of individuals in charge of its destinies before I took over. Now I am frequently given much false credit for making Cold Spring Harbor much more than a minor institution along a very pretty shore. But as this volume shows, already between 1900 and 1968 some extraordinary science was done here that profoundly affected modern biological thought. We shall remain long in debt to Jan Witkowski for so well assembling and commenting upon many splendid examples of science ahead of its time.

James D. Watson
2 August 1999
Cold Spring Harbor, New York

Preface

In the fall of 1999, Cold Spring Harbor will welcome the first graduate students who, upon completion of their research, will receive CSH doctorates instead of degrees from neighboring academic institutions. We believe that this is a most appropriate time to put together an *intellectual* history of the Laboratory through a selection of some of the many notable papers published by scientists working here. From these papers, entering students will see that our Laboratory has been intellectually exciting virtually from its inception. Not only have our scientists contributed to many biological areas, many of them have their actual origins through experiments done here.

This book contains 20 papers, published between 1903 and 1969. As might be expected, the research interests of Cold Spring Harbor scientists have mirrored contemporary trends, so we hope that the papers will be of interest to a wide audience, and not just to those who know Cold Spring Harbor. Some of these earlier research topics are likely to surprise scientists who think of Cold Spring Harbor Laboratory primarily as a center for research on genetics and cancer.

Each paper is preceded by a short essay. These are intended to fulfill four functions: First, to provide a brief description of the research; second, to set the research in the context of its time and so illustrate its significance; third, to provide biographical information about the scientists; and, fourth, to use the paper as a basis for telling a little of the history of research at Cold Spring Harbor. I hope that the essays make a sufficiently integrated series that anyone reading only them will learn the story of these years at Cold Spring Harbor.

It is difficult to know quite how to refer to the complex of buildings and institutions on the southwest shore of Cold Spring Harbor. For 60 years, there were two institutions here that were funded separately but shared directors and, usually, had common research goals. Their intertwined histories are shown in the charts printed on the inside covers. The first institution was the Biological Laboratory (1890), founded as a summer school to provide further training in biology for teachers. Research, such as it was, was concerned with the contemporary "hot" topic of evolution and was pursued by observations of nature and experimentally, through breeding of a variety of plants and animals. Just a few years later, the rediscovery of Mendel's laws put such research on a firmer footing, and in 1904 the Carnegie Institution of Washington was persuaded to establish the Station for Experimental Evolution at Cold Spring Harbor. The Station was built next door to the Biological Laboratory, but whereas the former benefited from strong funding from the Carnegie Institution, the latter's fortunes waxed and waned. The curious situation of having two institutions so closely linked yet independent was resolved in 1965 when they merged to form Cold Spring Harbor Laboratory.

Genetics had become the dominating theme of research at Cold Spring Harbor by 1904 and has remained so ever since. In this early period, the genetics of plants and human beings predominated, but soon the latter became transmogrified into eugenics and split off into the Eugenics Records Office, still based at Cold Spring Harbor. There are no papers on eugenics in this book. Although eugenics assuredly was an important part of Cold Spring Harbor's institutional history, the quality of research does not—and did not at the time—make the grade, motivated as it was by political rather than scientific goals.

Plant genetics provided the first major success of research here with George Shull's development of hybrid corn. Not long after, Blakeslee and Belling began a series of pioneering studies on the triploid mutants of *Datura*, correlating the characteristics of each mutant with a specific chromosome. Cytogenetics was also the forte of Barbara McClintock when she was hired by Demerec in 1942 to continue her work on maize. During the 1950s and 1960s, McClintock was the sole plant geneticist at Cold Spring Harbor as she tracked movable genes in maize.

In the 1920s, Cold Spring Harbor became a center for mouse geneticists with the C57BL inbred strain developed by Clarence Little for research on inherited susceptibility to cancer. His colleague, Carleton MacDowell, produced the C58BL strain that was susceptible to leukemia. *Drosophila* research at Cold Spring Harbor took off when it became the subject of research by Demerec and Kauffman, and by Calvin Bridges, a summer visitor. But the search for ever more tractable organisms continued and so, in the 1940s, *Drosophila* was superseded by bacteria and bacteriophage. This transition was promoted by Demerec and fueled by two summer visitors, Max Delbrück and Salvador Luria. From the early 1950s, microbial genetics came to dominate Cold Spring Harbor, with studies on gene structure by Demerec, pioneering work on DNA repair by Witkin, studies of phage by Hershey, and studies of chromosome structure by Cairns.

Even though genetics was the mainstay of research here, other trends were not ignored. The primary movement, following the example set by Jacques Loeb, was toward *quantitative* research in biology, as opposed to the merely observational. Here the Laboratory took a leading role following the appointment of Reginald Harris as director in 1924. Quantitative biology was exemplified by biophysics. As W.B. Hardy put it, "...the biologist's job is to take the findings of physics and chemistry and faithfully apply them to the riddle of this impossible elusive living slime... ." The biophysicists were interested especially in using physico-chemical techniques to investigate the colloidal nature of cytoplasm and the properties of the cell membrane that surrounded it. Radiation was also used as a tool to probe the cell and the gene, and the Laboratory's first biophysicist, Hugo Fricke, made important contributions to radiation biology as well as to studies of membrane properties.

From the turn of the century through the 1930s, research on internal secretions or hormones was of the highest priority. Their role of controlling metabolism ensured their interest to the physiologist and biochemist, and their clinical importance ensured that there would be funding for the research. Cold Spring Harbor was not left out, and adrenocortical trophic hormone and prolactin were discovered first here by Swingle and Pfiffner, and Riddle, respectively. This physiological-based research did not establish a footing, perhaps because Cold Spring Harbor was not associated with a hospital and was not engaged in other clinical research. However, Cold Spring Harbor made a significant contribution to medicine and the war effort when, in 1945, Demerec developed a high-yield strain of penicillin mold.

A collection like this cannot be comprehensive, and some areas of research are not represented by papers. For example, from the 1930s through the 1940s, population genetics was part of the research program. Bruce Wallace worked here, joined by Theodosius Dobzhansky and Ernst Mayr during the summers. The summer visitors deserve a volume to themselves. Their influx brought excitement and novelty to Cold Spring Harbor, swelling the meager number of year-round researchers. Their names and activities are not recorded consistently in the annual reports, but some of the research that they did was very significant. The most famous example is the work of Delbrück and Luria, who planned and carried out key experiments in phage genetics during the many summers they spent here. Others who worked here in the summers included Calvin Bridges, H.J. Muller, and Ed Lewis (*Drosophila* genetics), Sewall Wright (mammalian genetics), J.Z. Young and Kenneth Cole (electrophysiology), George Corner (hormones), H. Davson and J.F. Danielli (membrane permeability), and Marcus Rhoades (maize genetics).

Scientists here benefited also from the influx of meetings participants. Begun by Reginald Harris in 1933, under Demerec's guidance the Cold Spring Harbor Symposium on Quantitative

Biology quickly became the meeting place for molecular geneticists and molecular biologists from the world over. Everyone who was anyone came to present their latest findings, and the Symposium volumes—in their "cyanotic New England ox-blood red" covers (as John Cairns described them)—were primary sources of information. In 1969, at the end of the period covered in this book, there were 4 meetings and 5 courses, and in 1999, 20 meetings and 25 courses. Thus, Cold Spring Harbor played a key role in the professional lives of many thousands of biologists. Seymour Benzer, for example, recalls how just one day in the Phage course converted him from a physicist to a geneticist.

I have been fortunate in receiving much very useful and pertinent advice from many friends, none of whom let our friendship stand in the way of making critical comments. All of these have contributed immeasurably to whatever merits the introductory essays possess. These advisors are too many to list here, and their names can be found on the following page. At Cold Spring Harbor, Clare Bunce was indefatigable in searching the Archives for me, Inez Sialiano at the Press dealt with ever-changing copy, and Duncan Watson read the text with a layman's eye and made me clarify what was murky and obtuse.

Jan Witkowski
2 August 1999
Cold Spring Harbor, New York

Acknowledgments

The author wishes to thank the following colleagues for their invaluable assistance:

Garland Allen, Washington University

Michael Ashburner, University of Cambridge

Guiseppe Bertani, University of Southern California

Elizabeth Blackburn, University of California, San Francisco

Thomas Brock, University of Wisconsin

John Buntin, University of Wisconsin-Milwaukee

Elof Carlson, State University of New York, Stony Brook

Bayard Clarkson, Memorial Sloan-Kettering Cancer Center

Nina Fedoroff, Pennsylvania State University

Joe Felsenstein, University of Washington

Errol Friedberg, University of Texas Southwestern Medical School

Lester Gabrilove, Mt. Sinai School of Medicine

Murray Gardner, University of California, Davis

Stan Gartler, Washington University

Mel Green, University of California, Davis

Carol Greider, Johns Hopkins University

Benjamin Hall, University of Washington

Philip Hartman, Johns Hopkins University

Zlata Hartman, Johns Hopkins University

Kendall Lamkey, Iowa State University

Robert Martienssen, Cold Spring Harbor Laboratory

Herbert C. Morse III, National Institute of Allergy and Infectious Diseases

James Neel, University of Michigan

Karen Rader, Sarah Lawrence College

Philip Reilly, Eunice Kennedy Shriver Center, Waltham

Lianne Russell, Oak Ridge National Laboratory

Frank Stahl, University of Oregon

Jonathan Stoye, National Institute for Medical Research, London

Waclaw Szybalski, University of Wisconsin

Michael Yarmolinsky, National Institutes of Health

Prologue*

The first biological laboratory that we know about is that established by Aristotle, the Stagirite, on the Aegean Sea, where he made dissections of the marvelously varied forms that the sea shore yielded. Thus were laid the foundations of Comparative Anatomy, the science that in the later years has offered much of the strongest evidence for the blood relationship of animals. About 1790 the great comparative anatomist Georges Cuvier, of France, worked at the sea shore and dissected many marine animals, and thus was enabled to make a classification of the animal kingdom which, with many modifications, stands today. With the rapid development of biology in the latter part of the nineteenth century permanent laboratories became established at Arcachon in France in 1865; at Trieste at the head of the Adriatic Sea; at Naples in 1872; at Granton near Edinburgh in 1884; at Plymouth, England; Finisterre, France; Bergen; Trondhjem; Heligoland; Isle of Man; Banyul on the Mediterranean; Roscoff; Wimereaux near Bologne on the English Channel; near Christiania (now Oslo), and elsewhere.

When Louis Agassiz migrated to America he brought with him the ideal of a marine laboratory, but was not able to recognize it until a year before his death when he was offered use of the island of Penikese in Buzzards Bay for such a laboratory. Many young biologists studied there and among them was Franklin W. Hooper who later became Professor of Biology at Adelphi College, Brooklyn. Professor Hooper found in that city a foundation left by Augustus Graham that was being inadequately utilized and, with the assistance of local men of means, he built out of it the Brooklyn Institute of Arts and Sciences and became its first Director. Among the local men of means who were associated with him in the work of the Institute was Mr. Eugene G. Blackford, a leader in the fish marketing industry of Fulton Market, Manhattan. He had been appointed by the Governor, in 1879, Fish Commissioner of the State. Mr. Blackford wanted a fish hatchery on Long Island and engaged Mr. Fred Mather, a well-known pisciculturalist, connected with the United States Fish Commission, in the establishment of which he had played a part, to examine the waters of Long Island and report. Mr. Mather first established a temporary station at Roslyn and then moved to Cold Spring Harbor. He thereupon reported that Cold Spring Harbor was a fine place for both fresh and salt water fish culture and the Fish Hatchery was soon thereafter established. Mr. Blackford was much pleased with the location and told Professor Hooper about it. A group representing the Institute and its Department of Zoology (including Professor Hooper and Mr. Blackford) visited Cold Spring Harbor in February, 1890. After they had examined the Fish Hatchery plant and taken lunch with Mr. Mather, the foreman, it came to different members of the party that this would be a good place for a Biological Laboratory and plans were immediately laid for conducting the first session of such a Laboratory at the Fish Hatchery during the coming summer.

The Laboratory thus established has now continued through half a century, and it is its foundation and development that we celebrate today. The conditions at Cold Spring Harbor which led to its establishment and which have been largely responsible for its successful development there are partly geological, partly faunistic-floristic, and partly economic.

*Excerpts from the Long Island Biological Association Semi-Centennial of The Biological Laboratory, June 29, 1940, *How the Laboratory Was Planted at Cold Spring Harbor and Why It Grew There* by Robert Cushman Murphy, President, The Long Island Biological Association, Inc.

Long Island was formed during the cretaceous period as a bar resulting from the silts and sands borne by the rivers coming from the highlands of New England. During periods of flood, sands and gravels were deposited, during drier periods mud. The result was that Long Island was made up of alternating layers of gravel and clay which are responsible for the many springs in its valleys and the artesian wells that flow along the edges of the sea. This pure water has been indispensable to the Laboratory and to the extensive fresh water ponds nearby. After Long Island was established it became elevated and later dissected by streams, forming on the north side long and deep valleys. These have later been depressed and water-filled, forming fjords. It is at the head of such a fjord that our Laboratory was established

Where the streams flow into the harbor, silt has been deposited forming extensive mud flats and these constitute the homes of a large number of kinds of mollusks, worms and crustacea. At Cold Spring Harbor a sandy bar has been thrown up at the outer end of this mud flat, and this bar affords a population which in numbers is probably equal to the human population of North America. A strictly marine fauna is found in the harbor. On account of the secluded conditions in the harbor the eggs of young of the species that inhabit it are not washed out to sea, but form a veritable soup during the latter part of the summer. Thus the mass and variety of the surface fauna (called plankton) are extraordinarily great. Although the harbor is far removed from the open ocean still at times of easterly winds the floating forms of the high seas are carried to our shores. In the outer harbor and in the Sound are found numerous strictly marine animals including a worm-like form (Balanoglossus) which is regarded by most biologists as the nearest living representative of the oldest precursor of the vertebrate group.

The fresh water animal life is extraordinarily abundant and that of the upland can hardly be surpassed in numbers and variety. The marine and fresh water floras are also very rich and have been made the object of extensive researches. The flora of the upland is likewise varied.

The meterological conditions at Cold Spring Harbor have also contributed to the success of the Laboratory. The winters are comparatively mild so that work during this season is not interrupted. The summers do not have the extreme heats found in the interior, and the rural surroundings permit a dress suitable to the climate. Finally, Cold Spring Harbor is accessible to the metropolis especially by the fine automobile roads which bring it to within an hour's ride. Proximity to the city is of the greatest advantage because of its libraries, museums, laboratories, biological societies and an active group of biological workers greater than is to be found in any other center of the United States.

The place that was selected by Mr. Mather for the Laboratory was part of an estate of historical interest. It lies at the head of Cold Spring Harbor, an elongated body of water about a mile wide and five miles long, opening at its northern end into Long Island Sound. From this estate one looks through the axis of the harbor over to the Connecticut shore. At the head of the harbor, Nachaquatuck Creek enters. A sand spit about half a mile long serves as a natural breakwater to the inner harbor, whose bottom is exposed at low tide. Here the Matinicock Indians, who well appreciated natural advantages, had a village, and here Captain David Jones bought from Philip Youngs in 1785 a large plot of ground, including a house which he enlarged and which was the birthplace and home of some of the famous Jones family of Cold Spring Harbor. This land passed in 1807 to his nephew, John H. Jones, who established important business enterprises which were centered here, including a large whaling industry, and grist and woolen mills and a steamboat line making daily trips from Cold Spring Harbor to New York. He also planned to bring the railroad to Cold Spring Harbor, but actually did not build it beyond Syosset. The estate passed to his son, John D. Jones, who had been born in the old homestead and lived there until it was destroyed by fire in 1861; then he removed to the south side of the Island. Meanwhile the factories had been discontinued and the commercial activity of Cold Spring Harbor had become a thing of the past. It was under these circumstances that the Fish Hatchery was established in 1887 on land leased to it by John D. Jones.

Robert Cushman Murphy

Ecology, Adaptation, and Evolution:
Early Research at Cold Spring Harbor

Davenport C.B. 1903. **The Animal Ecology of the Cold Spring Sand Spit, with Remarks on the Theory of Adaptation.** (Reprinted, with permission, from *Decennial Publications University of Chicago* **X:** 157–176.)

Charles B. Davenport at about the time he moved to Cold Spring Harbor (circa 1900).

CHARLES DAVENPORT (1866–1944) BEGAN HIS professional career as an instructor in zoology at Harvard in the 1890s (MacDowell 1946 [reprinted in this volume, pp. 311–360]). This was a time of great change in biology, as there developed a movement away from descriptive natural history and toward an experimental, laboratory-based science (Allen 1975). The champion of this approach was Jacques Loeb. His research program was set out in the declaration that opened his book *The Mechanistic Conception of Life* (1912): "It is the object of this paper to discuss the question whether our present knowledge gives us any hope that ultimately life, i.e., the sum of all life phenomena, can be unequivocally explained in physico-chemical terms" (Loeb 1912; Pauly 1987). Davenport was one of those, along with T.H. Morgan, E.B. Wilson, E.G. Conklin, and R.G. Harrison, who embraced this new biology. He introduced a new course at Harvard, entitled "Experimental Morphology," that included physiology as well as statistical and experimental studies of variation and heredity (in 1896, this was, of course, pre-Mendelian heredity and was based on the statistical work of Galton and Pearson). Davenport's students carried out physico-chemical research on the effects of temperature and chemicals on cytoplasm, and, like Loeb, on the responses of animals to light and gravity (Davenport and Cannon 1897).

In 1898, Davenport was appointed Director of the Summer School at the Brooklyn Institute of Arts and Science's Biological Laboratory at Cold Spring Harbor. The Summer School provided additional training for biology teachers, and he intended to use the available local resources for both teaching and research on the origin of species. His intention was to determine the geographical distributions of different forms of organisms and to examine their adaptations to the differing environments. The paper reprinted here was his first (indeed, only) step in this research program.

The major portion of the paper is an ecological description of the organisms that Davenport found in the differing environments provided by the Cold Spring Harbor sand spit. These included the outer beach exposed to the force of the Long Island Sound; the tip where the current in a channel between inner and outer harbors was strong; and the inner, mud-coated beach. In the second part of the paper, Davenport turned to adaptation and discussed the relative importance of a species "selecting" an environment to which it has favorable adaptations and the role of selection in driving adaptive changes to a particular environment. Davenport believed that the former process had not received sufficient attention. One example he gave was of the mussels found in large numbers at the tip of the sand spit where their byssus threads, binding them tightly to rocks and stones, were essential for withstanding the strong currents. He argued that mussels must have had some form of thread—useful, perhaps, for some other function—that enabled them to colonize such environments. Subsequently, natural selection came into play, leading to the survival of variants with longer or stronger threads. Thus, adaptation resulted from a combination of "...the improvement of the organ to meet the requirements of the environment, through selection, and of improvement of situation to meet the abilities of the organism."

Evolutionary studies of this kind were not destined to become the research theme at the Biological Laboratory. Instead, Davenport became interested in the nature and origins of the variations themselves, a research program that came to the fore with the discovery of Mendel's work in 1900. Davenport was one of the first American scientists to write about Mendel (Davenport 1901), and he was soon breeding mice and other animals and studying the inheritance of traits in man (see page 23).

His opportunities to promote genetics research were increased greatly in 1904, when, after a 3-year campaign, he persuaded the Carnegie Institution of Washington to establish a Station for Experimental Evolution adjacent to the Biological Laboratory. This was a decisive moment in the history of Cold Spring Harbor, providing as it did a center for year-round research. During the 60 years of its existence, this institute—later renamed the Department of Genetics in acknowledgment of the changed nature of its research—was home to some of the world's most eminent geneticists, a fitting tribute to Davenport's hopes for experimental biology at Cold Spring Harbor.

Charles Benedict Davenport was born on June 1, 1866, at Davenport Ridge, near Stamford, Connecticut. Although Charles was a keen naturalist, his father insisted that he acquire a practical education, and Davenport graduated from Brooklyn Polytechnical College with a B.S. in civil engineering. His career as a civil engineer was short-lived, for in 1887 he entered Harvard to read Natural History. He graduated in 1889 and completed his Ph.D. in 1892. Davenport was an instructor at Harvard until he moved to the University of Chicago in 1893 and became director of the summer courses at the Biological Laboratory, Cold Spring Harbor, in 1898 and the director of the Station for Experimental Evolution in 1904. He established the Eugenics Records Office in 1910 and thus was responsible simultaneously for three institutions. Davenport was an indefatigable organizer—he was on the editorial boards of eight journals and was a member of 64 societies. He developed pneumonia while spending days outdoors in January, 1944, trying to prepare the skull of a killer whale that had beached at the eastern end of Long Island. He died on February 18, 1944.

Allen G. 1975. *Life science in the twentieth century.* John Wiley, New York.

Davenport C.B. 1901. Mendel's law of dichotomy in hybrids. *Biol. Bull.* **2:** 307–310.

————. The animal ecology of the Cold Spring sand spit, with remarks on the theory of adaptation. *Decennial Publications, University of Chicago* **X:** 157–176.

Davenport C.B. and Cannon W.B. 1897. On the determination of the direction and rate of movement of organisms by light. *J. Physiol.* **21:** 22–32.

Loeb J. 1912. *The mechanistic conception of life.* University of Chicago Press, Chicago, Illinois.

MacDowell E.C. 1946. Charles Benedict Davenport, 1866–1944: A study of conflicting influences. *Bios* **17:** 3–50.

Pauly P.J. 1987. *Controlling life: Jacques Loeb & the engineering ideal in biology.* Oxford University Press, New York.

THE ANIMAL ECOLOGY OF THE COLD SPRING SAND SPIT, WITH REMARKS ON THE THEORY OF ADAPTATION

C. B. DAVENPORT

COLD SPRING sand spit runs from the west shore of Cold Spring Harbor, Long Island, eastward to within 100 or 200 feet of the eastern shore of the harbor. The history of the formation of the spit is briefly this : Cold Spring Harbor (Fig. 1) is a fiord-like re-entrant about ten kilometers long, emptying at its lower or northern end into Long Island Sound about one-fifth of the way from Hell Gate at New York city to "The Race," south of New London. The Sound itself, 175 kilometers long by 35 kilometers broad at its widest point, and having a prevailing depth of about 30 meters, receives large streams of water from the north — the Connecticut, Housatonic, and Quinnipiac rivers, and many minor ones. These give it a low specific gravity, 1.020, and a muddy bottom. Long Island is covered over its northern part with glacial débris, forming hills that rise to a height of over 100 meters. Between these hills streams flowing north have cut valleys and, in the general sinking to which the whole

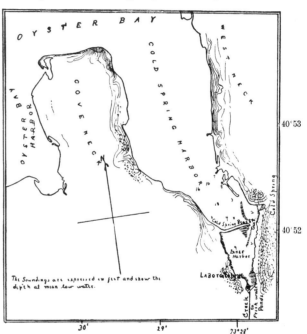

FIG. 1.—Map of Cold Spring Harbor, showing spit (Cold Spring Beach).

area has been subjected, these valleys have become drowned, forming long, straight but shallow harbors of which Cold Spring Harbor is an excellent type (Shaler, 1902). Into the head of the harbor a small stream flows with a summer discharge of not far from five cubic meters per minute. This stream is dammed thrice along its course of two miles, so that the deposit that it carries into the basin above the beach is only fine mud.

The effective winds at Cold Spring Harbor blow from the northeast and, striking with violence upon the bluffs on the west side of the harbor, tend to wear them away. The currents setting southward then carry this eroded material until it is dropped in the shallow and protected waters of the upper end of the harbor, where the sand spit or "beach" is now found. The inner harbor thus cut off is about 800 meters long by 600

157

wide, and at mean high tide has a prevailing depth of less than 2.5 meters.   Its sandy bottom is for the most part covered with half a meter of fine black mud.   This is largely exposed at low tide, and supports an abundant growth of "sea cabbage" (Ulva).   The specific gravity of the water at high tide varies from 1.018 near the entrance to the inner harbor to 1.005 at the surface near the inlet of the creek.   Near the "gut" where the tides rush in and out of the inner harbor the bottom is kept scoured and consists of gravel and stones.   Outside the sand spit the water has a pre-

FIG. 2.—View of Cold Spring Harbor at high tide looking north. The inner harbor is in the foreground; between it and the outer harbor is seen the sand spit. Close to the sand spit on the inside is a fringe of Spartina polystachya.

vailing specific gravity of 1.019 and is about 6 to 8 meters deep at high water.   A bar (Fig. 2), almost submerged at high tide and consisting largely of hard mud, has formed opposite to the outer entrance to the gut and affords a "shallow sea" fauna.

The sand spit itself (Figs. 2 and 3) is 660 meters long and varies in width from 80 meters at its western end to 15 meters at its eastern end at high water.   The prevailing width is about 40 meters.   The highest part of the spit is about one meter above mean high tide, and the average height of tide is 2.4 meters.   The slope is gradual on the outside (north side), it being 40 meters from high to low water at the western third of the spit.   The slope increases to 2.4 meters in 15 at the eastern end.   The material of which the spit is built is sand, which at the western end contains much gravel, apparently because of the greater carrying power of the surf, which is highest here.   At the gut the sand is very coarse on account of the rushing tide.   In the middle stretches of the beach the sand is finer.   On the inner face of the spit mud is deposited.   On account of the sandy beach and the currents and surf that are washing it all the time, no plants grow on the outside of the sand spit between average tides. On the inside of the spit, on the other hand, the marsh grass, Spartina, finds abundant foothold. All these facts have an important influence on the distribution of the animals of the sand spit. Indeed, for the purposes of this account of the fauna of the spit it will be necessary to treat separately its outer and inner margins and its tip, just because the conditions are so different in the three situations.

FIG. 3.—View of sand spit taken near western end looking east. Terrestrial border zone with Ammophyla in the foreground. The outer beach with the storm bluff, wreckage-strewn upper beach, and lower beach. At the right the inner beach, passing into mud flats at the edge of which Spartina polystachya is growing.

A. ANIMALS OF THE OUTER BEACH

The outer beach may, for our purposes, be divided into three zones which we may herein designate as : (1) the submerged zone; (2) the lower beach; and (3) the upper beach. The submerged zone includes all that portion of the beach that lies below mean low tide, but which may be exposed by the lowest spring tides combined with southerly winds (Fig. 4). It is a region that is normally covered with water; it is the very margin of the shallow sea. The lower beach is that zone which lies between mean low tide and mean high tide. It is the zone that is twice each day exposed to the air and submerged. It passes without any sharp break into the submerged zone (Figs. 4 and 5). The upper beach is limited on the one hand by the line of débris that marks the average high tide, and on the other by a bluff, half a meter high, that has been cut by storms (Fig. 5).

I. FAUNA OF THE SUBMERGED ZONE

The animals that live below the high-tide line may be considered under four heads : (*a*) sessile species, (*b*) crawling species, (*c*) burrowing species, and (*d*) swimming species.

a. *The sessile species of animals,* inhabiting the lower beach, include certain molluscs, especially bivalves, or Lamellibranchiata.   All the lamellibranchs feed on minute particles of organic matter : Algæ, Infusoria, and decaying bits of organisms.

FIG. 4.—Photograph of lower beach, north side of sand spit, near the eastern end, at very low tide. The gut is seen in the background. At the water's edge (submerged zone) is a dense growth of the alga Enteromorpha. The naked lower beach is thickly peopled with the mud snail, Nassa obsoleta, visible as dots in the photograph.

The silt brought down by the creek is rich in such material and the algæ thrive on the mud flats, consequently these flats and the shallow sea around the flats are especially favorable feeding-grounds for these molluscs.   Here occur oysters (Ostrea virginiana) in a semi-domesticated state.   They normally attach themselves to some solid object while still young, but those of Cold Spring Harbor have been mostly transplanted and lie loose in the shallow waters, but altogether incapable of movement or of any other defense than that provided by their two thick valves.   Here also occurs the scallop (Pecten irradians), which attaches itself while yet less than two weeks old, in the middle of August.   The attachment is chiefly to eel-grass or to small stones.   About the middle of September the scallops migrate into the inner harbor to live on the mud-flats like the oyster, but they never lose their power of free migration. The jingle shells (Anomia simplex) are permanently attached to stones or larger shells, such as Pecten, from early life. On the lower beach one also finds two species of Arca (Arca transversa and Arca pexata); the hen-clams, Mactra solidissima and Mactra lateralis; the hard clam, Venus mercenaria; and Liocardium mortoni—all lying on or imbedded in the muddy bottom. Here also are found certain species of lamellibranchs that burrow in the mud or sand and, to facilitate that burrowing, have become elongated; namely, the soft-shelled clam, Mya arenaria; the razor clam, Ensis americana, and Solenomya velum. This great group of bivalves represents then a society of animals that are fairly common because of favorable food conditions, but very helpless and much exposed to predaceous animals, were it not for their hard shell or their habit of burrowing into the mud.

  b. The crawling species belong chiefly to the three groups of Mollusca, Echinoderma, and Crustacea. The crawling molluscs are slow-moving snails (Gastropoda) which are there partly to feed on decaying animal and vegetable matter, partly to feed on the growing Ulva, and partly to prey upon such living animals, chiefly bivalves, as have no means of escape. The chief omnivorous and carrion snail is the mud snail, Nassa obsoleta (Fig 4), which is abundant everywhere and even remains exposed on the middle beach at low tide, if busy feeding on a dead oyster. The little snails, Anachis avara and Astyris lunata, feed on the Ulva, or sea lettuce. The carnivorous species are of larger size and include two Muricidæ, Eupleura caudata and Urosalpinx cinerea ("oyster-drill"), and the great whelks, Fulgur caniculatum and Fulgur carica. All these feed upon oysters, scallops, and other surface bivalves by drilling holes through the shell. Two species of Naticidæ, Neverita duplicata and Lunatia heros, seek out the burrowing lamellibranchs, and so we find them burrowing into the sand. Then, too, they find in the sand of the beach the proper material for their egg cases, which are made out of agglutinated sand molded in the shape of a spiral collar. The crawling echinoderms are chiefly the starfishes, which are here because of the oysters and other bivalves upon which they prey. They cannot bore through the oyster's shell, and so they smother it until it is forced to open its valves for fresh water. The crawling Crustacea, finally, feed on organic débris of all sorts. Here belong the crabs, such as the three spider crabs, Libinia canaliculata, Libinia emarginata, and Libinia dubia, of which the latter comes farthest in-shore. Here, too, are the three mud crabs, Panopeus depresus, Panopeus herbstii, and Panopeus sayi, of which three the latter is found nearest the sand spit. On the very edge of the submerged zone are found also the two hermit crabs; the small one, Eupagurus longicarpus, finds protection for its abdomen in the cast-off shells of the small gastropods Nassa and Anachis. The large species, Eupagurus pollicaris, occupies such large shells as those of Lunatia heros and Fulgur carica. These scavengers, carrying their borrowed shells behind them, travel quickly along, but just below, the edge of the water, seeking for dead fish and other organic matter that may be resting there. Finally, the horseshoe crab, Limulus polyphemus, the largest and most aberrant of our Crustacea, will be seen, especially during June, traveling over the shallow water and occasionally coming to land to lay its eggs in the sand.

c. *The burrowing animals* of the submerged zone constitute a remarkable fauna of, for the most part, elongated animals. We have already seen that many molluscs burrow. So do a few sea anemones, such as the white-armed sea anemone, Sagartia leucolena, and the flesh-colored or white Halocampa producta. These sea anemones seem to feed on bits of organic remains, of which the sand is full. The other burrowers are here for a similar purpose; the circumpolar sea cucumber, Synapta inhærens; the worm that shows affinities with vertebrates, Balanoglossus Kowalevskii; two nemertines, Cerebratulus leidyi and Cerebratulus lacteus; and some seventeen different kinds of jointed worms, Annelida. All these find shelter, moisture, and food in the sand and mud beneath the sea bottom. But their immunity from attack is not complete, for as the moles have followed the earthworms and insect larvæ into the subaerial ground, so have several predaceous Crustacea followed the annelids into the mud. These are: the mantis shrimp, Squilla empusa, which is only a little smaller than the lobster, and Callianassa stimpsoni and Gebia affinis, that are somewhat smaller than a large crayfish.

d. *The swimming animals* are partly scavengers and partly predaceous. To the first class belong the prawn Palæmonetes vulgaris, which scours the edge of the tide for floating débris; and also the swimming crabs, the blue crab, Callinectes hastatus, and the commoner "lady crab," Platyonichus ocellatus. Here, too, we may place the little killifishes, Fundulus, although these pick up many live shore snails. The majority of the fishes are predaceous, feeding on the crawling and even the burrowing species that I have enumerated. The "skates" that lie close to the bottom catch burrowing worms and molluscs and also the snail Lunatia heros. The sand sharks and dogfish (Carcharias littoralis and Mustelus canis) gather in the spider crabs, squillas, and hermit crabs. The flounders, likewise, living close to the bottom, get Gebia and the prawns. The toad-fish, which lays its eggs in old shoes or in tin cans or under stones, feeds on the mud snail, Nassa, on crabs, and on prawns. Thus we see that in the shallow sea each species that occurs is present on account of particular relations that it bears to other species or to the non-living environment. The presence of the sharks is determined by that of the squillas, the squillas by the worms, the worms by the decaying vegetation, this by the living vegetation, and this by the salts and the nitrogenous food brought down by the creek from the valley above. This microcosm of the submerged zone affects in turn the lower and the upper beaches.

II. THE FAUNA OF THE LOWER BEACH

As already stated, the lower beach is a zone where aquatic and terrestrial conditions alternate every day. On this account, and on account of the sand, it is an area devoid of all living vegetation excepting the unicellular algæ that grow upon the stones (Fig. 4). It is also a region where oxygenation is combined with abundant moisture, so that conditions peculiarly favorable for respiration would seem to be afforded.

A fine layer of silt is dropped over the whole surface with each flood tide, affording thus abundant but microscopic food. But, on the other hand, it is a region of great exposure to terrestrial animals; so that only the stratum a little below the surface offers great safety. Also the lower beach is a region of wave action which makes it difficult for animals to secure a permanent place on it. Finally, and most important of all, the waves and currents cause the sands to shift, and this adds to the difficulty of maintaining a foothold. Consequently there are but few animals living on the lower beach, and such as there are live a curious and very strenuous life.

All over the lower beach will be found, upon careful examination, large numbers of extremely minute and active insects belonging to the group of Thysanura. These are arctic forms of Collembola of the species Xenylla humicola, O. Fabr. and Isotoma besselsii, Pack., together with an occasional Anurida maritima, Guer.[1] These Collembola are feeding on the rich microscopic débris which has been dropped on the lower beach. Being insects, they are air breathers; and the question arises: What do they do when the tide comes in? To answer this question I made measurements of the area occupied by the Collembola at different stages of tide. At tides lower than one-half tide the upper limit of the Collombolan zone is about nineteen meters north of the storm-cut bluff, or about ten meters north (*i. e.*, seaward) of the mean high-tide line. The lower limit is about two meters from the momentary tide-line. As the tide retreats the lower limit follows, while the upper limit remains constant. Thus, on September 10, 1901, with falling tide, the following determinations were made:

HOUR	DISTANCE FROM STORM BLUFF	
	Upper Limit of Collembola	Lower Limit of Collembola
3:10 P. M.............................	17.7 m.	28.4 m.
3:30 P. M.............................	17.7 m.	29.5 m.
4:50 P. M. (low-tide).................	17.7 m.	31.1 m.

Note that the lower limit of the Collembola travels down *pari passu* with the tide.

As the tide rises, the Collembola tend to retreat before the edge of the water, so that they are even crowded together there. Thus, on September 10, at 7 A. M. (one-third tide, rising), the lower edge of the Collembolan zone was about three meters away from the water's edge. As the tide rises still higher they crawl into the sand, until, at high tide, most of the Collembola are under the sea. But not all of the Collembola are there. At high tide one finds some of them floating on the quiet water out

[1] For the determination of these species I am indebted to my friend, Dr. J. W. Folsom, of the University of Illinois. (*Cf.* Wahlgren, 1899; Folsom, 1901.)

at a distance of ten meters or more, and moving hither and thither upon the surface. We conclude then that the rising tide has caught up with certain of the little insects. They rest upon the surface of the water by virtue of numberless fine hairs with which they are covered, in which the air is entangled so that the bodies of the insects are prevented from getting wet These are chiefly Anurida.

A second species that occurs on the middle beach in great numbers during the latter part of June is the horseshoe crab, Limulus polyphemus. This occurs here because it lays its eggs in the sand of the beach, thereby reaping the advantage of the superior oxygenation afforded by this situation over sand constantly submerged. The eggs are laid in nests containing several hundred each. The eggs are oval and about two millimeters in diameter. Each is enveloped in a tough membrane so that the sand cannot injure it. The position of the nest may be detected by a slight depression in the surface over it, and by the absence of pebbles. Not all the middle beach is occupied by these nests, but only those regions where the sand is coarse enough to let water through readily, but not so gravelly as to make hard digging. The east end of the sand spit where the current flows swiftest affords the best conditions, and here the nests are crowded together. In June also one finds many carcasses of female horseshoe crabs that have died in consequence of oviposition; for, as in many other species, oviposition is accompanied by a great mortality. Most of these carcasses are eventually thrown up to the high-tide line, and their fate will be considered in connection with the fauna of the upper beach.

Finally, mention should be made of the great annelid, Nereis limbata, that occurs burrowing in the sand above low-water mark. This again is confined to the tip of the sand spit where oxygenation is best carried on.

III. THE FAUNA OF THE UPPER BEACH

The upper beach I shall define as the zone including the high-tide line and above to the storm bluff (Fig. 3 left, Fig. 5). This region is inhabited by a very few annual plants; its main characteristic, however, is the débris cast up by the sea (Fig. 5). All over the world the upper beach is the graveyard of the shallow sea. In this graveyard two sorts of remains are found: first, such as have been floating on the surface of the sea; and, second, such as have fallen to or were lying on the bottom of the shallow sea. The floating remains are carried in toward the shore by winds from the sea. If the sea is quiet, they are merely dropped at the time the tide begins to fall; consequently they mark the high-water line (Fig. 5). If the sea is heavy, the floating or drowned débris may be thrown against the upper part of the upper beach and even against the storm bluff. This flotsam and jetsam consists, in the first place, of such things as lumber and articles of wood and cork; fruits and seeds; bits of eel grass; stems of last year's marsh grass, Spartina; fronds of Ulva torn from the mud flats; jelly fishes; drowned insects, especially heavy-bodied beetles, which have probably been blown out to sea and been drowned or have fallen in during migration. The second class includes chiefly empty shells, whose inhabitants have perhaps met with a violent death through predaceous animals, or the smothering of a stirred up muddy bottom,[2] also the dead bodies of Crustacea, such as Limulus.   This débris is piled up at the lower edge of the middle beach and is renewed twice a day.   Especially, however, after a storm is the accumulation large.   At such times and at certain seasons of the year one may meet with particular species in large numbers.   Thus, early in September, 1901, as the young Pectens were swimming into the inner harbor, a combina-

FIG. 5.—Photograph of north side of sand spit, near the western end, at low tide. In the central foreground is the high-tide line, marked by a mass of débris. On the left is the gravelly lower beach; the middle beach and storm bluff are at the right.

tion of high tide and sea breeze left thousands of them stranded on the upper and even on the lower beach.[3]   The drowned insects are largely leaf eaters (Chrysomelidæ, including the Colorado potato beetle) and, especially in the early summer, ladybirds (Coccinellidæ) of various species.   All of these constitute a rich, frequently replenished food supply, the only disadvantage connected with it being the dangerous proximity of the sea, with its occasional very high tides and its storm-born breakers. As could have been anticipated, certain animals have come together to make use of this food material.   Some are herbivorous, others are scavengers, and others still are predaceous.

[2] In March, 1890, the levee gave way on the left bank of the Mississippi river above New Orleans. The waters pouring through Lake Ponchartrain into western Mississippi Sound so stirred up the muddy bottom that the great beds of oysters of this region were killed. (Smith, 1894.)

[3] During a visit to Santa Rosa island, outside of Pensacola Bay, Florida, in March, 1902, I found the beach covered with thousands of Portuguese men-of-war (Physalia), and the floating gastropod, Janthina fragilis, thrown up by the southerly storms.

Of the herbivorous feeders on the wreckage of the sea may be mentioned, first of all, the Amphipoda, marine Crustacea which are so adapted to a terrestrial life that they are rarely submerged.   At Cold Spring Harbor two species are found—the small, dark Orchestia agilis and the large, sand-colored Talorchestia longicornis.   Both may be seen in great numbers by turning over some of the cast-up Ulva fronds, under which they live and upon which they feed.   Here they dwell in a saturated atmosphere and so need no special modification of the respiratory apparatus to fit them for breathing air.   They both burrow, also, forming holes varying from three to five millimeters in diameter in the fine sand under or slightly above the line of wreckage. These holes enable the amphipods to reach moisture, they prevent them from being swept away by the sea, and they may serve as nests for eggs.   For some reason the Talorchestia only is found at the tip of the spit.   If it be asked why these Amphipods have left the water thus to assume a half-terrestrial life, I think it is a sufficient answer to say, first, that they find here abundant food; second, that they are here comparatively immune from the attacks of their greatest enemies—the fish; and, third, that their organization permits them readily to assume a semi-terrestrial life, as is shown by the fact that some of their allies have become even more terrestrial than they.   (Compare Talitrus platycheles, and Talitrus saltator, Semper, 1881, p. 188.) That the Talorchestia is no longer an aquatic animal is shown by the way it retreats before the tide, especially if abnormally high.

Secondly, rove-beetles (Staphylinidæ) of the genus Bledius are found in the débris.   This terrestrial insect is here found side by side with the marine Talorchestia, even burrowing into the sand.   It feeds upon decaying vegetable matter.   A third organism found under the débris is a minute white earthworm of world-wide distribution.   This is Enchytræus albidus Henle (Halodrilus littoralis of Verrill, 1873).

As the plant débris is being devoured by the amphipods, staphylinids, and Enchytræus, so the animal remains are being carried off by a number of scavengers. Among these the ants are the most important; there are two species of them.   The first, Formica rufa var. obscuriventris Mayr, is reddish brown and about four millimeters long (see Emery, 1893).   It digs holes in the sand in the upper part of the upper beach, the grains of sand being brought individually to the surface and deposited in a ring around the hole.   This ant also occurs under the shelter of boards and logs.   Immediately after the tide has begun to fall and dropped its burden of carcasses these ants sally forth in paths that run perpendicularly to the high-tide line and begin to seize and carry to their nests the drowned insects that have been left there stranded.   A second species of ant has its home somewhat farther out of reach of the tide; but its habits are quite similar.

There is, however, a larger carrion fauna to be utilized.   I have already referred to the dead horseshoe crabs and the dead molluscs that are left on the shore.   These soon attract great numbers of the flesh flies (Sarcophaga carnaria) which lay their eggs in the carrion.   A second fly with bronze abdomen (undetermined) is also found here. To test the quickness with which these flies accumulate I took a recently drowned turtle from the water at 3 p. m. July 7, and left it at the edge of the retreating tide. At 4:15 p. m. (wind south) I counted over thirty of the bronze flies upon it. It remains to be determined whether it is chiefly sight or smell that attracts them.

Underneath the carcasses of the horseshoe crabs one will find also large numbers of the carrion beetle, Necrophorus, and larvæ of the museum pests (Dermestidæ).

The rich fauna of carrion feeders and herbivorous and omnivorous species determines still another fauna, namely, a predaceous one. Thus, running spiders, almost as white as the sand, frequent the upper beach and sometimes pass down to the upper part of the lower beach. These belong to the species Lycosa cinerea — a species that occurs commonly in Europe also, and is by no means confined to the sand of *beaches*. I have seen them carrying off Orchestia toward the storm bluff. More powerful still are the robber flies (Asilidæ) which pounce upon carrion flies. Their path of flight is curved, and at the moment of alighting an offset is taken several centimeters to one side. Altogether the irregular flight seems well adapted to the end of putting the victim off its guard, like the curve in the path of the ball thrown by an expert baseball player. Tiger beetles also occur abundantly on the beach in the bright sunlight, especially in August and September. They are chiefly of the species Cicindela repanda Dej. They are the most rapid fliers among beetles and feed upon the other insects of the beach. Finally, all these insects must fall victim to the swift and powerful swallows which course up and down the beach, especially in the latter part of the afternoon.

On the upper part of the upper beach there occur also a few stragglers from the vegetation-grown top of the sand spit. Here one finds grasshoppers almost as white as the sand. The white color of this species seems partly determined by the color of the background, for placed in a cage on grass these spiders become darker, as experiments at the Biological Laboratory have repeatedly shown. Whether the grasshoppers feed upon the dead vegetation of the tide-line was not ascertained. Crickets, Gryllus abbreviatus, quite as black as those living in the grass, occur scatteringly on the upper beach; and under the waste lumber the sow-bug, Oniscus, which is only half adjusted to a terrestrial existence, finds a living on the water-soaked wood.

B. THE BEACH ON THE TIP OF THE SAND SPIT

The tip of the spit is transitional between the inner and the outer sides. Here the sandy beaches of the north side gradually pass over into the mud-flats of the south side with their thick growth of marsh grass, Spartina. Indeed, at the tip of the spit there is a constant struggle going on between the upbuilding tendencies of the Spartina, which tends not only to hold the mud in which it grows together, but also to accumulate additional silt, and the scouring away tendencies of the tide that rushes through the gut. Wherever a weak spot appears in the mass of Spartina there a channel becomes gradually worn through (Fig. 6). These channels gradually widen and may anastomose, and thus the Spartina be left on elevated hummocks, which would seem destined to become smaller and smaller until entirely washed away. But from this fate they are preserved by an interesting association. The current that rushes through the channels carries with it an abundant supply of microscopic food, such as lamellibranchs can make use of. This food is taken advantage of by the mussels which come to line the muddy banks on the channels, and form so close a wall that erosion is almost completely stopped (Fig. 6). Thus the mussels assist the Spartina in their constructive work. The mussels that line the banks are Modiola plicatula, Modiola

FIG. 6.—Photograph taken at the inside of the hook of the sand spit, looking north, at low tide, through one of the passages scoured out by the tidal currents. At the base of the Spartina patches, on the right, are seen some of the beds of mussels which protect the roots from exposure.

modiolus, and Mytilus edulis, the first-named being the most abundant. The channels have irregular bottoms in which shallow pools of water stand when the tide is out. Here the mud snails, Nassa obsoleta, aggregate, scarcely submerged. High up on the stems of the Spartinas, exposed to the air during perhaps half the day, are found clinging the Littorinas, whose lack of a siphon makes it necessary for them to keep out of the mud. Littorina rudis and Littorina palliata were still the prevailing species in 1898 (see Balch, 1899), but in 1901 Littorina littorea, which is rapidly advancing up the harbor, had gained a marked predominance.[4] The independence of the sea water that is exhibited at the tip of the sand spit is also illustrated in the marsh at the head of the harbor through which Cold Spring creek runs. On this marsh Littorina occurs

[4] This habit of clinging to rushes is even more exaggerated in the southern Littorina irrorata. For, in a visit to the Lagoon on Ship Island, Mississippi Sound, in March, 1902, I found nearly all of them living on the stems of the short marsh grass twenty to thirty centimeters above the water level and exposed to the sunlight. Cooke (1895, pp. 20, 93, 151, note) cites other cases of Littorina living out of water.

at places where it is submerged for only a short time at high tide and then under water that is nearly fresh. Littorina rudis behaves similarly in other parts of the world. Thus Fischer (1887, p. 182) states that at Trouville (Calvados) on the English Channel he has found L. rudis on rocks two meters above the other marine animals and moistened only by the highest tides. In fact, according to Simroth (1891, p. 84),

FIG. 7.—Photograph taken at half tide from the base of the sand spit, looking east, showing inner harbor, the hook of the spit, and the gut beyond.

species of Littorina pass the winter out of water with their gill chambers full of air. Littorina rudis, then, has evidently progressed far on the road toward adaptation to a terrestrial life—a road that the Pulmonata must have traveled long ago.

C. THE INNER EDGE OF THE SAND SPIT

Along the whole length of the inner beach, not far from the high-tide line, occur the holes of the fiddler crabs. These crabs belong to two species, namely, Gelasimus pugnax and Gelasimus pugilator. A remarkable thing in the distribution of these species is the fact that although their habitats are not markedly different their areas of distribution are so well defined that they hardly overlap. Both species occur at the edge of the Spartina. The pugnax is found all the way from the western, proximal end of the sand spit to about two-thirds of the way toward its eastern point. Then pugilator abruptly comes in. For a distance of a meter or so the two species occupy ground in common, and, so far as I could make out, peacefully. The pugnax alone occurs at the head of the inner harbor. It burrows in the banks at a level that is reached only by the high tides. Walking along the beach at high tide July 7, I found that many of the fiddlers had migrated to above the high-tide level. It is clear, I think, that they do not find submergence altogether agreeable, and it is probable that prolonged submergence would drown them as it does Ocypoda arenaria, the sand crab of the beaches south of Cape May. G. pugnax prefers the marshier ground and the higher water, and it is probably that preference which determines its spacial separation from pugilator on the sand spit.

D. THE TOP OF THE SAND SPIT (TERRESTRIAL BORDER ZONE)

Above the storm bluff on the outer beach, and at a less well-defined line on the inner beach, lies the zone of permanent vegetation in which certain shrubs have gained a foothold. Here the fauna at once assumes a strictly terrestrial aspect. No close ally, even, of a marine form occurs. On the contrary, the animals living on vegetation are precisely those species that occur in the fields, especially the plant feeders: the plant lice (Aphidæ), the leaf beetles (Chrysomelidæ), the bright-colored Buprestidæ, and the various blister beetles. On the sandy ground are sand-colored grasshoppers, sand-colored spiders, Lycosa cinerea, and also small black spiders (Lycosa communis; cf. Emerton, 1885), black crickets, and little red ants apparently identical with species that people the upper beach. Over the vegetation wandered, in early July, an abundance of the predaceous dragon-flies, a black wasp (Polistes), and an occasional dusk-flying butterfly (Hesperidæ)—quite the fauna of a meadow not far from water.

SUMMARY ON THE ANIMAL ECOLOGY OF COLD SPRING BEACH

The outer beach is a region of breakers where débris is thrown on the shore. The submerged zone is crowded with marine animals, some of which make their way out of the water and others of which contribute the débris with which the upper beach is strewn. The lower beach is covered with Collembola that feed upon microscopic organic débris and crawl into the sand at high tide. The line of débris is a rich feeding-ground for animals that live on vegetable matter and on carrion. The débris feeders—Amphipoda, staphylinids, earthworms, ants, carrion flies, necrophorous beetles, attract a predaceous fauna of spiders, robber flies, and tiger beetles. These predaceous species are fed upon, in turn, by the swallows.

The tip of the beach, where the marsh grass grows, is a region of swift currents which the lamellibranchs (mussels) find advantageous because of the food that the currents bring. The currents tend to wear away the spit, but the mussels grow so abundantly on the banks of the channels as in turn to protect these banks from further erosion.

The inside of the sand spit is a region of sedimentation. Plants grow here, and here the plant-feeding snails and fiddlers live. The organisms that are found on the beach are not accidentally there, nor is the fauna determined by causes remote and too complex to be unraveled. The fauna is determined by definite proximate causes of a simple sort that act, the world over, in the same way, and so give to a similar sea beach in other parts of the world a similar collection of animals — excepting that each species may be replaced by another.

COMPARISON OF COLD SPRING BEACH WITH THAT OF LAKE MICHIGAN NEAR CHICAGO

The question arises : How far is the fauna of the sea beach determined by the beach conditions of sand, sunlight, and proximity to a body of water with its strand zone of débris? Will beaches in general, whether of a fresh-water lake or of the sea, tend to have the same fauna? To test this question I have examined the fauna of the shore line of Lake Michigan, south of Jackson Park, Chicago. Here one finds a sandy beach essentially like that at Cold Spring Harbor. On one side extends a huge body of water which differs from that of the harbor chiefly in its lower specific gravity and in the absence of marked tides, but resembles it in its waves. We may recognize here a submerged beach and a terrestrial beach.

I. FAUNA OF THE SUBMERGED ZONE

This inundated part of the beach supports, as one would expect, a fauna the species of which are quite unlike those of the sea. Yet we may recognize a sessile fauna, a crawling fauna, and a swimming fauna.

a. The sessile fauna includes here, as in the sea, the lamellibranchs. These belong to two families, the Unionidæ and the Spheridæ. These, like their marine allies, feed on minute organic particles, chiefly algæ. The Unionidæ are the large forms and seem to take the place of the marine Mactra, Venus, and Mya. They occur in the streams and lakes of all parts of the northern hemisphere, but the group is best developed in North America. The Spheridæ are small, and take the place of the Nuculas of Cold Spring Harbor and Donax of our southern seashore. They are found in the streams and lakes of all countries.

b. Crawling animals belong chiefly to the groups of gastropod Mollusca and Crustacea — the Echinoderma being wholly absent. The snails are mostly small and seem to replace the Littorinas and the Rissoas of the seacoast. The principal crawling crustacean is the isopod Asellus communis, which lives on the wood and among the roots of the shore line. The group is, indeed, poorly represented here as compared with the sea. Burrowing animals seem to be almost entirely absent, possibly on account of the absence of such predaceous forms as occur in the sea.

c. The swimming animals are here, as in the sea, partly scavengers and partly predaceous. First of all we have in the water above the shore numerous Entomostraca. The prawns of the sea are replaced by a very closely related species, Palæ-monetes exilipes, which may have gained the great lakes by the way of the Missis-sippi river, in which it is abundant. The marine lobster is replaced by the crayfishes (Cambarus propinquus and C. virilis). The predaceous forms are the fishes which feed largely upon the snails and the Crustacea.

II.   FAUNA OF THE BEACH

In the tideless lake the lower and upper beaches are hardly to be distinguished. On the lake strand Collembola are found just as on the lower sea beach. In the line of débris that the waves deposit are found the wrecks of all the shallow-water forms of which I have spoken, and, in addition, the carcasses of vast numbers of insects that have fallen into the lake, have drowned, and are cast up by the waves. This wreck-age line affords, then, just the feeding-ground for inland species that the marine species find on the coast. What animals do we find here? The burrowing Orches-tidæ seem, indeed, to be absent, but there is a closely related species (Allorchestes) that lives in the shallow water. That it is not a beach burrower may be due to just these causes that have eliminated the burrowing habit in general. But under the débris rove-beetles and insect larvæ are to be found feeding on the vegetable matter. And small red ants build nests on the beach and visit the débris for the carcasses of insects. A similar carrion fly and carrion beetle (Necrophorous) occur. Feeding upon this fauna is a running spider (Lycosa cinerea) the same as that of the coast. Here, too, occur robber flies and tiger beetles, and even white grasshoppers. Thus the lake beach, having a similar strand zone of decaying vegetation and plant wreck-age with the sea, has, at a distance of nine hundred miles from the sea, excepting certain strictly marine species, practically the same fauna as the sea. The conclusion to be drawn from this fact is the immense importance of habitat (i. e., of environ-mental details) in determining similarity of fauna, or, in other words, the fauna of a point is, within limits, determined rather by the environmental conditions than by the geographical position of the point.

REMARKS ON THE THEORY OF ADAPTATION

Everyone must admit the fact of adaptation of the structures of animals to their environment. The generally accepted theory to account for this is that of Darwin and Wallace that a species coming into a new habitat gradually acquires a fitness to that habitat by the killing off of the less fit individuals born into the species. There are some cases, as, for instance, that of the leaf insects, that of the fungus beetle, and those of mimicry, that I can see no other explanation for but this, that an exter-nal condition existed first and a structure or coloration was acquired by the race that fitted or adjusted it to that external condition.

We must not, in accepting any theory as a true one, try to force it as a universal theory and become blinded to other possible theories. Now there is another and fundamentally different possible theory of adaptation, and this is that the structure existed first and a fitting environment was sought or fallen into by the species having the peculiar bodily condition. Thus the adaptive result is, on this theory, not due to a selection of structure fitting a given environment, but, on the contrary, a selection of an environment fitting a given structure. I shall now consider some special cases that are best explained on this theory. Thus, Eigenmann (1899) shows that the cave fishes, which in many points show an adaptation to the cave environment, are not to be thought of as having accidentally got into caves and as having subsequently gained a structure fitting them for that environment. But, on the contrary, as they all belong to one family, their getting into the caves was evidently not an accident. Moreover, this family includes species that are structurally especially fitted for cave life, even when they occur in regions where there are no caves and never have been any. They shun the light, and live in crevices and under stones. Their bodily conditions fit them for cave life and when, in their constant search for dark holes, some of them succeeded in getting into caves, they naturally thrived there.

Again, in many cases of parasitism among snails the radula is known to be absent altogether, and this has been accounted for by Cooke (1895) on the hypothesis that these snails lost their teeth through disuse. However, it is pointed out as a curious fact that the same absence of a radula occurs in species of Eulima known to be not parasitic. Cooke suggests the hypothesis that in cases like this the form must have derived from parasitic ancestors. It is equally probable that Eulima is a mollusc that will probably soon be driven to parasitism because it has no radula.

Now, that which is true for the cave animals, and probably true for edentulate snails, is illustrated time and time again in the animals of the beaches. We have seen that the Anurida are covered with fine hairs, which enable them to float upon the tide and thus keep them from drowning. Are not the fine hairs a remarkable adaptation to the necessities of the situation? They certainly are, but the probability is that the hairs were not developed to meet this situation at all; at least, such a coating of fine hairs is widespread among Collembola, and the hairs subserve a variety of functions. Thus, Schäffer (1898) finds that the long hairs protect the animal against the action of the sun's rays in the case of certain species that live on leaves; and the importance of such protection would seem to be great, for Absolon (1900) finds that certain cave forms of Collembola, which, so far as he describes them, seem to be scantily covered with hairs, are killed by a few minutes of exposure to sunlight. Hairs are, we may then say, common occurrences on the thin-skinned Collembola. The hairs are important to keep the thin skin from desiccation. Because the skin is thin, the Collembola favor damp or wet places; just because they are covered with hairs, they can float on the water; just because they can float on the water, they can live on the lower beach. Also, they find here their appropriate food. Having by some means got to the beach, they remain there, because they find the conditions existing on the beach peculiarly suitable.

This law is illustrated again by an inhabitant of the upper beach, Oniscus. Oniscus seems, on the whole, rather poorly adjusted to a terrestrial life. Its gills lack the well-developed tracheal chambers of the wood louse, Porcellio, and the pill bug, Armidillidium, its close relatives (Stoller, 1899). Correspondingly we find it only in moist situations, under logs, in cellars, in greenhouses, etc. On the other hand, Porcellio ratzburghii, Brandt, is a species in which all five pairs of outer gills possess tracheal chambers. As Stoller (1899) remarks: "This species lives in situations where the air is charged with moisture only in a moderate degree in excess of that of ordinary atmospheric air. Their habitat is under the bark of dead trees, and they may often be found a meter or more above the ground." Now, experiments that I have made show that a water-inhabiting isopod (Asellus), if taken out of the water, will go back into it if free to do so; and Porcellio, if put in water, will leave it for dry land, if free to do so. Similarly we may conclude that Oniscus chooses a situation that affords the requisite moisture. Shall we conclude that the reason why Oniscus has no tracheal chambers is the result of its living in a moist situation, or is it the cause? We find it where it is because it is hydrotactic. Now, is it hydrotactic because it has no lungs, or is it without lungs because it is hydrotactic? Certainly it would be rash to assert the latter, even though we cannot prove the former.

So likewise we find Nassa, which has a siphon to enable it to draw pure water from above the mud, living in the mud; whereas Littorina, which has no such siphon, clings to the stems of the marsh grass above the mud. Can we say that Littorina has no siphon because it clings to the marsh grass, or does it cling to the marsh grass because it has no siphon? I maintain that the latter is no less true than the former.

Let us consider still one other case. We have seen that the mussels cling to the banks of the channels in such numbers as to make a protecting wall. Of the advantageousness of that situation for the mussel as a lamellibranch there can be no doubt; abundant food and oxygen are brought to its doors every day. The wonder is that no other lamellibranchs than the mussel occupy this favorable situation. Why is this? It is because the mussels are the only species living about the sand spit that have a byssus in the adult stage. Lacking a byssus, the other species cannot attach themselves to the banks. Now, does Mytilus have a byssus because it tends to attach itself to banks, or, being provided with a byssus, was it led to take advantage of the favorable position offered? Did the situation or the organ precede? In this case we may see, I think, the necessity of the organ being well developed before the special habit (of attachment) could be exercised. However, it is quite likely that the byssus has been improved in the race by selection, and with every step of improvement the race has been able to take up a more and more advantageous habitat.

This brings us, indeed, to the most reasonable hypothesis of adaptation, namely, the combination of the improvement of the organ to meet the requirements of the environment, through selection, and of improvement of situation to meet the abilities of the organism. There have gone on, hand in hand, a selection of more appropriate organs and of more congenial habitats. Adaptation of organization to environment has been effected by the double process of selection by environment of the most appropriate organization and by the organism of the most congenial environment.

This hypothesis, I think, should be welcomed by those palæontologists who, like Osborn (1897), are led from their phylogenetic studies to conclude "that there are fundamental predispositions to vary in certain directions." They help out, too, I believe, the fundamentally important observations and experiment of DeVries (1901), who finds that race change is a series of steps, of mutations, that may often have no relation to adaptation; the adaptation comes later. For all those theories, in general, that assume that change of specific structure occurs independently of selection of the fittest, the hypothesis here proposed must be considered a welcome complement. It may be well to point out that the selection of a fitting environment is not confined to migratory animals. It is applicable to all organisms that have a means of dispersal. The seed that falls upon good ground—the race that gets into a favorable environment—will survive.

The theory may thus be summarized: The world contains numberless kinds of habitats, or environmental complexes, capable of supporting organisms. The number of kinds of organisms is very great; each lives in a habitat consonant with its structure. Each species is being widely dispersed, and, by chance, some members of a species get into an environment worse fitted for them; others into one better fitted. Those that get into the worse environment cannot compete with the species already present; those that get into a habitat that completely accords with their organization will probably thrive and may make room for themselves, even as the English sparrow has made room for itself in this country. This process may go on until the species is found only in the environment or environments suited to its organization. As Darwinism is called the theory of the survival of the fittest organisms, so this may be called the theory of segregation in the fittest environment.

In conclusion I repeat that the theory of segregation in the fittest environment does not replace that of survival of the fittest organism, but is complementary to it. It has this *raison d'être* that it shows how unadaptive mutations may become adaptive *if only they can find their proper place in nature.*

LITERATURE CITED

Absolon, P. C. K. "Einige Bemerkungen über mährische Hohlenfauna." *Zool. Anz.*, Bd. XXIII (1900), pp. 1–6.

Balch, F. N. "List of Marine Mollusca of Coldspring Harbor, Long Island, with Descriptions of One New Genus and Two New Species of Nudibranchs." *Proc. Boston Soc. Nat. Hist.*, Vol. XXIX (1899), pp. 133–62.

Cooke, A. H. *Mollusca* in "Cambridge Natural History," Vol. III (1895), pp. 459.

De Vries, H. *Die Mutationstheorie. Versuche und Beobachtungen über die Entstehungen von Arten im Pflanzenreich.* Leipzig: Veit & Co., 1901.

EIGENMANN, C. "Cave Animals: Their Character, Origin, and Their Evidence for or against the Transmission of Acquired Characters." *Proc. Amer. Assoc. Adv. of Sci.*, Forty-eighth Meeting, at Columbus, O., December, 1899, pp. 255.

EMERTON, J. H. "New England Lycosidæ." *Trans. Conn. Acad. Arts and Sciences*, Vol. VI (1885), pp. 481–517, Pls. XLVI–XLIX.

EMERY, C. "Beiträge zur Kenntniss der nordamerikanischen Ameisenfauna." *Zoologische Jahrbücher, Abth. f. Systematik*, Bd. VII (1893), pp. 633–82.

FISCHER, P. *Manuel de conchyliologie et de paléontologie conchyliologique.* Paris: Savy, 1887. Pp. 1,369; 22 plates.

FOLSOM, J. W. "The Distribution of Holarctic Collembola." *Psyche*, February, 1901.

OSBORN, H. F. "Organic Selection." *Science*, N. S., Vol. VI (1897), pp. 583–5.

SCHAFFER, C. "Die Collembola des Bismarck-Archipels nach der Ausbeute von Professor Dr. F. Dahl." *Arch. für Naturgeschichte*, Jahrg. 64 (1898), 1. Bd., pp. 393–425, Taf. XI–XII.

SEMPER, C. *Animal Life as Affected by the Natural Conditions of Existence.* International Scientific Series, Vol. XXX. New York: Appleton, 1881.

SHALER, N. S. "Report on Marshes and Swamps of Northern Long Island, Between Port Washington and Cold Spring Harbor," in *Report on Mosquitoes, North Shore Improvement Association* (W. T. Cox, Secretary), New York, 1902.

SIMROTH, H. *Die Entstehung der Landtiere.* Leipzig: Engelmann, 1891.

SMITH, E. A. "Report of the Geology of the Coastal Plain of Alabama." *Geological Survey of Alabama.* Montgomery, Ala., 1894.

STOLLER, J. H. "On the Organs of Respiration of the Oniscidæ " *Zoölogica*, Heft 25 (1899), pp. 31, 2 Taf.

VERRILL, A. E. (and S. I. SMITH). "Report upon the Invertebrate Animals of Vineyard Sound and the Adjacent Waters, with an Account of the Physical Characters of the Region." *Report of the U. S. Commission of Fish and Fisheries*, Part I, for 1871–72, (1873), pp. 295–778; 38 plates.

WAHLGREN, E. "Beitrag zur Kenntniss der Collembolafauna der äusseren Schären " *Entom. Tidskrift*, Vol. XX (1899), pp. 183–93.

The Early Applications of Mendel's Genetics to Human Beings

Davenport G.C. and C.B. Davenport. 1907. **Heredity of Eye-Color in Man.** (Reprinted, with permission, from *Science* **26:** 589–592 [©American Association for the Advancement of Science].)

Charles and Gertrude Davenport with their children, Millie (left), Charles, Jr.; and Jane (right) (circa 1914).

Two questions left unresolved by Darwin's great insight that species arise through natural selection acting on variations were the origin and the inheritance of the variations themselves. Although he wrote several books describing variation, Darwin could not account for its origins or why new variants were not lost through blending in subsequent generations. Mendel's work—rediscovered at the turn of the century—showed clearly that traits did not become blended but still left unresolved the question of how variations arise.

Davenport's plans for the Carnegie Institution's new Station for Experimental Evolution—vague as they were (MacDowell 1946 [reprinted in this volume, pp. 311–360])—included studies of variation, hybridizing, inbreeding, and cytology. Investigators bred a wide variety of organisms and Davenport published papers—none of which were of great significance—on heredity of traits in sheep, canaries, and chickens. More importantly, Davenport brought many fine geneticists to Cold Spring Harbor in the 30 years of his directorship, including Blakeslee, Shull, and Demerec (see later papers in this collection.)

Davenport also developed an early interest in applications of Mendel's findings to human beings. As he wrote in his 1909 report to the Carnegie Institution, although human genetics could not be experimental and hence fell outside the purview of the Station for Experimental Evolution strictly interpreted, nevertheless, "... the necessity of applying new knowledge to human affairs has been too evident to permit us to overlook it." This research was done in collaboration with his wife, Gertrude Davenport, and she was first author on the four "Heredity of ... in Man" papers that appeared between 1907 and 1910. The paper reprinted here, "Heredity of Eye-Color in Man," was the first of these papers and exemplifies the Davenports' approach.

Human genetics relies on studying the inheritance of traits within families, and the Davenports were able to collect information on eye color in three-generation families. The data were not collected by the Davenports themselves, but rather they enlisted "school principals and other friends" to do the work for them. Their failure to be more closely involved in the data collection led to difficulties in the subsequent analysis when they found that there was confusion on precisely what their color assignments meant. This became a much more serious problem when Davenport moved on to other traits that required even more subjective assessments. Nevertheless, the Davenports showed that blue is generally recessive to brown and that there is considerable variation within blue and brown coloration. In a later paper on skin pigmentation (Davenport and Davenport 1910), they made what appears to be the first reference to polygenic inheritance—a phenotypic trait influenced by two or more genes (Stent 1949).

This paper ends with a discussion of the "...practical applications of these results to human marriage..." Davenport quickly became more interested in these "practical applications" than in developing a science of human genetics (Rosenberg 1961). As early as 1909 he had published a paper entitled "Influence of Heredity on Human Society" that appeared not in a scientific journal but in the *Annals* of the American Academy of Political and Social Science, and in the next year, he published a short book, *Eugenics—The Science of Human Improvement by Better Breeding*.

The transformation of human genetics into eugenics at Cold Spring Harbor began in 1910 when Davenport established the Eugenics Records Office (ERO) (Kevles 1985). The ERO achieved public prominence through the indefatigable efforts of Harry Laughlin, its superintendent. Laughlin carried out surveys of prisons, mental institutions and orphanages, and concluded that American society was in danger of being overwhelmed by increasing numbers of "socially inadequate." Sterilization of the unfit was his solution and by 1932, 30 states had passed legislation authorizing sterilization of "degenerate" individuals and almost 4000—mainly women—were sterilized in that year (Reilly 1991). Laughlin believed also that his analyses revealed that "...the recent immigrants as a whole present a higher percentage of inborn socially inadequate qualities than do the older stocks." Continued entry of such people would undermine the character and moral fiber of the American people, and Laughlin was active in advising—indeed was seconded to—the House Committee that drafted the legislation for the 1924 Immigration Act.

By the late 1920s, however, increasing numbers of geneticists condemned the scientific basis of the ERO's work and deplored its social consequences. In 1925, T.H. Morgan wrote: "...the student of human heredity will do well to recommend more enlightenment on the social causes of deficiencies rather than more elimination in the present deplorable state of our ignorance as to the causes of mental differences" (Morgan 1925). The ERO became an embarrassment to the Carnegie Institution of Washington, and it was shut down in 1940. (See Kitcher 1997 for a discussion of the societal implications of modern human genetics.)

Biographical details on Davenport can be found on page 2.

Davenport C.B. 1909. Influence of heredity on human society. *Ann. Am. Acad. Pol. Soc. Sci.* **34:** 16–21.

———. 1910. *Eugenics—The science of human improvement by better breeding.* Henry Holt & Co., New York.

Davenport G.C. and C.B. Davenport. 1908. Heredity of hair-form in man. *Am. Nat.* **42:** 341–349.

———. 1909. Heredity of hair color in man. *Am. Nat.* **43:** 193–211.

———. 1910. Heredity of skin pigment in man. *Am. Nat.* **44:** 641–731.

Kevles D. 1985. *In the name of eugenics.* Alfred A. Knopf, New York.

Kitcher P. 1997. *The lives to come: The genetic revolution and human possibilities.* Touchstone Books, New York.

MacDowell E.C. 1946. Charles Benedict Davenport, 1866–1944: A study of conflicting influences. *Bios* **17:** 3–50.

Morgan T.H. 1925. *Evolution and genetics,* pp. 200–207. Princeton University Press, Princeton, New Jersey.

Olby R.C. 1966. *Origins of Mendelism.* Constable, London.

Reilly P.R. 1991. *The surgical solution: A history of involuntary sterilization in the United States.* Johns Hopkins University Press, Baltimore, Maryland.

Rosenberg C.E. 1961. Charles Benedict Davenport and the beginning of human genetics. *Bull. Hist. Med.* **35:** 266–276.

Stent C. 1949. *Principles of human genetics,* pp. 325–332. W.H. Freeman, San Francisco, California.

[*Reprinted from* SCIENCE, *N. S., Vol. XXVI., No. 670, Pages 589–592, November 1, 1907*]

HEREDITY OF EYE-COLOR IN MAN

IT has been known that eye-color in man is inherited as an alternative character. Alternative inheritance is usually associated with Mendelism. Is human eye-color inherited in Mendelian fashion? The importance of knowing whether it is depends on the fact that, if Mendelian, the result of any combination of eye-colors of the parents upon the eye-color of the offspring can be, within certain limits, predicted.

The data on which this study has been made were collected with the assistance of school principals and other friends. The records were made on blanks calling for the eye-color through three generations. The total number of cards—each giving the ancestry of one individual—is 132, of which 57 are single cards to a family while the remaining 75 are distributed in 20 families, an average of 3¾ children to a family.

Human eye-color falls into the main classes, blue and brown. The blue color of the iris is what is known as a structural color; no blue pigment is present, but there is a small quantity of scattered granules, reflection of the light from which gives a blue color exactly as reflection from suspended particles makes the air blue. The black pigment of the choroid coat gives a background that favors the reflection of light and prevents transmission; in albinos, who have no black choroid coat at the retina, light is reflected from the back of the eye and the iris appears reddish by transmitted light even as the sky is red at sunset. Brown eyes, on the contrary, contain melanic pigment, reflection from which yields black. Thus the blue eye is the absence of pigment. In addition to the two fundamental types we have black eyes, due to a greater quantity of pigment, and light (*i. e.,* dilute) brown eyes. In addition to black pigment the iris frequently contains more or less yellow in specks or patches. This is doubtless a fat-pigment or lipochrome. The combination of black and yellow pigment gives a green color as it does in the green canary, and such green and blue eyes are commonly called "gray." But "gray" is also used for blue eyes with some brown pigment in larger or smaller patches.

The nomenclature of eye-color which collaborators were requested to employ was as follows: Light blue, dark blue, blue-green or gray, hazel or dark gray, light brown, brown, dark brown, very dark brown or black. This nomenclature was generally followed and seemed to be understood except in the case of "hazel," which we suspect was employed in certain dark bluish-grays. The classification was probably too detailed and the three groups of blue, gray and brown would doubtless have sufficed. In the following summaries minor divisions of these three fundamental groups will frequently be neglected.

The first result which an analysis of the pedigree data reveals is that blue eye-color is recessive to brown. The first evidence of recessiveness is the purity of the germ cells of the recessive type, so that when two recessive individuals are mated *inter se* they throw only the recessive type. Of the offspring of two blue parents 69 are blue and 6 blue-gray or gray. Two additional cases of so-called "hazel" eyes we suspect to be of a blue type. Again, whenever in one family, both father and mother have blue eyes, all children have blue eyes. This is true in the Ge. and Sw. families of three children each, the Hur. family of 4 children and the Re. family of 6 children.

SCIENCE

Reference Letters	Children	Mother Father	Nature of Mating	Mother's Mother Father's Mother	Mother's Father Father's Father
Al.	1 Gray	Gray Blue	D × R	Br (gray?) Gray (blue?)	Gray Blue
Be.	5 Blue	Blue Blue	R × R	Dk Br (Blue?) Blue	Blue ——
Br.	1 Blue 4 Gray	Gray (blue?) Blue	DR × R	Gray Gray (blue?)	Dk Br (blue?) Blue
Bu.	1 Gray	Blue Gray (blue)	R × DR	Gray (blue?) Blk (gray?)	Dk Br (blue?) Blue
Do.	4 Br	Blue Dk Br (blue)	R × DR	Blue Dr Br	Blue Blue
Ge.	3 Blue	Blue Blue	R × R	Blue Blue	Gray (blue?) Blue
He.	1 Gray	Blue Gray	R × D	Blue "Blue"	Blue "Blue"
Huf.	3 Blue	Blue Blue	R × R	Blue Blue	Blue Blue
Hur.	4 Blue	Blue Blue	R × R	Blue Br (blue?)	Blue Gray (blue?)
La.	3 Br 2 Blue	Br (blue) Gray (blue)	DR × DR	Blue Gray	"Gray"?? Blue
Lu.	1 Gray	Gray (blue) "Blue-gray"	R × R	Blue Blue	Gray Blue
Ma.	1 "Blue-gray"	Blue Blue	R × R	Blue ——	Blue Blue
McB.	5 Dk Br	Dk Br Dk Br	D × D	Dk Br Dk Br	Dk Br Dk Br
McC.	1 Gray	Gray (blue) Blue	DR × R	Gray Blue	Blue "Blue-gray"
Mi.	2 Blue 4 Dk Br	Blue Dk Br (blue?)	DR × R	Gray (blue?) Dk Br	Blue Dk Br (blue?)
Oa.	6 Gray 2 Blue	Gray (blue?) Blue	DR × R	Gray Blue	—— Blue
Re.	6 Blue	Blue Blue	R × R	Gray (blue?) Blue	Blue Blue
Ri.	1 Gray	Gray Br (blue)	DR × R	Br (gray?) Br	Gray Blue
Sa₁.	2 Dk Br	Br Blue	D × R	Br Blue	Br Blue
Sa₂.	3 Br 2 Blue	Blue Br (blue)	R × DR	Blue Dk Br	Blue Blue
Sa₃.	2 Br 3 Blue	Br (blue) "Gray"	DR × R	Blue Gray	"Gray" ?? Blk (blue?)
St.	1 Gray	Gray (blue) Gray	DR × R	Blue Gray	Gray Gray
Sw.	3 Blue	Blue Blue	R × R	Blue Blue	Blue Blue
Th.	1 Gray	Gray Blue	D × R	Gray Gray (blue?)	Gray Gray (blue?)

SCIENCE   3

Reference Letters	Children	Mother Father	Nature of Mating	Mother's Mother Father's Mother	Mother's Father Father's Father
Va.	1 Gray	Gray (blue) Blue	DR × R	Br (gray?) Blue	Violet Blue
Vo.	1 Blue 2 Gray	Blue Gray (blue?)	DR × R	Br (blue?) Gray	Blue Br (blue?)
Wal.	1 Gray	Gray Blue	D × R	"Blue" Blue	"Blue" Blk (blue?)
War.	1 Blue 1 Br	Dk Br (blue) Blue	DR × R	Dk Br Blue	Blue Blue

Abbreviations: Br, brown; Dk Br, dark brown; Blk, black; D, dominant; DR, dominant and recessive (heterozygous); R, recessive.

Colors in parentheses are recessive; without a ? means observed, with a ? means hypothetical. Quotation marks means doubt if the term is used with precision. Double query, doubt as to correctness of color assigned.

The second criterion of recessiveness is the absence of offspring of the recessive type from parents one of which is of the recessive type and the other a homozygous dominant. The only family that seems to meet the conditions of having a homozygous dominant brown parent is a small one (Sa.) as follows:

Children	Parents	Grandparents
Boy, dark brown ⎤ ⎦ Girl, dark brown ⎰	⎧ light brown ⎨ ⎩ blue	⎧ brown ⎨ brown ⎧ blue ⎨ light blue

A third criterion is found in crosses of the R × DR type where a recessive is mated with a heterozygous dominant; in this case there should be an equal number of offspring of each type. Six matings of this sort give 16 dark-eyed to 9 light-eyed offspring—a deficiency of the light-eyed group which is probably due to the small numbers.

Since blue or absence of pigment is recessive we should expect to find some cases of two homogametous dominant browns which produce only brown-eyed offspring. We apparently have one such family (McB.) in which the four grandparents, two parents and five children have all dark brown eyes. The behavior of brown alone thus confirms that of browns when crossed with blues, and all results prove that black iris pigment is dominant over its absence.

It remains to consider the behavior of gray in inheritance. Upon tabulating the crosses of blue with gray we find that gray dominates over blue. This is true, for example, in the Al., Bu., He., Lu., McC., Ri., Va. and Wal. pedigrees given in the Appendix. In families where blue × gray parents have a blue-eyed child (Br., Oa., Vo. families) the gray is doubtless heterogametous, containing recessive blue. Again, when both parents are gray-eyed they have produced 9 gray-eyed to 2 blue-eyed children—indicating that both grays are DR (containing recessive blue) expectation being three gray to one blue. Consequently, gray or partial pigmentation is dominant over the pigmentless blue and the occasional enumeration (Ma. family) of descendant of two blue-eyed parents as "blue-gray" or "gray" is due to a slight inaccuracy of classification. On the other hand, gray is recessive to brown (La. family), *i. e.*, a slight pigmentation to an extensive one.

The facts brought out by these statistics show, first, that there are two principal classes of eye-color—brown and blue: that brown varies in intensity from black to light brown; that blue or absence of pigment varies from pale to deep; that blue is frequently imperfect owing to the presence of specks or patches of pigment—the "gray" or "hazel" color; that blue is recessive to gray and gray is recessive to brown.

The practical applications of these results

4   *SCIENCE*

to human marriage are as follows: Two blue-eyed parents will have only blue-eyed children; two gray-eyed parents will have only blue-eyed and gray-eyed but not brown- or black-eyed children; brown-eyed parents may have children with any of the colors of eyes. Gray and blue-eyed parents will tend to have either gray-eyed children only or an equal number of gray- and of blue-eyed children according as the gray-eyed parent is homozygous or heterozygous. When one parent has blue eyes and the other brown the children will be all brown-eyed, if the brown-eyed parent is homozygous— otherwise they will have eyes of various colors according to the gametic constitution of the brown-eyed parent. In case one parent has gray eyes and the other brown we may have the following cases in the offspring: all of them brown-eyed (dark parent homozygous); 50 per cent. gray and 50 per cent. brown (brown parent heterozygous in gray or blue); 25 per cent. blue, 25 per cent. gray and 50 per cent. brown (both parents containing recessive blue germ-cells).

GERTRUDE C. DAVENPORT
CHARLES B. DAVENPORT

The Development of Hybrid Corn at Cold Spring Harbor

Shull G.H. 1908. **The Composition of a Field of Maize.**
(Reprinted from *Proc. Am. Breeders Assoc.* **4:** 296–301.)

Shull G.H. 1909. **A Pure-line Method in Corn Breeding.**
(Reprinted from *Proc. Am. Breeders Assoc.* **5:** 51–59.)

A young George Shull.

THE REDISCOVERY OF MENDEL'S WORK IN 1900 excited both scientists interested in heredity and evolution and breeders who wanted to use genetics to improve traits such as milk yield in cows and the number of kernels on a corncob. These two groups had different goals and worked in different environments, although on occasion they found common ground. When George Shull began his experiments on corn at the Station for Experimental Evolution, he was motivated by the attractiveness of corn for genetic analysis and not by practical goals, yet his research contributed to one of the greatest revolutions in agricultural history.

George Shull (1874–1954) became interested in the biometrics of Galton and Pearson while he was a graduate student at the University of Chicago (Mangelsdorf 1955). It was there that he met Davenport, and when Davenport founded the Station for Experimental Evolution, Shull joined the staff at Cold Spring Harbor to continue his analysis of variation in plants, using the evening primrose and shepherd's purse. In 1905, he added maize to his studies and used the number of rows of kernels on a cob as an easily scorable trait (Shull 1952). Shull propagated plants by self- and cross-fertilization and found that trait variation within the lines of selfed plants decreased as contrasting forms of each trait became established; in the terminology of the day, each contrasting form represented a particular biotype. His conclusion was that a maize plant in "...an ordinary cornfield is a series of very complex hybrids produced by the combination of numerous elementary species" (Shull 1908). These hybrids are maintained by cross-hybridization of the plants, and selfing could reveal and isolate the constituent biotypes.

Shull found also, as others had before him, that there was a loss of vigor in the purebred lines. He suggested that whereas inbreeding with selection for desirable traits at first was successful in eliminating "inferior components" not contributing to the trait, prolonged selection then led to the "...loss of one after another of the component biotypes which had added to the physiological vigor of the strain..." These purebred plants could be reinvigorated by cross breeding with each other.

29

This phenomenon of hybrid vigor, or heterosis as Shull called it, was well known, but Shull proposed that it, together with the preparation of purebred lines, should be the basis for a new breeding strategy. In essence, he proposed making purebred lines to reveal the biotypes "hiding" in the corn plants growing in a field and then crossing these purebred lines to determine which combinations of biotypes led to hybrid vigor of those traits of interest to the farmer. Furthermore, because the F_1 hybrids are identical, a much greater uniformity of the plants would be expected.

In his second paper, Shull detailed a breeding scheme based on a single cross of carefully selected purebred lines (Shull 1909). However, the yield of seed from the inbred lines used to produce the F_1 hybrids was low, and because the seed was very expensive, Shull's plan was not adopted. The preferred breeding scheme was that devised by East and Jones at the Connecticut Agricultural Station in 1918 (East and Jones 1919). In this double-cross method, two F_1 hybrids were crossed and their seed was used. The extra cost of having to maintain four purebred lines and perform two crosses was more than offset by the greatly increased yield of seed. (In the 1960s, there was a shift to the single-cross method as farmers found that the benefits of using this seed outweighed its extra cost.)

It was a further decade before suitable double-cross hybrids were developed, but, beginning about 1930, hybrid corn was rapidly adopted by farmers so that by 1950, almost all corn grown in the United States was hybrid corn (Mangelsdorf 1974). However, farmers had to buy new seed each year from seed merchants because they could not afford to produce the purebred lines and test the F_1s for hybrid vigor. Indeed, it has been argued that the economic self-interest of the seed merchants was a significant factor in the adoption of hybrid corn (Paul and Kimmelman 1988; Fitzgerald 1990). Be that as it may, hybrid corn led to a tripling of the yield of corn per acre in the USA between 1930 and 1980 (Russell 1993).

George Harrison Shull was born April 15, 1874, in Clark County, Ohio. Despite minimal formal schooling, he entered Antioch College in 1892 and graduated in 1901. He moved to the University of Chicago to do a Ph.D. in botany but spent time working for the U.S. National Herbarium and the U.S. Bureau of Plant Industry. Shull became interested in genetics and submitted for his Ph.D. thesis research on variation in wild asters that he had carried out in his final year at Antioch. Shull moved to the Station for Experimental Evolution in 1904 even before the building had been finished. Shull left Cold Spring Harbor in 1915 to become Professor of Biology at Princeton University, where he remained until his retirement in 1942. It is estimated that he grew more than 1,000,000 shepherd's purse and 750,000 specimens of evening primrose in his career. Shull was the Founding editor of Genetics *and a Charter Member of the Genetics Society. He was elected to the National Academy of Sciences in 1948 and won the Academy's Public Welfare Medal in that same year. Shull died on September 29, 1954.*

East E.M. and D.F. Jones. 1919. *Inbreeding and outbreeding: Their genetic and sociological significance.* J.B. Lippincott Company, Philadelphia, Pennsylvania.

Fitzgerald D. 1990. *The business of breeding: Hybrid corn in Illinois, 1890–1940.* Cornell University Press, Ithaca, New York.

Mangelsdorf P.C. 1955. George Harrison Shull. *Genetics* **40:** 1–4.

————. 1974. *Corn: Its origin, evolution, and improvement.* Belknap Press and Harvard University Press, Cambridge, Massachusetts.

Paul D.B. and B.A. Kimmelman. 1988. Mendel in America: Theory and practice, 1900–1919. In *The American development of biology* (ed. R. Rainger et al.), pp. 281–310. University of Pennsylvania Press, Philadelphia, Pennsylvania.

Russell W.A. 1993. Achievements of maize breeders in North America. In *International crop science I* (ed. D.R. Buxton et al.), ch. 29, pp. 225–233. Crop Science Society of America, Inc. Madison, Wisconsin.

Shull G.H. 1908. The composition of a field of maize. *Proc. Am. Breeders Assoc.* **4:** 296–301.

————. 1909. A pure-line method in corn breeding. *Proc. Am. Breeders Assoc.* **5:** 51–59.

————. 1952. Beginnings of the heterosis concept. In *Heterosis: A record of researches directed toward explaining the vigor of hybrids* (ed. J.W. Gowen), pp. 14–48. Iowa State College Press, Ames, Iowa.

THE COMPOSITION OF A FIELD OF MAIZE.

By GEORGE H. SHULL, *Cold Spring Harbor, N. Y.*

While most of the newer scientific results show the theoretical importance of isolation methods, and practical breeders have demonstrated the value of the same in the improvement of many varieties, the attempt to employ them in the breeding of Indian corn has met with peculiar difficulties, owing to the fact that self-fertilization, or even inbreeding between much wider than individual limits, results in deterioration.

The cause of such a result is wholly unknown at present. The old hypothesis which sought an explanation of the deleterious effects of inbreeding in the inharmonious or unbalanced constitution produced by the accumulation of disadvantageous individual variations, can hardly stand in the face of the fact that a very large number of plants normally self-fertilize and a noteworthy few have even given up sexual reproduction entirely, without in the least degree lessening their physiological vigor and evident chances of success in competition with sexually produced plants. The dandelion is propagated parthenogenetically, i. e., its seeds are produced without fertilization, but only the advocate of an unwarrantable theory will maintain that this plant is on that account undergoing a process of deterioration which threatens it with summary extinction. Many species of violets produce most of their seeds from flowers that never open, and one of the most vigorous forms of the small-petaled evening primrose (*Oenothera cruciata*) does the same. In the breeding of tobacco, it is now well known that cross pollination within the limits of a single strain produces inferior offspring and only self-fertilization gives offspring of the highest degree of vigor, though hybrids between distinct strains of tobacco often display a vigor superior to that of either parental strain. Examples could be continued indefinitely, but even one instance in which long-continued inbreeding results in no injurious effects would be sufficient to disredit the old hypothesis.[a]

Some results of the pedigree-breeding of maize at the Cold Springs Harbor Station for Experimental Evolution have suggested a different explanation of the deterioration which has been universally observed in self-fertilized maize. For several years a series of investigations on Indian corn has been in progress at the Station, which involved parallel cultures of cross-pollinated and self-fer-

[a] For a good discussion of inbreeding see A. D. Shamel, on "The effect of inbreeding in plants," Yearbook U. S. D. A., 1905, pp. 377–392.

tilized lines of as nearly equivalent parentage as possible. Although a study of the injurious effects of self-fertilization was not the aim of the investigation, it was immediately apparent in the smaller, weaker stalks, fewer and smaller ears, and the much greater susceptibility to the attacks of the corn-smut (*Ustilago maydis*). These results were almost as marked when the chosen parents were above the average quality as when they were below it, which in itself refutes the idea that the injurious effect is due to the accumulation of deficiencies possessed by the chosen parents.

All the cross-bred rows were similar in structure, vigor, variability, etc., but each self-fertilized row could be seen to differ from other self-fertilized rows in ways capable of description in fairly definite terms. Without entering into a description of all the different self-fertilized rows, a comparison between the two rows showing the greatest contrast will suffice to illustrate and serve as a basis for the conclusion to be arrived at. Following the method everywhere known among breeders as the "ear-row test," I planted parallel rows from ears having given numbers of rows of grains, one self-fertilized and one cross-fertilized ear form each row-class, i. e., one row was planted from a self-fertilized ear and beside it a row from a cross-fertilized ear having 10 rows of grains each; one row each was planted from two ears having 12 rows of grains; and so on.

Taking for comparison the row produced from a self-fertilized ear having 12 rows of grains, and the row produced from a self-fertilized ear having 14 rows of grains, the following differences were observed. In each row the variability was slight and the different qualities noted were characteristic of the entire row. The characters are given in contrasted pairs (designated as a and b), the qualities of the row originating from a 12-rowed ear (a) being given first in each pair: (a) Average height, $6\frac{1}{4}$ feet, (b) average height, $8\frac{1}{2}$ feet; (a) stalks moderately stocky, (b) stalks slender; (a) strong tendency to sucker near the ground, (b) no suckers; (a) leaves broad, dark green, and spreading, (b) leaves rather narrow, light green. not strongly spreading; (a) ears rather strongly diverging on long shanks, the latter usually as long as the internodes or longer, (b) ears erect on short shanks which are usually not over half the length of the internode; (a) husks with well-marked leafy appendages, (b) husks without appendages; (a) grains flinty, (b) grains starchy; (a) most common number of rows to the ear, 10, (b) most common number of rows, 14.

Most of the features here contrasted differ more or less in both cases from the cross-fertilized rows derived from the same original stock. If self-fertilization is assumed to be the direct cause of any of the above characteristics of the one row, it is obviously illogical to attribute the opposite characteristic possessed by the other row to self-fertilization as a direct result. We come to the conclusion therefore that the observed differences between these rows are not *directly* attributable to self-fertilization, but must be due to an indirect effect. The distinguishing characters of these two rows are permanent inheritable qualities and each therefore represents what is known as an elementary species or biotype, as Johannsen has appropriately named the elementary form-group. Self-fertilization has simply isolated the two described forms by separating them from their hybrid combinations with other elementary species.

By rearing under different conditions or by selecting in different directions, it is possible to get a number of somewhat different strains within the same biotype; but, if we do not distinguish clearly between biotypes and these strains which differ only because of the different treatment they have received, only confusion can result. The difference between biotypes and the different strains of a single biotype lies in the nature of the characters which they possess as regards their heritability. Two biotypes will remain distinct from each other without resort to selection as long as they are kept pure-bred and are grown under like conditions, provided those conditions are sufficiently favorable generally to allow such characters as each possesses to develop normally. Two strains within the same biotype may be just as distinct from each other as some biotypes are, but when they are grown under the same conditions, constant selection will be required to keep them distinct, and if selection is omitted the distinguishing characteristics quickly disappear, usually within several generations. Such characters as can only be retained in a pure-bred race by constant selection or by culture under a particular set of external conditions are called fluctuating characters. Now, the inheritance of the fluctuations in any character follow a well-known law, usually known as Galton's law, whose essential feature is the lagging of the average value of single characters of the offspring behind the average of the parents with respect to the same characters. In other words, the average condition of children with respect to any fluctuating character, stands between the average condition of the parents and the average condition of the biotype to which they belong.

It follows from this law that when a given degree of a fluctuating condition is continuously selected under fairly constant cultural conditions, the ideal which is followed in the selection marks the theoretical limit of progress which will take place in the direction of that ideal, and there will always be some lagging back of the average condition, which lagging becomes less and less, the longer the selection is continued. To be specific, in the strain of maize with which I started the most frequent number of rows of grains per ear was 14. According to theory, if we are dealing here with a fluctuating character of a single biotype, we should never expect to be able to exceed 20 rows on the average, by continuously selecting 20-rowed ears for seed, and in like manner we could never hope to get a strain whose average number of rows is less than 12, by continuously selecting 12, provided no change in the conditions of the culture tended to increase the number of rows generally in the former instance or to decrease them generally in the latter. Not the least significant contrast therefore between the two self-fertilized strains above described is that which deals with the number of rows of grains on the ears. In the case of the selection to 14 rows, the result shows a considerable predominance of 14 among the ears of this year's (1907) crop, nearly 40 per cent falling into that single class as compared with 38 per cent in the same class among the unselected population with which the experiments were begun four years ago. As 14 rows was the original prevailing class or "modal" class as it is called, it is quite what we would expect, to find that continued selection of this modal number has simply increased the relative value of class 14. In the case of selection to 12 rows on the other hand we are met with a surprise, for instead of the average number of rows being between 12 and 14 as the supposition that we are dealing with the fluctuations of a single elementary species whose normal mode is 14 rows, would lead us to expect on theoretical grounds, we find the prevailing class to be 10, with nearly 39 per cent in that class. Knowing that this row was the result of continued selection of 12-rowed ears, one would infer from the data of this year's crop that the original condition of the population— the normal condition for the race to which this row belongs—is probably 10 rows or possibly even 8 rows as the modal class, instead of the 14 rows possessed by the original stock from which all my cultures came. It is demonstrated therefore that these two rows belong to distinct races or elementary species of corn, though the original stock appeared to be fairly homogeneous. Most of the other self-fertilized rows showed by various marks that they were likewise to be considered members of distinct biotypes, instead of fluctuant parts of a single biotype as I believed they were when I began my investigation.

The obvious conclusion to be reached is that an ordinary cornfield is a series of very complex hybrids produced by the combination of numerous elementary species. Self-fertilization soon eliminates the hybrid elements and reduces the strain to its elementary components. In the comparison between a self-fertilized strain and a cross-fertilized strain of the same origin, we are not dealing, then, with the effects of cross and self-fertilization *as such*, but with the relative vigor of biotypes and their hybrids. The greater vigor of the cross-fertilized rows is thus immediately brought into harmony with the almost universal observation that hybrids between nearly related forms are more vigorous than either parent.

The components of a hybrid strain may be separated by means of cross-fertilization just as surely as by self-fertilization, if the parents of the cross are rigidly selected, generation after generation, for definite characteristics; but the process of segregation will be in this case much slower, because in each cross some of the elements which were eliminated from the mother will be reintroduced by the father, and *vice versa.* For this reason the deterioration which comes from close inbreeding coupled with cross-fertilization should not be as rapid though just as sure as by self-fertilization. This again is in accord with such observations as are on record.

As most of the important characteristics for which the corn breeder strives are closely related to the question of physiological vigor the fundamental problem in breeding this plant is the development and maintenance of that *hybrid combination* which possesses the greatest vigor. Up to a certain point the common empirical method of selection will mostly eliminate only those components which do not contribute to the best possible result, and the more rigid the selection during this period the more rapid will be the improvement of the selected strain; but if the selection is continued in the same rigid manner after these inferior components are eliminated, it may lead to the loss of one after another of the component biotypes which had added to the physiological vigor of the strain and there will then be a resultant deterioration, especially if among the characteristics which guide the selection are some which are unrelated to vigorous growth. The fundamental defect in every empirical scheme of corn-breeding which simulates the isolation methods of the breeder of small grains, lies in the fact that there is no intelligent attempt in these methods to determine the relative value of the several biotypes *in hybrid combination,* but only in the pure state.

In the present state of our knowledge it is impossible to predict from a study of two pure strains what will be the relative vigor of their hybrid offspring. That is an important relation which future investigations must unlock for us. The problem of getting the seed-corn that shall produce the record crop, or which shall have any specific desirable characteristic combined with the greatest vigor, may possibly find a solution, at least in certain cases, similar to that reached by Mr. Q. I. Simpson in the breeding of hogs by the combination of two strains which are only at the highest quality in the first generation, thus making it necessary to go back each year to the original combination, instead of selecting from among the hybrid offspring the stock for continued breeding. That is, it may be found that the desirable combination of elementary species of Indian corn will be best attained by separating and recombining in some definite manner the different elementary species, or on the other hand it may be found that selection according to the empirical methods now most approved can be carried to a point at which the most efficient combination has been isolated from the less efficient components and may then be maintained only by a relaxation of the rigid selection.

Such questions as these cannot be settled in the study, but only in the field by means of carefully conducted experimentation. I hope that those experiment stations which are dealing with the problems of the improvement of maize will undertake the solution of these fundamental problems and that as a consequence the technique of corn-breeding will find a basis in scientific knowledge quite different from the present more or less blind conflict between empirical selection and the little understood injurious effects of inbreeding.

In conclusion I wish to say that the idea that in breeding maize we are dealing with a large number of distinct elementary species or biotypes is not presented here as a new idea, for De Vries, in his little book on "Plant breeding" presents this view, and Dr. East in a recent bulletin from the Connecticut Station has indicated the great complexity of the corn breeder's problems owing to the concurrence of these elementary species and fluctuating variations. I have aimed simply to point out how my own experience in corn-breeding supports the same view. I think, however, that the suggestion here made, that continuous hybridization instead of the isolation of pure strains is perhaps the proper aim of the corn breeder, is new and it is this view that I wish to submit for your consideration.

A PURE-LINE METHOD IN CORN BREEDING.

By Dr. George Harrison Shull
Cold Spring Harbor, N. Y.

Last year I described[1] a series of experiments with Indian corn which led me to the conclusions: (1) that in an ordinary field of corn the individuals are generally very complex hybrids; (2) that the deterioration which takes place as a result of self-fertilization is due to the

[1]The Composition of a Field of Maize. Report American Breeders' Association, 4:296-301, 1908.

gradual reduction of the strain to a homozygous condition; and (3) that the object of the corn-breeder should not be to find the best pure-line, but to find and maintain the best hybrid combination.

The continuation of these studies during the past year have given still further proof of the correctness of the first two of these propositions, and besides has given unexpected suggestions for a new method of corn breeding by which the essential feature of the third proposition may be realized. It is my purpose to discuss this new method briefly in the following pages. I will first, however, describe the results of the past year's experiments in so far as they bear upon the points in which we are interested here. For convenience I will refer to the two self-fertilized families contrasted in my paper last year as "Strain A" and "Strain B." It will be remembered that these two families resulted from the self-fertilization of different, apparently equal, individuals; but that notwithstanding this fact, they differed from each other in height and stockiness of stems, width and greenness of the leaves, length of shank of the ears, appendages of the husks, quality of the grains, and the number of rows of grains on the ears. (See fig 1.)

In addition to the parallel cultures of self-fertilized and cross-fertilized families which have been continued from the beginning of these experiments in 1904, I had during the past season the F_1 offspring of a cross between two sibs in Strain A, and two families representing reciprocal crosses between Strain A and Strain B. It was observed that every one of the mentioned characteristics which distinguished Strains A and B, remained constant distinguishing features in the pure-bred families, but in regard to the number of rows on the ears, it is now obvious that Strain A has the normal mean number 8, as compared with 14 in Strain B, for in this year 89 per cent of the ears produced by Strain A had only 8 rows of grains, though the selection of ears for seed in this strain during three years was for 12 rows on the ear, and only in the last year was an 8-rowed ear used because a suitable 12-rowed ear was not available. This result is a striking confirmation of the suggestion made last year that according to the law of regression the occurrence of a mean number of rows less than 12 in Strain A indicated that the normal number of rows for this strain is 10 or possibly only 8.

The cross between two sibs in Strain A was grown beside the self-fertilized family belonging to the same strain, and these two families were so similar during the entire period of their development that they were considered identical, but at the end of the season it was found that the cross-bred family was a trifle taller and produced over 30 per

AMERICAN BREEDERS' ASSOCIATION   53

FIG. 1. TYPICAL SPECIMENS OF STRAIN A (AT RIGHT) AND STRAIN B (LEFT),
SHOWING CONTRAST OF VEGETATIVE CHARACTERS. DRAWN BY J. MARION
SHULL FROM A PHOTOGRAPH.

54   A PURE-LINE METHOD IN CORN BREEDING

cent more grain by weight than the self-fertilized family. In the self-fertilized family, 73 ears were produced, weighing 12 lbs., and in the cross between sibs the 78 ears weighed 16½ lbs. There was also a striking difference between these two families as regards variability in the number of the rows on the ear, as may be seen in this table:

NUMBER OF ROWS ON THE EARS	8	10	12	14
Self-fertilized	65	6	2	0
Cross-fertilized	8	50	19	1

Unfortunately the parents of these two families were not identical in the number of rows, the mother of the self-fertilized family having 8 rows and that of the cross-fertilized family 10. The greater height, greater weight of grain produced, the higher number of rows on the ears, and the greater variability in the number of rows, in the cross-fertilized family, all point to the same conclusion, namely, that my self-fertilized Strain A was not yet reduced completely to a homozygous condition, and that the parents, or at least one of them, of my cross-bred family was heterozygous.

The two families which were the product of reciprocal crosses between Strain A and Strain B, have proved of great interest, for although the individuals of both Strain A and Strain B were small and weak, and the self-fertilized families of these produced respectively only 12 lbs. and 13 lbs. of ear-corn, the hybrid family in which Strain A supplied the mother and Strain B the father, produced 92 ears weighing 48 lbs., and the reciprocal cross produced 100 ears weighing 55 lbs. Typical ears of Strain A and Strain B, and of their reciprocal hybrids, may be compared in fig. 2. If we reduce these results to bushels per acre on the basis of 10,000 ears per acre and 70 lbs. per bushel, it is found that Cross A×B has produced 74.4 bushels per acre and Cross B×A has produced 78.6 bushels per acre, the average for the two families being nearly 77 bushels per acre. The two families which I have kept continuously cross-bred during the period in which these experiments have been in progress, and which have been likewise continually selected to 12 and 14 rows of grains, may be properly taken as controls. These two families together produced 203 ears weighing 107½ lbs., or at the rate of 75 bushels per acre, and when the comparison is extended so as to include my other continuously crossed families—8 families in all—it is found that these produced collectively at the rate of a little less than 75 bushels per acre.

My farmer friends, especially here in the heart of the corn country, will not be greatly impressed with these yields of 75—78 bushels per acre, but I must call attention to the facts that the light gravelly soil of Long Island bears a very unfavorable comparison with Mississippi valley alluvium for the production of Indian corn, and further that the summer of 1908 was notable for one of the longest periods without rain that has ever been experienced there. The important point will

FIG. 2. TYPICAL EARS OF STRAIN A (AT RIGHT) AND STRAIN B (LEFT) AND OF THEIR RECIPROCAL HYBRIDS. EACH HYBRID STANDS NEAREST ITS MOTHER-STRAIN.

not be missed however that the crosses between two self-fertilized strains yielded a little more grain than those strains which had been kept carefully cross-fertilized by hand. To be sure, the difference is not great enough to seem of any particular significance in itself, but it must be remembered that the two self-fertilized strains, A and B, have been essentially unselected, being simply those two strains which have first approached the pure homozygous state as a result of self-fertilization. It is scarcely conceivable that other pure strains crossed together should not give in certain combinations considerably greater yields than those produced by the combination of Strains A and B. At any rate the result is sufficiently striking to suggest that the method of separating and recombining definite pure-lines may perhaps give results quite worth striving for.

This suggestion will be more readily appreciated perhaps if I discuss briefly the theoretical aspect of this method of pure-line breeding as compared with the method now in use among the most careful corn-breeders. In the light of my results, the constant precautions that are taken in the method now in use to prevent in-breeding, have for their real object the retention of the most efficient degree of heterozygosis or hybridity, and it is obvious that the selection of the most vigorous individuals for seed, really picks out those individuals which have this most efficient degree of hybridity. While I have not investigated the inheritance of the various characteristics of the pure lines of maize and am not in position to say that they all follow Mendel's law, many investigations of particular· characteristics in corn have shown that those characteristics are Mendelian. Even if some of the differentiating characteristics of corn should not prove to be Mendelian, it seems not improper to discuss the two methods on the Mendelian basis.

In the method which selects for seed the most heterozygous individuals, the characteristic splitting and recombination of unit-characters must produce an offspring of quite various degrees of heterozygosis. Some individuals will be as complex as the selected parents, others will have many of the same units in the homozygous condition, and thus be less complex and consequently less vigorous. According to the laws of chance a few individuals in the field may be expected to be almost or quite completely homozygous, and as a result will be very inferior in vigor and will produce but little grain. The result of such a process must always be to give a crop of lower average yield than the average of the selected seed. Moreover, these different combinations of unit-characters and different degrees of hybridity in the offspring of a complex hybrid must introduce a certain amount of heterogeneity into the crop which will have the effect to also lower the average quality with respect to any other desirable points which have been used as guides in the selection of the seed-corn, and efforts at the attainment of homogeneity by the method now in use tend to lessen physiological vigor, and therefore lessen the yield, owing to the fact that such homogeneity in the offspring of hybrids is to be attained only through homozygosis in respect to all those characteristics which affect the form and size of the ear, width, depth, shape, and composition of the grains, and any other feature in which homogeneity may be desired. This is doubtless the explanation of the interesting experience related by Mr. Joseph I. Wing at the meeting of the American Breeders' Association in Columbus two years ago. His father had selected a very fine deep-grained variety of corn in which great uniformity had been attained but only at the expense of decreased yield.

In the pure-line method outlined below all individuals in the field will be F_1 hybrids between the same two homozygous strains, and there are theoretical grounds for expecting that both in yield and uniformity superior results should be secured. Thus, every individual will be as complex as every other one and should produce an equal yield of grain if given an equal environmental opportunity, so that in so far as hereditary influences are concerned the vigor of the entire crop should be equal to the best plants produced by the methods now in use. This would seem to result necessarily in a larger yield than can be produced by the present method. But not only will all the plants in the field have the same degree of complexity, but they will all be made up of the same combination of hereditary elements, and consequently there must result such uniformity as is at present unknown in corn.

With such a prospect as this, I believe we will be sufficiently interested to make the discussion of the method by which such results are to be attained worth while. The question naturally arises as to whether the technique of the new method will be sufficiently simple to make it practicable. To this question I believe I can safely answer that the pure-line method will be considered simpler than the elaborate one now in use among the most careful breeders, e. g., those at the Illinois, Connecticut, and Ohio State Experiment Stations. The process may be considered under two heads: (1) Finding the best pure-lines; and (2) The practical use of the pure-lines in the production of seed-corn.

(1) In finding the best pure-lines it will be necessary to make as many self-fertilizations as practicable, and to continue these year after year until the homozygous state is nearly or quite attained. Then all possible crosses are to be made among these different pure strains and the F_1 plants coming from each such cross are to be grown in the form of an ear-to-the-row test, each row being the product of a different cross. These cross-bred rows are then studied as to yield and the possession of other desirable qualities. One combination will be best suited for one purpose, another for another purpose. Thus, if the self-fertilized strains be designated by the letters of the alphabet, it may

58   A PURE-LINE METHOD IN CORN BREEDING

be found that Cross C×H will give 120 bushels per acre of high-protein corn, that F×L produces a similar yield of low-protein corn that K×C gives the highest oil-content accompanied by high yield, and so on. Moreover, it seems not improbable that different combinations may be found to give the best results in different localities and on different types of soils.   The exchange of pure-bred strains among the various experiment stations would greatly increase the number of different possible hybrid combinations and facilitate the finding of the best combination for each locality and condition.

(2) After having found the right pair of pure strains for the attainment of any desired result in the way of yield and quality, the method of producing seed-corn for the general crop is a very simple though

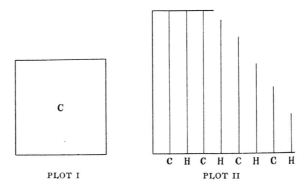

PLOT I   PLOT II

FIG. 3. ARRANGEMENT OF THE TWO ISOLATED PLOTS FOR THE PURE-LINE
METHOD OF CORN BREEDING.

somewhat costly process. Two isolated plots will be necessary, to which I may conveniently refer as Plot I and Plot II. (See fig. 3.) In Plot I will be grown year after year only that pure strain which investigation has proved to be the best mother-strain for the attainment of the desired end. Thus, if it has been found, as in the example already cited, that Cross C×H gives the desired result, Plot I will be occupied by Strain C. This will require no attention from the breeder's point of view except that any exceptionally vigorous or aberrant individuals should be eliminated, as such plants might be safely assumed to be the result of foreign pollinations. In Plot II Strain C and Strain H are to be planted in alternate rows, and all of Strain C is to be detasseled at the appropriate time. All the grain gathered from the detasseled rows will be seed-corn for the general field-crop, and that gathered from the tasseled rows will be pure-bred Strain H to be used again the following year in the same way. Here again in pure Strain H all exceptionally vigorous or aberrant individuals should be discarded as being probably due to the entrance of foreign pollen.

I am not prepared at present to say what will be the probable cost of seed-corn when produced by this method, but have reason to suppose that it would be more expensive than by the present method; nor can I surmise what relation this increased cost will bear to the increased yield that will be produced. These are practical questions which lie wholly outside my own field of experimentation, but I am hoping that the Agricultural Experiment Stations in the corn-belt will undertake some experiments calculated to test the practical value of the pure-line method here outlined.

Chromosomal Changes and Heredity

Blakeslee A.F., Belling J., and Farnham M.E. 1920. **Chromosomal Duplication and Mendelian Phenomena in Datura Mutants.** (Reprinted, with permission, from *Science* **52:** 388–390 [©American Association for the Advancement of Science].)

Department of Genetics, 1920, outside what was then the Animal House and is now McClintock Laboratory. John Belling and Albert Blakeslee are in the middle of the second row. Charles Davenport is standing behind Blakeslee.

THE MAIN TENETS OF THE CHROMOSOMAL BASIS of heredity were set out clearly and explicitly by Sutton in 1903, and a new field of research—*cytogenetics*—was created from a fusion of Mendelian genetics and cytology—the study of chromosomes (Wilson 1925; Darlington 1932). Analysis of the morphology and behavior of chromosomes was used to explain Mendelian inheritance, or, more often, to explain apparent exceptions to Mendelian inheritance. It was a powerful tool, and three of its most adept exponents—Albert Blakeslee, John Belling, and Barbara McClintock—worked in the Carnegie Institution of Washington's Station for Experimental Evolution.

Blakeslee came to Cold Spring Harbor in 1915 to take the place of George Shull, who had moved to Princeton University (Sinnott 1959). He brought with him his own experimental material—the plant *Datura stramonium*, known as jimsonweed. Blakeslee had found a form in which the seed capsule was smooth rather than spiny and had used this for teaching genetics in his botany classes at the Connecticut Agricultural College in Storrs. At Cold Spring Harbor, Blakeslee continued to use *Datura* and in 1915 found a plant in which the seed capsule was spherical—the mutation, *Globe*—and inheritance of this trait was an exception to Mendel's Laws. By 1920, he had found a total of 12 mutants, clearly distinguishable from each other in a number of traits and all showing the same unusual pattern of inheritance (see figure, p. 52). In that year, Blakeslee was joined by John Belling, a brilliant cytogeneticist (Babcock 1933), and the two of them set about determining the chromosomal basis of the twelve *Datura* mutants. The paper reprinted here is the preliminary report of their findings.

Belling had developed a much improved method for making chromosomal preparations and staining them with his iron aceto-carmine stain (Belling 1921, 1926). He and Blakeslee distinguished 12 pairs of chromosomes in normal *Datura* plants, for a total of 24 chromosomes. In contrast, each mutant had 25 chromosomes—the plants had 2 each of 11 chromosomes but 3 of the 12th. The key observation was that this trisomic chromosome was different in each mutant, leading them to conclude that the characteristics of each mutant were caused by genes on the extra chromosome. They expected to be able to determine what genes were carried on each chromosome by simply examining the traits of each of the mutant plants and seeing which was the extra chromosome. They were able to show, for example, that *Globe* had three copies of the S chromosome whereas *Poinsettia* had three copies of the m chromosome. However, in this first paper, they simply showed that their findings accounted for the peculiar inheritance patterns of the mutants.

At the height of his research program, Blakeslee was growing 70,000 plants each summer and generating large amounts of data which he summarized in 1931. (Belling, who suffered from ill health, had moved to California in 1927, where he continued his cytogenetic studies.) By then, Blakeslee had identified all manner of abnormal chromosomes—some induced by treatment with radiation and chemicals—and he was able to conclude that "Extra chromosomal material thus brings about distinct and specific changes in the appearance of the plant in which it is present" and it does so by virtue of the "...genetic factors which the extra material contains." Blakeslee emphasized that *Datura* was unlikely to be unique in having variable numbers of chromosomes and that the desirable traits in crop plants might be due to extra chromosomes as well as to single mutations. He hoped that such extra chromosomes would be "...consciously utilized as a source of desirable variations in plants of economic importance."

Blakeslee was a major figure who made many contributions to genetics. Continuing his interest in abnormal chromosome numbers (polyploidy), Blakeslee attempted to make tetraploid plants at will by treating the plants with extremes of heat and cold or by irradiation. These were largely unsuccessful, but in 1937, at the suggestion of Dr. Eigsti, one of his colleagues, he began to treat seeds with colchicine, an alkaloid poison. He found that it induced chromosome doubling with high frequency (Blakeslee and Avery 1937). Colchicine treatment was later adopted by the animal cell cytologists and played a key role in the first correct determination of the number of human chromosomes (Hsu 1979). Blakeslee also became interested in human genetics, especially the inheritance of sensory traits such as smell and taste (see page 95).

> *Albert Francis Blakeslee was born on November 9, 1874, in Geneseo, New York. He was an undergraduate at Wesleyan University (B.S., 1896) and went to Harvard University where he discovered mating types in* Mucor. *He was awarded a Ph.D. in 1904 and then studied in Germany for two years. From 1907 to 1915, Blakeslee was Professor of Botany at Connecticut Agricultural College, Storrs, before moving as a Staff Scientist to the Station for Experimental Evolution. In 1934 he was appointed director of what was now the Department of Genetics, a post he held until his retirement in 1941. This was followed by 12 years as a Visiting Professor at Smith College where he established a Genetics Experiment Station and continued his research, publishing no fewer than 38 papers in his "retirement." Blakeslee was elected a Member of the National Academy of Sciences in 1929. He died on November 16, 1954.*

Babcock E.B. 1933. John Belling. *J. Hered.* **24:** 297–300.

Belling J. 1921. On counting chromosomes in pollen-mother cells. *Am. Nat.* **60:** 573–574.

———. 1926. The iron-acetocarmine method of fixing and staining chromosomes. *Biol. Bull.* **50:** 160–162.

Blakeslee A.F. 1931. Extra chromosomes as a source of variation in the Jimson Weed. *Smithson. Rep. 1930* 431–450.

Blakeslee A.F. and Avery A.G. 1937. Methods of inducing doubling of chromosomes in plants. *J. Hered.* **28:** 393–411.

Darlington C.D. 1932. *Recent advances in cytology.* Blakiston, Philadelphia, Pennsylvania.

Hsu T.C. 1979. *Human and mammalian cytogenetics: An historical perspective.* Springer-Verlag, New York.

Sinnott E.W. 1959. Albert Francis Blakeslee. *Biogr. Mem. Natl. Acad. Sci.* **33:** 1–38.

Wilson E.B. 1925. *The cell in development and heredity.* The Macmillan Company, New York.

SPECIAL ARTICLES

CHROMOSOMAL DUPLICATION AND MENDELIAN PHENOMENA IN DATURA MUTANTS

There are 12 separate and distinct mutants of the Jimson weed (*Datura Stramonium*) which have recurred with more or less fre-

OCTOBER 22, 1920]   *SCIENCE*   389

quency in our cultures of this species during the past six years. The majority of these 12 mutants have been already briefly described or figured elsewhere.[1]

The twelve have certain characteristics which distinguish them from the normal stock from which they arose. They are of feebler growth than normals and have a relatively high degree of pollen sterility, while pollen from normals is relatively good with less than 5 per cent. obviously imperfect grains when examined in unstained condition. The breeding behavior of the twelve is peculiar in that the mutant character is transmitted almost entirely through the female sex. Usually about one quarter or less of the offspring only from a given mutant reproduce the parental mutant type. The pollen entirely fails to transmit the mutant character, or transmits it only to a small percentage of its offspring. This is concluded from the fact that normal female plants crossed with mutant pollen produce no mutant offspring or only a small per-centage, and from the fact that the pollen of any of the 12 mutants seems to be no more potent in reproducing the mutants than pollen from normals.

Another type of mutant, provisionally called "New Species" because of the difficulty or impossibility of crossing it with normals has relatively good pollen and breeds true.

A study has been begun by the present authors of the relationship which exists in *Datura* between the cytological condition and the related phenomena of mutation and Mendelian inheritance. The cytological findings are based on counts of over 350 groups of chromosomes. We can confirm the report of others as to the presence of 12 pairs of chromosomes in the somatic cells of normal jimsons. The somatic number is accordingly twenty-four in contrast to the gametic number twelve. Chromosomal counts from the first division of pollen mother cells show that the gametic number in all the 12 mutants is

[1] Blakeslee, A. F., and Avery, B. T., "Mutations in the Jimson Weed," *Jour. of Heredity*, X., 111–120, Figs. 5–15, March, 1919.

apparently 12 and 13 giving a calculated somatic number of 25 instead of the 24 found in normals. Whereas in normals all the gametes have 12 chromosomes, in our dozen mutants presumably half the gametes have 12 and half have 13 chromosomes. Apparently in the 13-chromosome gamete the extra chromosome is brought in by a duplication of one of the regular twelve.

The suggestion lies near at hand that each of our 12 mutants is associated with, if not actually determined by, the duplication of a different individual chromosome to make up the calculated total of 25 characteristic of their somatic cells.

If each of our dozen mutants is characterized by the presence of an additional chromosome in a definite one of the 12 chromosome sets, it should be possible by breeding tests to identify the mutant which has as its extra chromosome the one which carries the gene for any particular Mendelian character. This we apparently have been able to do for two of the twelve sets.

The mutant *Poinsettia* (1) which appears to be caused by a duplication of one of the chromosomes carrying determiners for purple or white flower color will serve as an example. *Poinsettia* plants have 2 chromosomes in all the sets except in the one carrying the gene for flower pigmentation, which has three. Considering only the latter, we may have *Poinsettia* mutants, as regards their purple pigment, either triplex PPP, duplex PPp, simplex Ppp or nulliplex ppp.

A duplex purple *Poinsettia* with the formula PPp should, if the chromosomes assort at random, be expected to form egg cells of the following types: $2P + p + pp + 2Pp$. The pollen grains should have the same constitution; but, since the *Poinsettia* character fails to be carried by the pollen to any significant extent, the effective male gametes are $2P + p$. Combining male and female gametes in selfing we expect the following zygotes: $4PP + 4Pp + pp + 2PPP + 5PPp + 2Ppp$. The zygotes with 2 chromosomes in the set are normals, the zygotes with 3 chromosomes are *Poinsettia* mutants. We should have therefore among the normals 8 purples to 1 white, and among the Poinsettias 9 purples to no whites. The expectation of an equal number of normals and mutants is practically never realized, probably because of differential mortality in early stages favoring the normals.

A simplex purple heterozygote with the formula Ppp should have the following female gametic formula: $P + 2p + 2Pp + pp$. Its effective male gametes should be $P + 2p$ Selfing a simplex purple heterozygote therefore should give offspring showing a ratio of purples to whites in normals of 5:4 and in the *Poinsettias* of 7:2. Several *Poinsettia* plants of these two heterozygous purple types have been selfed and found to give color ratios in their offspring in close agreement with the calculated values above. When *Poinsettia* mutants are made heterozygous for the other known Mendelian factors, segregation occurs in normal manner giving the customary 3:1 ratio for the characters involved, in both normals and *Poinsettias*.

Two of the 12 mutants have each a single varietal type, which may be due to factors modifying the expression of the more typical complex. In addition two new mutant forms have arisen each of which in appearance seems to be a combination of two of the typical 12 recurrent mutants. It has not been possible as yet to count their chromosomes nor to study their breeding behavior.

We have discussed the duplication of a single chromosome from only one of the 12 sets, producing mutants with 25 somatic chromosomes, with 3 chromosomes in one set and 2 chromosomes in the other 11. We have obtained in addition the duplication of a single chromosome from each of the 12 sets producing a mutant triploid for all the 12 homologous sets.

The duplication may bring about a doubling of all the chromosomes, producing Gigas-like tetraploid mutants—the "New Species" type already mentioned. Such tetraploid plants have presumably 48 chromosomes in somatic cells and 24 in the gametes. From a study of the color ratios in over eight thousand offspring from tetraploid plants, it is possible to assert with some confidence that independent assortment of the chromosomes in the homologous sets of such tetraploid mutants is the rule. Selfed duplex purple heterozygotes throw 35 purples to 1 white, while the back-cross gives a ratio of 5:1. Simplex purple heterozygotes on the other hand give 3:1 ratios when selfed and 1:1 ratios when back-crossed.

Evidence is at hand which indicates that we may have plants with other of the theoretically possible combinations of chromosomes than those mentioned in the present paper.

The significance of the findings in *Datura* in relation to the peculiarities in inheritance in *Gigas* and other mutant types in *Œnothera* will be pointed out later. It is hoped that it may be possible to publish in the near future a series of more detailed papers on the phenomena of chromosomal duplication in the *Daturas*. The present preliminary publication will suffice to emphasize the distinction which must be kept in mind between chromosomal mutations and mutations affecting only single genes.

Albert F. Blakeslee.
John Belling,
M. E. Farnham

Carnegie Station for
    Experimental Evolution

Smithsonian Report, 1930.—Blakeslee PLATE 8

Normal

Globe   Poinsettia   Cocklebur   Ilex

Echinus   Rolled   Reduced   Buckling

Glossy   Microcarpic   Elongate   Spinach

CAPSULES OF THE 12 PRIMARY (2N+1) TYPES IN THE JIMSON WEED WITH A
CAPSULE OF A NORMAL PLANT ABOVE

(Modified, with permission, from Blakeslee A.F. and Belling J. 1924–1925. Chromosomal mutations in the Jimson weed, *Datura stramonium. J. Hered.* 15–16: 195–206 [©Oxford University Press].)

Mouse Headquarters, Cold Spring Harbor

Little C.C. 1920. **The Heredity of Susceptibility to a Transplantable Sarcoma (J.W.B.) of the Japanese Waltzing Mouse.** (Reprinted, with permission, from *Science* **51**: 467–468 [©American Association for the Advancement of Science].)

Richter M.N. and MacDowell E.C. 1930. **Studies on Leukemia in Mice: I: The Experimental Transmission of Leukemia.** (Reprinted, with permission, from *J. Exp. Med.* **51**: 659–673 [©Rockefeller University Press].)

Clarence Little

Carleton MacDowell examining mice in the 1930s.

AMONG THE MANY LEARNED SOCIETIES THAT claimed Charles Darwin as a member, it is likely that none brought him as much pleasure as his participation in the Columbarain and Philoperisteron Clubs. These were the haunts of pigeon fanciers, and Darwin mixed pleasure with work as he sought evidence for the power of selection, in this case of breeders creating, if not yet new species, at least varieties that were strikingly different. His association with pigeon fanciers brought him a certain notoriety among his friends, for the fanciers were not of his social class, but it is clear that Darwin became an enthusiast and "...kept every breed which I could obtain or purchase." Furthermore, he found what he required—the chapter on artificial selection opens *The Origin of Species*, and pigeons occupy more than six pages (Browne 1995).

Some 50 years after Darwin was admiring his pigeons, geneticists turned again to the world of breeders, but this time of mice (Morse 1981). The dealer whose mice were the source of many well-known laboratory strains was Abbie Lathrop, who lived in Granby, Massachusetts. She had been a teacher but was forced to retire because of pernicious anemia. Instead, she turned to breeding and selling fancy mice. Her business prospered although the clientele changed—eventually Lathrop was supplying more mice to laboratories than to mouse fanciers (Morse 1978a).

One of her early clients was William E. Castle, director of the Bussey Institution of Harvard

University (Keeler 1978; Morse 1985; Rader 1998). He had been a student of Davenport when the latter was an instructor at Harvard and, like Davenport, Castle became an enthusiastic Mendelian (Dunn 1965). Castle seized the opportunity provided by his appointment to begin work on mammalian genetics. He had many distinguished colleagues and students, including L.C. Dunn, Sewall Wright, George Snell, Gregory Pincus, L.C. Strong, E. Carleton MacDowell, R. Snyder, Lloyd Law, and Clarence Little, some of whom later turned to human genetics. (Essays by some of the pioneers of mouse genetics can be found in Morse 1978b.)

A major use of mice was in the study of cancer. Of particular interest at the time was the inheritance of susceptibility to cancer tested by, for example, transplanting pieces of tumor between mice. An early period of "...highly naive optimism..." faded when it was found that the results were so inconsistent that "...an investigator could not even verify his own observations" (Strong 1978). It was realized that it was the mice that were genetically "inconsistent," and the first efforts to make inbred strains began. One of the first inbred lines was DBA, developed by Little in 1909.

Davenport, too, was breeding mice at Cold Spring Harbor. As early as 1900, he noted that the Biological Laboratory had a colony of 50 mice, and he was studying the inheritance of color and other measurable characteristics (Davenport 1900). However, even though the Station for Experimental Evolution in the first year of operation was breeding jungle fowl, domestic fowl, ducks, pigeons, canaries, goldfinches, goats, sheep, cattle, cats, numerous invertebrates, and even more varieties of plants, mice were not included. Castle was mentioned in 1909 as an "associate" of the Station, but it was not until 1914 that former Bussey Institution scientist E. Carleton MacDowell (1887–1973) moved to Cold Spring Harbor (Rader 1997). Even then, mice did not figure in research at the Station, because MacDowell began research on the possible mutagenic effects on their offspring of alcohol given to pregnant rats. This project lasted over eight years and ended in disappointment; the effects he observed were due to the physiological effects of the alcohol on the mothers rather than on the genes of the fetuses.

In 1919 Clarence Little (1888–1971) came to Cold Spring Harbor following his service in the First World War (Snell 1975; Russell 1978; Rader 1997). Prior to the war, he had developed inbred mouse strains and, with Tyzzer, had carried out experiments on transplantation of tumors in mice (Little and Tyzzer 1916). At Cold Spring Harbor, Little published findings on coat color in dogs and sex-linked lethals in mice and man, but his main research continued to be on cancer, and the paper published here is typical of his research (Little 1920). Little had originally postulated, by analogy with inheritance of coat color in mice, that 12 to 14 "Mendelizing factors" were necessary to account for the inheritance pattern of susceptibility to the J.W.A. carcinoma. Here he describes findings with the J.W.B. sarcoma that indicate only 4 factors are involved, none of which are carried on the X-chromosome. It was while at Cold Spring Harbor that Little developed the C57BL inbred strain. This was derived from female 57 mated with male 52, both of which probably originated from Lathrop. All did not go well with Little's mice at Cold Spring Harbor. His colony of DBA mice was wiped out by mouse typhoid and it was only Strong's heroic efforts that led to the re-establishment of the strain from a single breeding pair (Strong 1978).

Little became assistant director to Davenport in 1921 but left in July, 1921, apparently following disagreements with Davenport over his (Little's) managerial style (Rader 1997). He went on to become the president of the University of Maine, but after only three years he moved to the University of Michigan as its president. However, his time there was short. A "stormy term" and "brilliant but tactless" are the phrases used in the official university history to describe Little's presidency. In 1929, he seems to have created something of a scandal when he divorced his wife of eighteen years and left her and their three children (Snell 1975). In 1929 also, he moved to Bar Harbor, Maine, to establish what became the Jackson Laboratory.

By 1922, MacDowell had abandoned both his studies of the mutagenic effects of alcohol on rats and a project begun two years earlier to examine inheritance of behavior in dogs. That year, he considered using mice to study blood group inheritance, but he failed to develop satisfactory tests and abandoned that project as well. For the next few years, MacDowell used mice for various studies, but a significant change occurred in 1928 when, for the first time in annual reports of the Department of Genetics, there was a description of new mutants in the mouse colony. There was also a brief reference to MacDowell's beginning a genetic analysis of leukemia in mice. It is clear that mouse genetics then became MacDowell's major preoccupation to the exclusion of other projects, such that by 1930, his work on leukemia in mice takes three pages in the annual report. In the paper reprinted here (Richter and MacDowell 1930), he describes the incidence of leukemia in an inbred strain of mice, C58. This was developed by MacDowell from crossing male 52 (the same used by Little to establish C57BL) with female 58. Almost 90% of C58 mice developed leukemia. Furthermore, Richter and he showed that the tumors could be transferred by cell-grafts within the strain but not to other strains of mice.

Similar inbred strains with high cancer incidence were developed by others—for example, C3H and A (mammary tumors) by Strong, and Ak (leukemia) by Furth. Genetic analysis showed that the "inheritance" of these susceptibilities to cancer was not in accord with Mendelian laws, reinforcing the suspicion that a virus might be to blame (Gross 1983a,b). However, MacDowell made many attempts to transmit leukemia with cell-free extracts, but these failed, as did Furth's experiments using his Ak line. Finally, Ludwik Gross injected cell-free extracts of Ak mouse tissue into newborn C3H mice; 12 of 14 injected mice developed leukemia (Gross 1951). Mammary tumors were also shown to be caused by a virus. Breeding experiments using DBA (high incidence of mammary tumors) and C57 (low incidence) mice showed that the susceptibility to inherit tumors depended on the mother being a DBA mouse. Further careful work by Bittner revealed that the causative agent (known first as "milk factor" and later as mouse mammary tumor virus) was transmitted in the mother's milk.

Keeler (1978) wrote that "If mouse genetics was born at the Bussey, it was cradled at Cold Spring Harbor ... ." To understand why this was so, we must look beyond the published record of mouse research at Cold Spring Harbor which, while laudable, does not appear to justify Keeler's remark. (For a detailed description of Cold Spring Harbor's role, see Rader 1997. There are also many charming anecdotes in the historical essays in Morse 1978b.) Both MacDowell and Little made Cold Spring Harbor a hub of informal exchanges and interactions between mouse geneticists. In particular, colleagues were invited to come as summer researchers; for example, L.C. Strong assisted Little in the summers of 1919 and 1920. Little instigated the "Mouse Club of America" whose membership eventually reached about 100. The mouse geneticists, like their maize and *Drosophila* counterparts, began an informal newsletter that fulfilled much the same purposes, listing strains and providing practical tips on maintaining mouse colonies. The address of the newsletter was given as "Mouse Headquarters, The Animal House, Cold Spring Harbor, New York!"

By 1922, Cold Spring Harbor was becoming increasingly important to the mouse world. Davenport wrote that because the Department had "...come to be the gathering place of many mouse geneticists during the summer...", it had been asked to assume the role of a central agency for maintaining mutant strains. In 1922, MacDowell—now feeling free of Little—asked Davenport for extra funds to expand the mouse colony at the Department of Genetics, writing that the research on mice was benefiting from having arisen in "...a cooperative spirit that I wish to foster in every possible way." Ironically, a few years later, after he had discovered the high level of leukemia in C58, MacDowell became notorious for not being cooperative, refusing to share C58 with other researchers (MacDowell 1932). It is said—perhaps unkindly—that he hoped that C58 would be useful in studies of alcoholism, a topic of special interest to the wealthy neighbors who supported his research (Ardevant quoted in Morse 1981). Furth was obliged to make his own high leukemic Ak line as a consequence. MacDowell remained at Cold Spring Harbor for the rest of his life. His last two papers on mouse leukemia—numbers XV and XVI in the series—were published in 1955, the year that he retired.

Clarence Cook Little was born in Brookline, Massachusetts on October 6, 1888. He was an undergraduate at Harvard University, 1906–1910, and was awarded his D.Sc.in 1914, following research with William E. Castle at the Bussey Institute. Little joined the U.S. Army for World War I and attained the rank of major. In 1918, he accepted an appointment at the Carnegie Institution of Washington's Department of Genetics at Cold Spring Harbor. In 1922 he became President of the University of Maine but in 1925 moved to become President of the University of Michigan. A group of wealthy Detroit industrialists offered to support a mammalian genetics research institute at Bar Harbor and what became The Roscoe B. Jackson Laboratory was founded in 1929. Little retired in 1955 but accepted a highly controversial post as the first scientific director of the Tobacco Industrial Research Committee, despite having been a two-term President of the American Cancer Society. Little was also President of the American Eugenics Society, the American Birth Control League, and the American Euthanasia Society. He was elected a member of the National Academy of Sciences in 1945. He died on December 22, 1971.

E. Carleton MacDowell was born in 1887. He was an undergraduate at Swarthmore College, graduating in 1909, and then gained his Ph.D. in 1912 at the Bussey Institute, Harvard, where he was a fellow student with Little. After teaching biology at Dartmouth and Yale, MacDowell came to the Department of Genetics at Cold Spring Harbor in 1914. He remained at Cold Spring Harbor for the rest of his career. MacDowell died on November 7, 1973.

Browne J. 1995. *Charles Darwin: Voyaging*, pp. 521–525. Alfred A. Knopf, New York

Davenport C.B. 1900. Investigations at Cold Spring Harbor. *Science* **12**: 371–373.

Dunn L.C. 1965. William Earnest Castle. *Biogr. Mem. Natl. Acad. Sci.* **38**: 34–80.

Gross L. 1951. Pathogenic properties, and "vertical" transmission of the mouse leukemia agent. *Proc. Soc. Exp. Biol. Med.* **78**: 342.

————. 1983a. The development of inbred strains of mice, and its impact on experimental cancer research. In *Oncogenic viruses*, 3rd edition, vol. 1, pp. 253–261. Pergamon Press, London.

————. 1983b. Mouse leukemia. In *Oncogenic viruses*, 3rd edition, vol. 1, pp. 305–458. Pergamon Press, London.

Keeler C. 1978. How it began. In *Origins of inbred mice* (ed. H.C. Morse III), pp. 179–193. Academic Press, New York.

Little C.C. 1920. The heredity of susceptibility to a transplantable sarcoma (J.W.B.) of the Japanese waltzing mouse. *Science* **51**: 467–468.

Little C.C. and Tyzzer E.E. 1916. Further experimental studies on the inheritance of susceptibility to a transplantable carcinoma (JWA) of the Japanese waltzing mouse. *J. Med. Res.* **33**: 393–427.

MacDowell E.C. 1932. Letter to C.B. Davenport in regard to supplying mice to Whitney, dated 9/15/32. Cold Spring Harbor Laboratory Archives, Cold Spring Harbor, New York.

Morse H.C. III, ed. 1978a. Introduction. In *Origins of inbred mice*, pp. 3–21. Academic Press, New York.

————, ed. 1978b. *Origins of inbred mice*. Academic Press, New York.

————. 1981. The laboratory mouse–A historical perspective. In *The mouse in biomedical research* (ed. H.L. Foster et al.), vol. 1, pp. 1–16. Academic Press, New York.

————. 1985. The Bussey Institute and the early days of mammalian genetics. *Immunogenetics* **21**: 109–116.

Rader K.A. 1997. The origins of mouse genetics—Beyond the Bussey Institution. I. Cold Spring Harbor: The Station for Experimental Evolution and the "Mouse Club of America". *Mamm. Genome* **8**: 464–466.

————. 1998. "The mouse people": Murine genetics work at the Bussey Institution, 1909–1936. *J. Hist. Biol.* **31**: 327–354.

Richter M.N. and MacDowell E.C. 1930. Studies on leukemia in mice: I: The experimental transmission of leukemia. *J. Exp. Med.* **51**: 659–673.

Russell E.R. 1978. Origins and history of mouse inbred strains: Contributions of Clarence Cook Little. In *Origins of inbred mice* (ed. H.C. Morse III), pp. 33–43. Academic Press, New York.

Snell G.D. 1975. Clarence Cook Little. *Biogr. Mem. Natl. Acad. Sci.* **46**: 241–263.

Strong L.C. 1978. Inbred mice in science. In *Origins of inbred mice* (ed. H.C. Morse III), pp. 45–67. Academic Press, New York.

[*Reprinted from* Science, *Vol. LI, No 1323, Pages 467–468, May 7, 1920*]

THE HEREDITY OF SUSCEPTIBIL-ITY TO A TRANSPLANTABLE SARCOMA (J. W. B.) OF THE JAPANESE WALTZING MOUSE

2

In 1916[1] the writer in collaboration with Tyzzer reported on the inheritance of susceptibility to a transplantable carcinoma (J. W. A.) of the Japanese waltzing mouse. This tumor grew in one hundred per cent. of the Japanese waltzing mice inoculated and in zero per cent. of the common non-waltzing mice. When these two races were crossed, the F_1 generation hybrids showed sixty-one out of sixty-two mice to be susceptible. In these mice growth was as rapid if not more so than in the Japanese waltzing mice themselves. The one exception may well have been due to faulty technique for a reinoculation test was not made.

The F_2 generation gave a very interesting result—only three out of 183 mice grew the tumor. At that time the results were explained on the basis of multiple Mendelizing factors[2] whose number was estimated at from twelve to fourteen. *Simultaneous presence* of these factors, themselves introduced by the Japanese waltzing race, was considered necessary for progressive growth of the tumor. The analogy between this case and that of coat color in wild mice, dependent upon the simultaneous presence of at least five known Mendelizing factors was at that time pointed out.

Later[3] while working with a transplantable sarcoma (J. W. B.) of the Japanese waltzing mouse, results were obtained which showed what semed to be a somewhat simpler quantitative condition of the same process. In this case, the parent races and F_1 hybrids behaved as before, but the F_2 hybrids gave a total of twenty-three susceptible, to sixty-six non-susceptible animals. It was previously estimated that from five to seven factors were involved. In order to determine more closely the number of factors, new experiments were devised as follows: F_1 hybrid mice themselves susceptible were crossed back with the non-susceptible parent race. This has recently given a back cross generation whose susceptibility

[1] Little, C. C., and Tyzzer, E. E., 1916, *Jour. Med. Research*, 33: 393.

[2] Little, C. C., Science, N. S., 1914, 40, 904.

[3] Tyzzer, E. E., and Little, C. C., 1916, *Jour. Cancer Research*, 1: 387, 388.

3

would depend upon the factors introduced through the gametes received from their F_1 parent. If one factor was involved, the ratio of gametes containing it formed by the F_1 animal, to those lacking it would be 1:1, if two factors, 1:3; if three factors 1:7; if four factors, 1:15; if five factors, 1:31; if six factors, 1:63; and if seven factors, 1:127. Susceptible and non-susceptible *individuals* would occur in the back cross generation in similar proportions.

The actual numbers obtained were twenty one susceptible to 208 non-susceptible. This result may be compared with expectations on three, four, five, and seven factor hypotheses, as follows:

	Susceptible	Non-susceptible	Ratio
Expected 3 factor....	28	201	1:7
Observed	21	208	1:99
Expected 4 factor....	14	215	1:15
Expected 5 factor....	7	222	1:31
Expected 7 factor....	1.8	227.2	1:127

The observed figures fall between the three and four factor hypothesis. The numbers are not large enough to give a definite test, but the F_2 generation already mentioned is interesting as a supporting line of evidence. If we compare this with the expectation, we find that the observed figures lie between the

	Susceptible	Non-susceptible	Ratio
Expected 3 factor....	39	50	1:1.3
Expected 4 factor....	29	60	1:2.1
Observed	23	66	1:2.8
Expected 5 factor....	21	68	1:3.2

four and five factor hypothesis. In both cases the four factor hypothesis figures are close and the three and five factor hypothesis are to be still considered as possibilities, though not probabilities. The six and seven factor hypotheses appear to be definitely eliminated.

The non-susceptible back cross animals which should by the multiple factor hypothesis contain in many cases part, but not all, of the factors for susceptibility are being tested by breeding back with the F_1 animals. If four factors are involved, as seems likely, of every

4

fifteen such back cross animals approximately four or 26.6 per cent. should have three; six or 40 per cent. two; four or 26.6 per cent. one; and one or 6.6 per cent. none of the four factors necessary for continued growth of the tumor. When crossed with F_1 animals these back cross types should give the following ratios of susceptible to non-susceptible animals in their progeny.

Type of Back Cross	Ratio of Susceptible to Non-Susceptible Progeny
Having three factors	1 : 3.7
" two factors	1 : 6.1
" one factor	1 : 9.7
" zero factors	1 : 15

The first two categories should be easily recognizable and together form 66.7 per cent. of the back cross animals. Such tests have now been begun.

The sex chromosome has been eliminated as a probable carrier of any of the four factors as follows. If mice like other mammals have the female XX and the male XY in formula, the use of susceptible Japanese waltzing males to form the F_1 animals used, gives daughters carrying his X, and sons his Y chromosome. If now his *sons* only are used to produce the back cross generation by mating with common non-susceptible females, all the X chromosomes in the resulting animals will be derived from common non-susceptible mice. Unless therefore, crossing over between the X and Y chromosomes occurs frequently, any susceptibility factor borne in the X chromosomes of the original Japanese waltzing males used, has been eliminated.

While further investigations are in progress, we may conclude provisionally that:

1. From three to five factors—probably four—are involved in determining susceptibility to the mouse sarcoma J. W. B.

2. That for susceptibility the simultaneous presence of these factors is necessary.

3. That none of these factors is carried in the sex (X) chromosome.

4. That these factors Mendelize independently of one another.   C. C. LITTLE

[Reprinted from The Journal of Experimental Medicine, April 1, 1930,
Vol. 51, No. 4, pp. 659–673]

STUDIES ON LEUKEMIA IN MICE*

I. The Experimental Transmission of Leukemia

By MAURICE N. RICHTER, M.D., and E. C. MacDOWELL, S.D.

(*From the Department of Pathology of the College of Physicians and Surgeons,
Columbia University, New York, and the Department of Genetics, Carnegie
Institution of Washington, Cold Spring Harbor*)

(Received for publication, February 6, 1930)

In a strain of mice designated hereafter as C58, which has been inbred by brother-sister matings since 1921, it was observed that a considerable number of animals that lived more than 8 months had lymphatic leukemia.** The present cooperative investigation was undertaken early in 1928, since which time the mice have regularly been autopsied and the lesions examined microscopically.

In a preliminary report (1), we recorded the existence of this leukemic strain, and the fact that the leukemia is transmissible to other mice of the same strain by inoculation with tissue emulsions, at an earlier age than that at which leukemia occurs spontaneously.

LITERATURE

A complete account of the recorded attempts to produce or to reproduce the lesions of leukemia is not within the scope of this paper. Briefly, it may be stated that attempts to transmit leukemia from man to man, from man to lower animals, or from animals of one species to those of another, have been unsuccessful. These experiments have been reviewed recently by Opie (2).

The first and most widely known transmissible leukemia is that which occurs spontaneously in the fowl, and which Ellermann and Bang (3) and later Hirschfeld and Jacoby (4) and Schmeisser (5) transmitted to other fowl by inoculation with organ emulsions. Ellermann and Bang also transmitted the disease by inoculation

---

* This investigation was supported by a grant from the Carnegie Corporation and an appropriation for technical assistance from the Research Fund of Columbia University.
** The diagnosis was first made by Dr. Alwin M. Pappenheimer in 1926.

660   LEUKEMIA IN MICE.   I

with cell free filtrates.   Both lymphoid and myeloid leukemia were thus transmitted, and change of type was noted in some of Ellermann's experiments.   Fowl
paralysis (neurolymphomatosis gallinarum), although not identical with leukemia,
has points in common, and is transmissible (6).

The first report of successful transmission of mammalian leukemia is that of
Snijders (7), who inoculated guinea pigs with tissue emulsions of a guinea pig which
had spontaneous leukemia.   The lesions produced were comparable to those of the
spontaneous case, and were transmissible by subsequent transfers.   Not only
were typical leukemias encountered, but also cases of aleukemic lymphadenosis,
leukosarcomatosis, and several almost pure tumors.   Snijders considered his
results as evidence favoring the neoplastic theory of leukemia.   Mettam (8)
has found conditions resembling leukemia in cattle (East coast fever and snot-
siekte) to be transmissible.

In mice both lymphoid and myeloid leukemia have been observed to occur
spontaneously, notably by Tyzzer (9), Haaland (10), Levaditi (11), Slye (12),
Simonds (13), and Hill (14).   Tyzzer and Haaland attempted to transmit the
disease by inoculation, but without success.

Since we reported the transmission of leukemia in Strain C58, Korteweg (15)
has recorded the transmission of "leukosarcomatosis" in mice.   Korteweg's
results are very similar to our own and, as will be seen, the "leukosarcomatoses"
and "leukemias" which occurred in the course of our experiments are interpreted
as varieties of the same disease.   Korteweg found leukosarcomatosis to behave, on
transfer, like a true tumor.   His material differs from ours, however, in that the
mice used for inoculation were not from a single strain, and were thus very different
genetically.   Furthermore, all of his material came originally from the tumor of
a single mouse.

Spontaneous Leukemia in Strain C58

The cases of lymphatic leukemia and related conditions which
have been observed in Strain C58 are similar in their anatomical and
microscopical features to those reported by other observers.   The
varieties observed by Simonds (13) in the Slye stock have also been
found in Strain C58.   It will be necessary, however, to note certain
general features of the disease as it has occurred in this strain.

Incidence.—A large proportion of mice of both sexes have had leukemia.   Of 155 mice that died since May, 1928, 137 had leukemia.
This, however, is not a final result as many mice are still alive.   There
are wide variations in the age of incidence and some animals have
died at relatively early ages.   The figures should not, therefore, be
read in terms of percentage.   An account of the history of leukemia
in Strain C58 will be published in a subsequent report.

MAURICE N. RICHTER AND E. C. MacDOWELL   661

Age Incidence.—In 137 cases of spontaneous leukemia that died since May, 1928, the age at death is given in Table I.   The youngest animal in this series was 186 days old.   104 or 76 per cent of the cases occurred between the 8th and 12th months; 18 or 13 per cent were less than 8 months old, and 15 or 11 per cent were over 1 year.

During the same interval, 18 other mice died and 4 were killed.   Of these, 6 were diagnosed as "doubtful" (usually because of extreme post-mortem decomposition), and 16 as "not leukemia."

Occurrence of Other Diseases.—In the early history of the strain, many animals were discarded before the age at which neoplasms are commonly found.   Since March, 1928, at which time discarding of animals was discontinued and necropsies regularly performed, a few neoplasms have been found other than those of the leukemic group under discussion.   These include one case of epithelioma of the skin,

TABLE I

	Age in months													
	6	7	8	9	10	11	12	13	14	15	16	17	18	19
No. of animals...............	5	13	19	35	24	26	5	2	2	2	2	1	0	1

one of osteogenic sarcoma and two of carcinoma of the lung, all of which were associated with leukemia.   There were also one case of chondro-osteosarcoma and one salivary gland tumor.   Of other diseases, none has been constant.   External parasites have occasionally appeared in the colony, and a few cestodes have been found.

Diagnosis.—The final diagnosis has, in each case, been made by gross and microscopic examination of the organs at autopsy.   In typical cases the lesions are sufficiently characteristic to enable accurate diagnoses to be made by gross findings alone, although sections have been regularly examined.   In only two cases were diagnoses of leukemia made at autopsy changed after examinations of the sections, but a few unsuspected cases of leukemia in early stages were found microscopically.

It is convenient to recognize the disease during life, and this we have found possible in a very large number of cases.   In general, enlargement of the spleen is the first symptom noted, and can be recognized by palpation of its lower free end and usually distinct edges. The number of cases detected in early stages has been roughly proportional to the regularity of periodic examinations of spleens in the colony. Although splenomegaly is also found in conditions other than leukemia, we have encountered but few such instances in these mice. There is also emaciation, causing the pelvic bones to become prominent, the nose sharp, and the vertebrae of the tail distinct. The abdomen becomes distended and feels firm and heavy in spite of the emaciation. The animal is less active and its movements slow. The presence of large intrathoracic tumors such as occur in the "leukosarcomatoses" has frequently been diagnosed by the occurrence of labored

TABLE II

Leucocyte Counts on 250 Normal Mice of Strain C58

White cell count	No. of counts
5–10,000 per cmm.	81
10–15,000 " "	110
15–20,000 " "	64
20–25,000 " "	27
25–30,000 " "	5
30–35,000 " "	2
35–40,000 " "	1

breathing. Blood examinations often give valuable confirmatory evidence of the presence of leukemia, but the absence of a typical blood picture does not necessarily indicate the absence of leukemic changes in the tissues. The data in Table II of leucocyte counts in mice of Strain C58, compiled from 290 counts on 250 normal, uninoculated mice between the ages of 1 and 4 months may be used for comparison with the counts on cases of spontaneous and transmitted leukemia.

All of these mice were subsequently used for inoculation. From the table it will be seen that in only 3 of the 290 examinations were there more than 30,000 white blood cells per cubic millimeter. Blood smears were not examined in every instance, but neither in those which were examined nor in the counting chamber when the leucocytes were counted were abnormal cells observed. In addition. smears of the blood of 175 mice less than 6 months of age, which were made in the spring of 1928, failed to show abnormal cells resembling those in the typical leukemic cases.   Many of these animals later developed leukemia spontaneously.

In 131 cases of spontaneous leukemia, the distribution of the leucocyte count is given in Table III, the highest count for each mouse being used.

Of these, 6 were over 300,000 and 2 over 600,000.   Thus in 22.9 per cent of these cases, the leucocyte count was never found to be above 30,000 per cubic millimeter, a figure occasionally reached in normal mice.   In several cases with counts below 30,000, abnormal cells were seen in smears or in the counting chamber.

Of the 100 mice that at some time had leucocyte counts of 30,000

TABLE III

Leucocyte Counts in Spontaneous Leukemia

White cell count	No. of animals
under 30,000 per cmm.	31
30–50,000 " "	35
50–100,000 " "	35
over 100,000 " "	30

or more, 33 subsequently had diminution of the leucocytes, which in 18 cases were then less than 30,000.   A few of these changes are, for example: from 66,000 to 11,000; 75,000 to 25,000; 354,000 to 31,900; 184,000 to 60,000; 229,000 to 3800 with subsequent rise to 46,000.

The abnormal cells which appear in the blood in typical cases of leukemia are larger than the normal mouse lymphocyte.   The nuclei are large, usually round, and leptochromatic.   When stained by one of the Romanowsky methods a nuclear network of chromatin is demonstrated, which is somewhat less dense than in the normal lymphocytes, though in both of these types of cells the structure is slightly more coarse than in the corresponding human leucocytes. Several nucleoli may be present, but are not constant.   The cytoplasm is basophilic, frequently with small clear areas near the nucleus, or with small vacuoles.   Specific granules are not present, but azure

664 LEUKEMIA IN MICE. I

granules are occasionally found. Mitoses have been observed. Many cells are intermediate in type between these immature forms and the normal lymphocytes.

Course.—No instance of recovery from spontaneous leukemia has been observed in this strain. The disease is progressive, but variations have been observed in the size of enlarged lymph nodes. There is increasing enlargement of the spleen, weakness, emaciation, and frequently marked anemia. In several cases the course was protracted, lasting in one instance 10 months after enlargement of the spleen was first noticed.

Varieties of Manifestation, and Terminology.—It is known that both myelogenous and lymphatic leukemia occur in mice, as well as related lymphoid hyperplastic and neoplastic conditions, which offer difficulties in classification. Some of the latter correspond to the "pseudoleukemias," "leukosarcomatoses" and related conditions as observed in man. Hodgkin's disease appears to be an exception, for but five cases have been reported in mice (13, 16).

In Strain C58 the cases of leukemia which have been examined since the spring of 1928 have been only of the lymphatic type. Among these are mice showing the typical lesions of lymphatic leukemia including a characteristic blood picture, and also, in smaller numbers, those which would correspond to cases of "leukosarcomatosis" or "pseudoleukemia," as described by Simonds (13). In all cases, however, the type of cell comprising the infiltrations is of lymphoid nature. The similarity of these infiltrations, the variations in the blood picture and the occurrence of cases intermediate in type between leukemia and leukosarcomatosis or lymphosarcoma, give one the impression that these different anatomical varieties are but variations in the distribution of lesions rather than differences of fundamental nature.

The distinction between "leukemia" and "pseudoleukemia," the latter term referring to Cohnheim's designation of that condition in which the gross and microscopic lesions are characteristic of leukemia, but in which the blood picture does not show leukemic changes, should not, in our opinion, be made in mice of this strain. Repeated blood examinations in mice with palpably enlarged spleens have shown that:

(1). Splenic enlargement frequently precedes the appearance of a leukemic blood picture;

MAURICE N. RICHTER AND E. C. MacDOWELL   665

(2) A leukemic blood picture usually occurs at some time during the course of the disease;

(3) The blood changes may be a very late manifestation;

(4) A leukemic blood picture is not always permanent, but may approach or reach normal without recovery or corresponding clinical changes.

In view of these variations in the blood picture, it is evident that the classification of a particular case as "leukemia" or "pseudoleukemia" depends on the stage at which death occurred, and that the blood picture is an inconstant symptom rather than an indication of fundamental differences. Likewise there are cases with lesions the distribution of which is intermediate between the diffuse changes of leukemia and the local growths of "leukosarcomatosis" or "lymphosarcoma." We therefore prefer to classify in one group all of those conditions which we believe to be fundamentally of the same nature. For this group the term "leukemia" is selected, not because it is literally descriptive, but because it is widely used and generally understood. "Pseudoleukemia," "leukosarcomatosis" and similar terms represent, as far as these mice are concerned, unimportant though interesting variations in the extent or distribution of the lesions.

Experimental Transmission of Leukemia in Strain C58

The first of our experiments on the transmissibility of leukemia in this strain was performed as follows:

A female mouse (50420) not previously used in any experiment was first observed to have a large spleen at the age of 7½ months. At 10 months the spleen was larger, and the circulating white blood cells numbered 694,000 per cubic millimeter, of which the great majority were lymphoid cells of immature type. The mouse was killed, the spleen removed aseptically, and a portion of it minced in sterile saline. Examination of the organs at autopsy showed enlargement of the spleen and liver and of the cervical, axillary, inguinal and mesenteric lymph nodes. The diagnosis of lymphatic leukemia was later confirmed microscopically.

The saline emulsion of spleen was immediately inoculated intraperitoneally into 8 mice of Strain C58 which were 4 weeks of age.

Following the inoculation, the leucocytes were counted at intervals and were found in each case to increase in number, reaching figures higher than are found in many cases of leukemia. The highest count found in this experiment was 262,000 about 3 weeks after inoculation. The increase in the number of white cells

666   LEUKEMIA IN MICE.   I

appeared to be due to the presence in the blood stream of immature lymphocytes comparable to those of the spontaneous case.

Two of the 8 mice were killed at 17 and 21 days respectively after inoculation, while the rest were allowed to live until death occurred spontaneously in 32 to 35 days after inoculation.   The gross appearance of Mouse 53983 shows the type of lesion found in this particular experiment, though it is not characteristic of all:

The spleen measured 1.9 x 0.5 x 0.3 cm.   It was of normal color, the follicles were distinct but not larger than normal.

At the site of inoculation in the anterior abdominal wall was a tumor about 1 x 0.8 cm. in diameter.   Near it and on the opposite side of the abdomen, were smaller nodules in the abdominal wall.   The cervical nodes were enlarged, slightly greater on the right side.   Axillary and inguinal nodes were but slightly enlarged. In the mesentery were many nodules, each less than 0.2 cm. in diameter.   A mass of lymphoid tissue was behind the pyloric portion of the stomach, between it and the liver.   There was also infiltration of the perirenal fat, forming tumor-like masses on either side of the vertebral column, mesial to the kidneys.

The liver, lungs, kidneys, thymus and mediastinal lymph nodes were normal.

Microscopic examination showed the lesions to be composed of lymphoid cells of abnormal type, with many mitoses.   Smears of the tissues show that these are the same type of cell described above in the blood of leukemic mice.

It is significant that in this and subsequent experiments, the lesions of leukemia were obtained in mice dying of the disease at an earlier age than that at which any mouse has been observed to have leukemia spontaneously, either in this strain or in any other recorded in the literature.

From one of the animals in this experiment, the spleen was removed aseptically, a saline emulsion prepared, and other mice of Strain C58* inoculated.   In this manner, by the inoculation of tissue emulsions** the disease has been repeatedly transferred, and is now in its 30th passage.   The experimental leukemia of this series, originating from Mouse 50420 is designated as "Line A," the first transfer as A1, the second as A2, etc.

* We have not been successful in transmitting the disease to mice of other strains.

** Our usual procedure has been to inoculate intraperitoneally with emulsion of spleen or lymph node.   Positive results have also been obtained after the inoculation of ascitic fluid, which in some experiments has been abundant, with heart-blood and with liver.   Subcutaneous inoculation is also successful, but the distribution of the lesions is somewhat different.

MAURICE N. RICHTER AND E. C. MacDOWELL   667

The experiment was repeated, using tissues of other mice with spontaneous leukemia, and the experimental leukemias thus obtained are designated as "Line B," "Line C," etc.   In all, 10 spontaneous cases have been used as donors, with successful transmission in each. Some of these lines have been allowed to die out. Four lines are now used for continued transmission and the current experiment in each is A30, E10, H10 and I9.

The lesions found in Experiment A1 are fundamentally the same as those of subsequent experiments, but variations have been observed in the distribution of the lesions, the degree to which individual organs are involved, and the occurrence of leukemic blood pictures. These variations are comparable to similar differences in distribution of the lesions observed in the spontaneous cases. In each of the experimental lines now being studied, minor differences in the distribution of the lesions, the type of reaction and the interval between inoculation and death are observed which, though not perfectly constant, are sufficiently characteristic to indicate the presence of differences in the inocula. Thus, after intraperitoneal inoculation, Line A is now characterized by the nearly constant presence of ascites; Line E by the relatively frequent occurrence of chest tumors with fluid in the pleural cavity, but infrequent ascites; Line H by somewhat higher leucocyte counts and marked infiltrations in the liver, without chest tumors and only rarely with ascites; and Line I by marked infiltrations in the kidneys and liver, with high leucocyte counts, but without chest tumors or ascites.

In the later transfers of each experimental line there has been a decrease in the average interval between inoculation and death, and, at the same time, less variation among animals inoculated with the same material. This enables us to obtain, at present, fairly uniform results after inoculation, with each experimental line presenting its individual characteristics. An abstract of these results is appended.

Microscopically the lesions found in different experiments are similar, in that they all consist of the presence of numerous large cells of lymphoid type with frequent mitoses. These cells appear in the blood stream and infiltrate distant organs or tissues. In some experiments the cells have infiltrated the liver, spleen, kidneys and pancreas, producing the same picture as in spontaneous leukemia. In

668   LEUKEMIA IN MICE.   I

Experiment	No. of mice	Positive results	Negative results	Average period of survival (days)*
		Line A		
A1	8	8	0	33.5 (6)
A2	4	1	3	(a)
A3	25	20	5	28.19 (16)
A4a (b)	8	8	0	19.3 (6)
A4b	8	8	0	21.6 (7)
A4c	6	5	1	30.25 (4)
A5a (c)	6	6	0	12.6 (5)
A5b (c)	3	3	0	23.0 (2)
A6a (d)	4	4	0	16.25 (4)
A6b (e)	4	4	0	16.7 (3)
A6c (e)	4	3	1	22.0 (2)
A7a (f)	4	4	0	15.5 (4)
A7b (g)	4	4	0	20.3 (3)
A8	4	4	0	13.7 (3)
A9	4	2	2	(h)
A10	4	4 (i)	0	29.5 (2)
A11	4	2	2 (j)	20.0 (1)
A12	4	4	0	10.5 (2)
A13a	4	4	0	9.0 (3)
A13b	4	4	0	10.0 (4)
A14 (k)	8	8	0	10.8 (7)
A15	8	8	0	12.0 (7)
A16	4	4	0	15.7 (3)
A17	4	4	0	10.0 (3)
A18	8	8	0	9.4 (7)

* The figures in parentheses indicate the number of positive cases from which the average is calculated.   Animals that were killed are not included.

(a) The one positive case was killed 46 days after inoculation.   One of the negatives died at 4 days.

(b) Three mice of A3 were killed and used as donors, hence the subdivisions of A4.   Similarly for A5, A6, A7 and A13.

(c) Donor from A4a.

(d) Donor from A5a.

(e) Donor from A5b.

(f) Donor from A6b.

(g) Donor from A6c.

(h) One killed at 15 days was positive.   One died at 141 days (age 206 days), may be a spontaneous case.

(i) One died at 147 days (age 214 days) may be a spontaneous case.   Not counted in average duration.

(j) One died 5 days after inoculation with negative result.

(k) Donor from A13a.

Experiment	No. of mice	Positive results	Negative results	Average period of survival (days)
		Line A—Concluded		
A19	12	10	2	13.4 (9)
A20	12	12	0	10.5 (11)
A21	12	12	0	10.5 (11)
A22	12	12	0	13.0 (11)
A23	8	8	0	13.0 (7)
A24	5	5	0	7.25 (4)
A25	8	7	1 (l)	11.6 (5)
A26	8	6	2	10.75 (5)
A27	8	8	0	13.7 (7)
A28	6	6	0	20.0 (4)
A29	8	8	0	14.1 (6)
A30	4	4	0	9.0 (2)
		Line B		
B1	12	5	7 (m)	54.5 (4)
B2	8	8	0 (n)	12.7 (6)
B3	8	2	6 (o)	17.0 (1)
B4	5	5 (p)	0	9.0 (3)
		Discontinued		
		Line C		
C1	7	6	1	58.1 (6)
		Discontinued		
		Line D		
D1	12	3	9	28.0 (2)
D2	5	4	1	17.3 (3)
D3	6	6	0	18.3 (6)
		Discontinued		

(l) The negative died three days after inoculation.

(m) Four negatives died in less than 4 days.

(n) In one case diagnosis is not certain.   Nodes and spleen were enlarged, but animal was decomposed and partly eaten by other mice.

(o) The 6 negatives died 1 day after inoculation.

(p) One case doubtful because of post mortem decomposition.

670   LEUKEMIA IN MICE.   I

Experiment	No. of mice	Positive results	Negative results	Average period of survival (days)	
Line E					
E1	9	8	1 (q)	66.8	(6)
E2	8	7	1 (r)	57.3	(6)
E3a	25	9	16	36.3	(8)
E3b	6	6	0	50.7	(6)
E4 (s)	4	4	0	51.3	(3)
E5	4	4	0	32.3	(3)
E6	4	4	0	25.0	(3)
E7	8	8	0	22.0	(7)
E8	8	8	0	26.8	(7)
E9	6	6	0	21.6	(5)
E10	8	8	0	19.5	(7)
Line F					
F1	5	4	1	15.3	(3)
F2	3	3	0	21.0	(3)
Discontinued					
Line G					
G1	8	8	0	48.25	(8)
Discontinued					
Line H					
H1	2	1	1 (t)	35.0	(1)
H2	6	6	0	30.5	(4)
H3	4	3	1 (u)	20.5	(2)
H4	7	7	0	16.1	(6)
H5	4	4	0	17.7	(3)
H6	4	4	0	18.7	(3)
H7	8	8	0	15.0	(7)
H8	8	8	0	14.0	(7)
H9	6	6	0	17.5	(5)
H10	6	6	0	18.6	(5)

(q) The negative case died 36 days after inoculation.   Others became ill not less than 49 days.   Two killed at 109 and 121 days were positive.

(r) Negative case died 1 day after inoculation.

(s) Donor from E3a.

(t) The negative case had a large spleen but was too much decomposed to diagnose microscopically.

(u) The negative case died 2 days after inoculation.

MAURICE N. RICHTER AND E. C. MacDOWELL   671

Experiment	No. of mice	Positive results	Negative results	Average period of survival (days)
		Line I		
I1	4	4	0	55.0 (2)
I2a	4	1	3	(v)
I2b	6	6	0	29.0 (6)
I3	4	4	0	36.0 (3)
I4	4	4	0	20.0 (3)
I5	6	6	0	29.6 (5)
I6	8	8	0	25.7 (7)
I7	6	6	0	19.8 (5)
I8	6	6	0	13.4 (5)
I9	6	6	0	10.75 (5)

Line J

(Inoculated at Cold Spring Harbor)

J1	4	3	1	22.0 (3)
		Discontinued		

(v) The one positive result was in an animal killed 14 days after inoculation and used as donor for I3. Emulsion kept on ice 48 hours before use. Two animals died 4 days after inoculation.

other cases the infiltration of these organs may be minimal or absent, yet the mesentery and omentum may have numerous small nodules, the fat tissue of the abdomen may be infiltrated and ascites may be present. These lesions will be reported in detail in a separate paper.

The appended protocols include only those mice of Strain C58 which were inoculated with emulsions of tissues not subjected to any preliminary treatment other than mincing in saline, with the exception of Experiment I3, the material for which was kept in a refrigerator for 48 hours before use. In calculating the average period of survival, only those mice that died of transmitted leukemia are included. The figure in parenthesis after each average indicates the number of mice from which the average is calculated. Except where noted, the inoculations were made in New York.

In addition to these C58 animals, there were inoculated 107 mice from the Storrs-Little strain, 19 of Dilute Brown, 16 Bagg albinos, and 3 of Strain 85, all with negative results.

672 LEUKEMIA IN MICE. I

A series of experiments performed in the laboratory at Cold Spring Harbor has given results similar to the above obtained in New York. A mouse from Experiment A17 and one from A25 were sent to Cold Spring Harbor and used as the donors in subsequent experiments, and the series of transmissions obtained therefrom correspond to those in this line reported in the protocols.

Of 78 mice thus inoculated, 73 developed leukemia, and 5 died in less than 4 days after inoculation.

SUMMARY

Lymphatic leukemia has occurred with great frequency in a particular strain of mice which have been inbred by brother-sister matings since 1921. In addition to typical cases of leukemia are others which, because of the absence of leukemic changes in the blood, correspond to "pseudoleukemia" and others which, by the presence of unusually great enlargement of certain lymph node groups resemble the "leukosarcomatoses" as observed in man.

Examinations of the blood of leukemic mice have shown that leukemic blood pictures are not necessarily early in their appearance, nor are they constant. The blood picture may not, therefore, be used as a criterion for the separation of the two diseases (leukemia and pseudoleukemia) but merely indicates different phases of the same condition. Likewise, cases with lesions intermediate between the local growths of "leukosarcomatosis" and the more general lymphatic enlargements of leukemia suggest that these conditions differ only in the distribution of lesions but not in their nature.

Lymphatic leukemia occurring spontaneously in this strain may be transmitted to other mice of the same strain, and carried, apparently, for an unlimited number of transfers in animals at an earlier age than that at which leukemia occurs spontaneously. In each of 10 such experiments transmissions were obtained. The lesions produced by inoculation correspond to those of spontaneous cases, in that they consist of growths of abnormal lymphoid cells which infiltrate tissues and organs and often appear in the circulating blood. Only minor differences have occurred, some of which are characteristic of certain experimental lines. After repeated transfers, the disease tends to run a more acute course.

MAURICE N. RICHTER AND E. C. MacDOWELL   673

Among the cases in which transmissions occurred, are some without leukemic changes in the blood, and many with local growths at the site of inoculation or in certain node groups. The differences in the blood pictures and distribution of lesions (which latter may be influenced to some extent by the method of inoculation) correspond to similar differences which are sometimes observed in the spontaneous cases.

REFERENCES

1. Richter, M. N., and MacDowell, E. C., *Proc. Soc. Exp. Biol. and Med.*, 1929, **26,** 362.
2. Opie, E. L., *Medicine*, 1928, **7,** 31.
3. Ellermann, V., and Bang, O., *Centralbl. f. Bakteriol.*, 1908, **46,** part 1, 4 and 395.
4. Hirschfeld, H., and Jacoby, M., *Berl. Klin. Wchnschr.*, 1909, **46,** 159.
5. Schmeisser, H. C., *Johns Hopkins Hosp. Repts., Monograph No. 8,* 1915.
6. Pappenheimer, A. M., Dunn, L. C., and Cone, V., *Jr. Exp. Med.*, 1929, **49,** 63. Pappenheimer, A. M., Dunn, L. C., and Seidlin, S. M., *Jr. Exp. Med.*, 1929, **49,** 87.
7. Snijders, E. P., *Nederl. Tijdschr. v. Geneesk.*, 1926, **70,** 1256.
8. Mettam, *9th and 10th Reports, Director of Veterinary Education and Research, Union of S. Africa.* (Quoted by De Kock, G., *Jr. S. African Vet. Med. Ass'n.*, 1928, **1,** 73.)
9. Tyzzer, E. E., *Jr. Med. Res.*, 1907, **17,** 137; 1909, **21,** 479.
10. Haaland, M., *Sci. Rep. Imp. Cancer Res. Fund.*, 1911, No. 4.
11. Levaditi, C., *Compt. Rend. Soc. de Biol.*, 1914, **77,** 258.
12. Slye, M., *Jr. Cancer Res.*, 1926, **10,** 15. Also paper read before the Amer. Ass'n. for Cancer Res., 1928 (to appear in *Jr. Cancer Res.*).
13. Simonds, J. P., *Jr. Cancer Res.*, 1925, **9,** 329.
14. Hill, F. McC., *Australian Jr. Exp. Biol. and Med. Sci.*, 1928, **5,** 89.
15. Korteweg, R., *Ztschr. f. Krebsforsch.*, 1929, **29,** 455.
16. Jobling, J. W., Monograph No. 1, Rockefeller Inst. for Med. Res., 1910, p. 81.

Unstable Genes in *Drosophila*

Demerec M. 1926. **Reddish–A Frequently "Mutating" Character in *Drosophila virilis*.**
(Reprinted, with permission, from *Proc. Natl. Acad. Sci.* **12:** 11–16.)

Milislav Demerec as a young man.

MILISLAV DEMEREC (1895–1966) SPENT 37 YEARS at Cold Spring Harbor. For 19 of these, he was Director of the Biological Laboratory and the Carnegie Institution's Department of Genetics, during a period that was key in the development of Cold Spring Harbor Laboratory. During his tenure, genetics was established definitively as the foundation of research at the Laboratory. This led to Cold Spring Harbor's playing a crucial role in promoting the fledgling field of molecular biology.

Demerec was born in 1895 in Kostajnica in Croatia (Glass 1971; Wallace 1971 [reprinted in this volume, pp. 369–371]). His interest in genetics began immediately following his graduation from the College of Agriculture, Krizevci, when he began work on plant breeding at the Krizevci Experiment Station. His research became dedicated to genetics. Following the First World War, he went to the College of Agriculture in Grignon, France. In 1919, he became a student of R.A. Emerson at the Department of Genetics at Cornell University. Emerson was one of the leaders in the field of corn genetics, and Demerec's first publications were on variegation in maize that Emerson ascribed to mutations in "unstable" genes. Much of Demerec's career focused on searching for the causes of mutations and understanding their nature.

While visiting Woods Hole, Demerec was introduced to *Drosophila* by C.W. Metz, then at the Carnegie Institution's Department of Genetics at Cold Spring Harbor, who brought Demerec to the Department of Genetics in 1923. It was a characteristic of Demerec's research that he was quick to turn to new organisms and tools if they offered him better opportunities for genetic analysis. Thus, Demerec analyzed unstable mutations in *Drosophila virilis* and the larkspur *Delphinium ajacis* when he came to Cold Spring Harbor. Later, in the 1930s, he turned to the study of mutations induced by X-rays in *Drosophila melanogaster*. It was X-rays that were used to produce a mutant strain of *Penicillium* that gave a high yield of penicillin (page 141). Finally, in the late 1940s and throughout his retirement, he worked in the field of bacterial genetics, especially fine structure analysis of the bacterial chromosome using transduction (Hartman 1988) (page 223).

The observations Demerec reported in the paper reprinted here (Demerec 1926a) dealt with the peculiar behavior of a gene, *reddish*, affecting body color in *Drosophila virilis*, a relation of *Drosophila melanogaster*, the standard fly for genetic analysis. *Reddish* was a sex-linked recessive—so carried on the X-chromosome—and Demerec thought that it might be a different version of another mutant with a similar appearance, *yellow*; the fact that *yellow/reddish* heterozygotes are yellow seemed to confirm that. As a further test he crossed these heterozygotes. If *yellow* and *reddish* were different versions of the same gene, no wild-type flies were to be expected. However, to his astonishment, normal colored flies appeared from these crosses. Demerec then did what all good fly geneticists would do—he tried to locate *reddish* on the genetic map. But the data were inconsistent and inexplicable—he could not find its position. Demerec realized that all his data could be accounted for if he assumed that *reddish* was a novel sort of mutation, one that was unstable, mutating at high frequency to give the wild type. Later, Demerec found two other genes that behaved similarly—*miniature-alpha* (Demerec 1926b) and *magenta-alpha* (Demerec 1927)—but he had no explanation for their behavior. It is said that all his *virilis* mutants were lost when he was attending the 9th International Genetics Congress in Edinburgh in 1939, and research on unstable mutants in *Drosophila* lapsed for almost a quarter of a century.

One problem was that no one had found unstable genes in *Drosophila melanogaster*. Demerec had been asked about this in the discussion to his review paper presented at the 1941 Symposium (Demerec 1941). He answered: "I have looked for many years but have found none." It was not until 1965 that Mel Green succeeded where Demerec had failed and found unstable mutants in *D. melanogaster* (Green 1967, 1986). While studying X-ray-induced reversion of mutants known to revert spontaneously, Green discovered an allele of the white gene, w^c, that was highly unstable. A detailed genetic analysis led Green to propose that he was studying a phenomenon similar to McClintock's *Ac-Ds* system in maize (Green 1967, 1969). Molecular analysis in Gerry Rubin's laboratory (Collins and Rubin 1982; Levis et al. 1982) showed that Green's conjecture was correct and, thus, the anomalous behavior of Demerec's "unstable genes" was solved—56 years later!

Quite apart from his research, Demerec showed a flair for organization and he became Director of both institutions on the Cold Spring Harbor site. Two of his most significant recruits were Barbara McClintock and Alfred Hershey. He was also skilled in recognizing new research of great potential, a skill put to good use in selecting topics for the Symposia. Those on the gene (1941, 1946, 1951, 1953, 1956, and 1958) are classics, attended by everyone who was anyone. Participants in the 1953 Symposium included Burnet, Delbrück, Dulbecco, Gajudsek, Hershey, Jacob, Luria, Lwoff, McClintock, Stanley, Tatum, and Watson—all then or future Nobel Laureates.

Demerec made significant contributions to genetics through the *Drosophila Information Service* (*DIS*), based on the *Maize Cooperative Newsletter* established by Marcus Rhoades in 1932. Founded by Demerec and Calvin Bridges in 1934, the *DIS* served as an informal source of information on strains and other resources for *Drosophila* geneticists and enhanced a sense of purpose and community. These were the inspiration for *The Worm Breeder's Gazette* of the *C. elegans* researchers, founded in the mid-1970s. Demerec was also the founding editor of *Advances in Genetics*, which began in 1947, and for 20 years was the only source for reviews of topics in modern genetics.

Finally, Demerec was an enthusiastic supporter of phage genetics, developed initially by Delbrück and Luria (Cairns et al. 1966; Fischer and Lipson 1988). Delbrück and Luria first worked together at Cold Spring Harbor in the summer of 1941 and annually after that. In 1945 they began a series of courses that promoted genetic analysis using phage. The Phage Course continued for 26 years and trained many key players in molecular biology (Susman 1995).

Milislav Demerec was born on January 11, 1895, in Kostajnica in Croatia. He attended the College of Agriculture, Krizevci, graduating in 1916. After three years at the Krizevci Experiment Station, Demerec moved to the College of Agriculture in Grignon, France in 1919 before crossing the Atlantic to Cornell University. Here he completed a Ph.D. in genetics (awarded 1923) and was then recruited to Cold Spring Harbor by Charles Davenport. Demerec was an investigator in the Department of Genetics from 1923, as well as Assistant Director, 1936–1941. On Blakeslee's retirement, Demerec took over as Acting Director until appointed Director in 1943, a post he held until his retirement in 1960. Additionally, Demerec took on directorship of the Biological Laboratory following Eric Ponder in 1941. Demerec continued research as Senior Geneticist at Brookhaven National Laboratory, 1960–1965, and then as Research Professor at C.W. Post College, Long Island University. Among many honors, Demerec was elected a Member of the National Academy of Sciences in 1946 and a Member of the American Philosophical Society in 1952. He played a major role in the Genetics Society of America, notably as President in 1939. Demerec died on April 12, 1966.

Cairns J., Stent G.S., and Watson J.D. 1966. *Phage and the origins of molecular biology.* Cold Spring Harbor Laboratory of Quantitative Biology, Cold Spring Harbor, New York.

Collins M. and Rubin G.M. 1982. Structure of the *Drosophila* mutable allele, *white-crimson*, and its *white-ivory* and wild-type derivatives. *Cell* **30:** 71–79.

Demerec M. 1926a. *Reddish*—A frequently "mutating" character in *Drosophila virilis. Proc. Natl. Acad. Sci.* **12:** 11–16.

———. 1926b. *Miniature-alpha*—A second frequently "mutating" character in *Drosophila virilis. Proc. Natl. Acad. Sci.* **12:** 687–690.

———. 1927. *Magenta-alpha*—A third frequently "mutating" character in *Drosophila virilis. Proc. Natl. Acad. Sci.* **13:** 249–253.

———. 1941. Unstable genes in *Drosophila. Cold Spring Harbor Symp. Quant. Biol.* **9:** 145–150.

Fischer E.P. and Lipson C. 1988. *Thinking about science. Max Delbrück and the origins of molecular biology.* W.W. Norton, New York.

Glass B. 1971. Milislav Demerec. *Biogr. Mem. Natl. Acad. Sci.* **42:** 1–27.

Green M.M. 1967. The genetics of a mutable gene at the white locus of *Drosophila melanogaster. Genetics* **56:** 467–482.

———. 1969. Controlling element mediated transpositions of the *white* gene in *Drosophila melanogaster. Genetics* **61:** 429–441.

———. 1986. Mobile DNA elements in *Drosophila melanogaster:* A retrospective on serendipity. *BioEssays* **4:** 79–82.

Hartman P.E. 1988. Between Novembers: Demerec, Cold Spring Harbor and the gene. *Genetics* **120:** 615–619.

Levis R., Collins M., and Rubin G.M. 1982. FB elements are the common basis of the w^{DZL} and w^c *Drosophila* mutations. *Cell* **30:** 551–565.

Susman M. 1995. The Cold Spring Harbor Phage Course (1965–1970): A 50th anniversary remembrance. *Genetics* **139:** 1101–1106.

Wallace B. 1971. Milislav Demerec 1895–1966. *Genet.* **67:** 1–3. (A collection of memoirs of Demerec was reprinted in *Advance in Genetics* [1971] **16.**)

Reprinted from the Proceedings of the NATIONAL ACADEMY OF SCIENCES,
Vol. 12, No. 1.   January. 1926.

*REDDISH—A FREQUENTLY "MUTATING" CHARACTER IN
DROSOPHILA VIRILIS*

BY M. DEMEREC

DEPARTMENT OF GENETICS, CARNEGIE INSTITUTION OF WASHINGTON, COLD SPRING
HARBOR, N. Y.

Communicated November 17, 1925

The body-color character "reddish" used in the experiments described below was first found in one of five identical pair matings which were back crosses involving several autosomal characters. Half of the males and none of the females from this pair mating were reddish, indicating that reddish is a sex-linked recessive character and that the parent female was heterozygous for it. Later experiments confirmed this interpretation.

Behavior of Reddish in Crosses with Yellow.—In appearance, reddish is very similar to the sex-linked character yellow body-color which has been known for a long time in Drosophila virilis.[1] Reddish, however, can be easily distinguished from yellow by its brighter color and still better by the color of hairs and bristles which is gray on yellow flies and yellowish on reddish flies.

Because of its similarity to yellow the first cross made with reddish was the cross between reddish and yellow, for the purpose of testing the possibility of reddish being an allelomorph of yellow. When it was found that F$_1$ females from such a cross were phenotypically yellow, it was assumed that reddish was a new allelomorph of yellow. Fortunately, however, an F$_2$ generation was bred from that cross. In this F$_2$ generation of 271 flies, 112 were reddish, 129 yellow and 30 were wild-type. These results made the assumption of allelomorphism very questionable. From a cross between two allelomorphs, which behave regularly, wild-type flies should not be obtained. Since there is no precedent in Drosophila, or as far as I am aware, in any other organism, for such behavior of two recessive non-allelomorphic characters, the case of reddish indicated either a novel complementary behavior of two recessive factors in heterozygous condition or an exceptional behavior of an allelomorph of yellow.

A Map of the "Left" End of the X-Chromosome.—Since the factors located in the "left" end of the X-chromosome were extensively used in the crosses considered here, the following map of that region of the chromosome will be an aid in interpreting the data.

PILOSE	SEPIA	YELLOW	SCUTE	VERMILION
0	0.18	2.18	2.88	21.88

The map of the "left" end of the X-chromosome.

Cross between Reddish and Sepia Yellow Scute Vermilion.—Since the crosses described above indicated that reddish is an allelomorph of yellow or is located in the region close to yellow, a cross was made between reddish and sepia yellow scute vermilion—involving a majority of known factors located in that region of the X-chromosome. In the course of the experiments this same cross has been repeated over and over for other purposes, thus giving a large amount of comparable data. Table 1 gives the summary of these data obtained from 31 independent experiments involving 168 F$_1$ females.

In these crosses again, in addition to reddish and yellow classes non-reddish and non-yellow flies were obtained. In an ordinary case they would represent cross-overs between the reddish and yellow loci. Closer analysis of the data, however, shows that in the present case they cannot be accounted for in that way. If an assumption were made that they are crossovers it would mean that reddish is located about 3.65 units from yellow, because in a total of 9259 flies 169 cross-overs were obtained.[2] The reddish locus, therefore, should be either to the left of pilose or between scute and vermilion. The first difficulty in the analysis was met in trying to determine which of these positions was held by the locus of reddish. By taking into account all classes obtained from the crosses between sepia yellow scute vermilion and reddish it can be seen that there is no place in that region of the chromosome where the reddish locus could be placed. An analysis of the data in table 1 shows this clearly. Here,

VOL. 12, 1926   *GENETICS: M. DEMEREC*   13

TABLE 1

GROUPING OF MALE CLASSES FROM THE CROSS $\frac{\text{se y sc v}}{\text{re}}$ ♀ × re OR se y sc v ♂, IN REGARD TO CROSSING-OVER IN DIFFERENT REGIONS, ASSUMING DIFFERENT POSITIONS FOR THE LOCUS OF REDDISH

ASSUMED ORDER OF LOCI	re AND y CLASSES								NON-re AND NON-y CLASSES				
FREQUENCY	se y sc v	re	y sc v	se re	se y	re sc v	se y sc	re v	y sc	re sc	+	v	sc v
	3363	3767	71	80	36	24	827	918	3	1	116	30	23
$\frac{\text{se y sc v}}{\text{re}}$	0³	0	1–2	1–2	3	3	4	4	1–2–4	3–4	1	1–4	1–3
$\frac{\text{se} \mid \text{y sc v}}{\text{re}}$	0	0	1	1	3	3	4	4	1–4	3–4	1–2	1–4	1–2–3
$\frac{\text{se y} \mid \text{sc v}}{\text{re}}$	0	0	1	1	3	3	4	4	1–4	3–4	1–2	1–2–4	1–2–3
$\frac{\text{se y sc} \mid \text{v}}{\text{re}}$	0	0	1	1	2–3–4	2–3–4	4	4	1–4	2–3	3–4	3	2

assuming different positions for reddish, it is shown in what region the crossing over should have occurred to give the observed classes. From that table it can be seen that reddish does not fit at all in any of the four possible positions. In which ever position reddish is assumed to be, some of the double cross-over classes are disproportionally high. In this cross, the occurrence of any of the double cross-over classes, except those involving the fourth region, would be an exception in itself. According to the published data[4,5] and those of Dr. C. W. Metz and Miss M. S. Moses not yet published, the smallest distance in which double cross-overs have been observed is more than fifteen units in length. In this case, the distance is less than five units long. The occurrence of any double cross-over in so small a distance would be very abnormal and the occurrence of so many double cross-overs as were observed is highly improbable.

Analysis as given in table 1 does not exhaust all possible recombinations because those which would bring yellow and reddish in the same chromosome are left out. There were two reasons for doing this. The first was that those cross-over classes probably do not occur at all, as will be shown later in the discussion of the cross between pilose scute and reddish, and the second is that nothing more than a guess could be made as to the phenotype of reddish yellow flies.

Cross between Reddish and Pilose Scute.—This cross brings out more clearly the points which were obscured in the cross between reddish and sepia yellow scute vermilion because in this case all characters involved are easily distinguishable whether alone or combined. In this cross, as in the previous one, no place for reddish could be found in the part of the chromosome where reddish should be located. As may be seen from table 2 all of the three possible positions for reddish are eliminated after a comparison is made between different cross-over classes. In this cross,

TABLE 2

GROUPING OF MALE CLASSES FROM THE CROSS $\frac{\text{pl sc}}{\text{re}}$ ♀ × re OR pl sc ♂, IN REGARD TO CROSSING OVER IN THE DIFFERENT REGIONS, ASSUMING DIFFERENT POSITIONS FOR THE REDDISH LOCUS

ASSUMED ORDER OF LOCI FREQUENCY	re 1280	pl sc 1119	re sc 5	pl 9	re pl 13	sc 22	+ 62
$\frac{\text{pl sc}}{\text{re}}$	0	0	2	2	1-2	1-2	1
$\frac{\text{pl sc}}{\text{re}}$	0	0	2	2	1	1	1-2
$\frac{\text{pl sc}}{\text{re}}$	0	0	1-2	1-2	1	1	2

again, the assumed double cross-over classes are large notwithstanding the fact that the distance between the farthest loci is less than five units. They

Vol. 12, 1926   *GENETICS: M. DEMEREC*   15

are so large in comparison with the single cross-over classes as to make it highly improbable that reddish can be located in any of these three positions.

One of the outstanding features of this cross is the failure to obtain the reciprocal class for the wild-type class which is the largest of the recombination classes. The reciprocal class would be pilose reddish scute. By crossing pilose reddish with scute flies it was possible to obtain pilose reddish scute flies which eliminated the possibility that that combination is not viable. The other possibility—i.e., that of low viability of pilose reddish scute flies—is also eliminated, because in the cross between pilose reddish scute and sepia yellow scute vermilion 116 flies of first parental class and 100 flies of second parental class were obtained, which shows that pilose reddish scute flies are at least as viable as sepia yellow scute vermilion flies, which are considered to have fairly good viability.

The Hypothesis.—When the results obtained from the cross between reddish and yellow were discussed, two possible interpretations were suggested, i.e., (1) that reddish is not an allelomorph of yellow, but only behaves uniquely in the crosses with yellow giving yellow F_1 generation or (2) that reddish is an allelomorph of yellow, which behaves exceptionally because in F_2 of the cross, reddish × yellow, wild type flies are found.

The first hypothesis appears to be untenable, because if reddish is not an allelomorph of yellow, but is an independent sex-linked character, it would be expected that it would behave in crosses as other mendelian characters do. The results from the crosses between reddish and sepia yellow scute vermilion, and between reddish and pilose scute, show clearly that that is not the case. The F_2 data from these crosses were so exceptional that it was not possible to "locate" reddish in the chromosome in the usual manner.

On the second hypothesis (i.e., that reddish is an allelomorph of yellow which behaves exceptionally) it is necessary to find an explanation for the occurrence of wild-type flies observed in the F_2 generation of crosses between reddish and yellow. Closer examination of the data from the cross between reddish and sepia yellow scute vermilion (table 1) shows that 116 out of 169 wild-type[6] males had the reddish chromosome from which reddish has been removed, and the rest of them had the same modified chromosome which had crossed-over in the yellow scute or scute vermilion region with the sepia yellow scute vermilion chromosome. Apparently all wild-type males obtained in this cross had the reddish chromosome in which the wild allelomorph was substituted for the reddish. Crossing over is the usual way by which one factor is substituted in a chromosome for the other one. In this case, however, as already indicated, crossing over cannot explain this substitution of wild for reddish. The mechanism by means of which this substitution is made is not known as yet and the word

16 *GENETICS: M. DEMEREC* Proc. N. A. S.

"mutation," used in its broadest sense, will be applied in describing it. It is assumed that reddish frequently "mutates" to wild-type.

Test of the Hypothesis.—The results of all experiments made so far, can be explained by the assumption that reddish is an allelomorph of yellow and that it mutates frequently to wild-type. When that assumption is made it is to be expected that the F_1 generation from a cross between reddish and yellow will be yellow or nearly so, and also some wild-type flies will be expected in the F_2 from the same cross. This hypothesis, also accounts for all irregularities observed in the results obtained from crosses between reddish and sepia yellow scute vermilion and between reddish and pilose scute. All wild-type classes obtained in the two crosses are results of mutation of reddish to wild-type. That separates them from the regular cross-over classes and eliminates all irregularities in behavior encountered otherwise in analysis of results.

[1] Metz, C. W., *Genetics,* **1,** 1916 (591–607).

[2] This would represent only half of the total cross-overs because no reciprocal cross-over class is represented. This feature is discussed in later paragraphs.

[3] In this and in the following table "0" signifies non-cross-over classes, "1" classes resulting from crossing over in the first region, etc.

[4] Metz, C. W., M. S. Moses and E. D. Mason, *Washington, Carnegie Institute Publ.,* **328,** 1923.

[5] Weinstein, A., these Proceedings, **6,** 1920 (623–639).

[6] Wild-type in this case refers to the body color only.

Quantitative Biology Comes to the Biological Laboratory: Isolating Hormones

Rowntree L.G., Greene C.H., Swingle W.W., and Pfiffner J.J. 1930. **The Treatment of Patients with Addison's Disease with the "Cortical Hormone" of Swingle and Pfiffner.** (Reprinted, with permission, from *Science* **72:** 482–483 [© American Association for the Advancement of Science].)

Alice and Wilbur Swingle at the cocktail party, probably associated with the 1937 Cold Spring Harbor Symposium "Internal Secretions."

IN 1924, TWO SIGNIFICANT CHANGES TOOK PLACE at the Biological Laboratory, affecting both its status and its research. The Brooklyn Institute of the Arts and Sciences decided that its relationship to the Biological Laboratory was too tenuous for it to continue its support. Davenport overcame this crisis by persuading prominent (and wealthy) members of the local North Shore community to contribute funds to a new entity—the Long Island Biological Association—that assumed management of the Biological Laboratory. The second significant change was the appointment of Reginald G. Harris as director of the Laboratory (Davenport remained as director of the Department of Genetics and the Eugenics Records Office).

Harris made many changes, but none so important as initiating a new line of research at the Laboratory and establishing the Cold Spring Harbor Symposia on Quantitative Biology. Harris (1933) wrote in the Introduction to the first volume of the Symposia series: "...The primary motive of the conference symposia is to consider a given biological problem from its chemical, physical and mathematical, as well as from its biological aspects." It was this *quantitative* biology that Harris wanted done at the Laboratory. This was not a new idea and, indeed, the Biological Laboratory was late in joining a movement that had been under way since the beginning of the century. Nevertheless, Harris embarked on the new program with great vigor, so that within a few years the Biological Laboratory was working on biophysics, endocrinology, and pharmacology, as well as genetics.

If genetics—particularly human genetics—is the hot topic in biomedical research at the end of the twentieth century, research on "internal secretions" occupied that position at the end of the nineteenth century. Addison (1855) recognized an association between degeneration of the adrenal glands and a disorder now known as Addison's disease. Brown-Séquard (1856) showed that surgical removal of the adrenal glands led to death of the animal, although his major contribution was to promote interest in the field through his claim that injections of extracts of testis led to his own rejuvenation. Be that as it may, by the end of the nineteenth century, adrenaline and vasopressin had been detected, and by the 1920s, secretin, insulin, and thyroxine had been isolated. It is not surprising, then, that Harris would have wanted endocrinology as part of his research portfolio, and, in 1924, he announced that the appointment of Wilbur Willis Swingle (1891–1975) to the Biological Laboratory's summer research staff was the "…most important single step in providing for an increase in investigation at the Laboratory."

As Harris had wished, Swingle's presence at the Biological Laboratory led to endocrinology's becoming an important part of its research program. In 1924, Swingle began to teach a course on endocrinology that continued until 1931 when it became subsumed in the General Physiology course, to reappear again in 1937, the year that the Symposium was devoted to internal secretions. George W. Corner first came to the Laboratory in 1929—the same year in which he discovered progesterone—and taught a course called "Surgical Methods in Experimental Biology" (1929–1930; 1934–1937). Reginald Harris himself turned to endocrinology research and began a program described in the Annual Reports as the "Physiology of Reproduction," searching for ovarian hormonal factors involved in the maintenance of pregnancy.

Swingle's first research report appeared in the Annual Report for 1927, itself notable as the first Annual Report to contain such reports—a clear signal of the increasing emphasis on research in the Biological Laboratory. Swingle had shown that the outer part of the adrenal gland—the adrenal cortex—was essential for the proper functioning of the kidneys. Now he was trying to isolate the active factor in extracts of adrenal cortex. Swingle took a year's leave of absence in 1928, but he returned for the summer of 1929. J.J. Pfiffner, a chemist working at Parke, Davis Company, had been appointed to the year-round staff of the Laboratory to work with Harris in his experiments on hormonal factors in the ovary that affect pregnancy. He also helped Swingle in extracting an active factor from the adrenal cortex. The situation was complicated by Swingle's moving from Iowa to Princeton, but Harris allowed Pfiffner to work at Princeton so that the work could continue uninterrupted. By this stage, they had an extract that had been depleted of adrenaline (also found in the adrenal gland) and that kept experimental animals without their adrenal glands alive for as long as 50 days. By 1930, they had eliminated adrenaline and had developed a water-soluble form that could be given by injection (Swingle and Pfiffner 1930a,b).

The paper reprinted here describes the first clinical application of their "cortical hormone," given to a patient with Addison's disease, caused by degeneration or destruction of the adrenal gland. The patient, under the care of Rowntree and Green at the Mayo Clinic, was in a state of "complete collapse…The outlook seemed hopeless…" but Rowntree and Green sent a telegram to Swingle and Pfiffner asking them for some of their extract. This was sent by airmail, and the extract restored the patient to activity within 2 to 3 days. Thus encouraged, the physicians, L.G. Rowntree and C.H. Green, administered the extract to three other patients with similar results. It is not surprising that pharmaceutical companies were interested in their results. We think of royalty and licensing incomes accruing to research institutions as a rather recent development, but the Biological Laboratory's annual report for 1930 tells us that Parke, Davis and Company were to pay the Laboratory no less than $10,000 per year over 3 years. This was a very large sum of money, amounting to about 15% of the Biological Laboratory's income in 1930.

The hunt for the active component(s) in extracts of adrenal cortex was taken up by Tadeus Reichstein (who had synthesized vitamin C) in Basel (Reichstein and Shoppe 1945) and by Edward C. Kendall (who had isolated thyroxine) at the Mayo Clinic, Minnesota (Kendall 1937). Using heroic extraction methods (beginning with up to 1000 pounds of bovine adrenal glands), they isolated corticosterone and 17-hydroxycorticosterone, as well as many other steroids. Reichstein and Kendall shared the 1950 Nobel Prize for Physiology or Medicine with Philip Hench, who had demonstrated the usefulness of cortisone in treating arthritis.

Pfiffner left the Laboratory in 1931 and Swingle did not return after 1933, and, after Harris's death in 1934, there was no full-time endocrinology research at the Biological Laboratory. A number of summer visitors worked on endocrinology, most notably Roger Gaunt from New York University who spent seven summers doing research on the adrenal glands and lactation. The 1937 Symposium on *Internal Secretions* provided an important stimulus, and Gaunt together with Haterius began a new endocrinology course. This was followed by a brief resurgence of endocrinology research in 1938, when there were five summer researchers whose work justified individual reports in the 1938 Annual Report, together with eight graduates or assistants. However, the Biological Laboratory was in decline and, by 1940, there was no one carrying out research in endocrinology.

Joseph John Pfiffner was born June 24, 1903, in Dubuque, Iowa. He took his BS (1925), MS (1926), and Ph.D. (1928), in biochemistry at the University of Iowa. He remained at Iowa as assistant biochemist and research associate until 1929 when he moved to the Biological Laboratory at Cold Spring Harbor. Pfiffner was a Research Associate at Princeton University, 1929–1934, and on the staff of College of Physicians & Surgeons, Columbia University, 1934–1936. For the next 22 years, until 1958, Pfiffner was a Research Biochemist at the pharmaceutical company Parke, Davis & Co. He returned to academia in 1958 as a professor in the Department of Physiology and Pharmacology, Wayne State University, and remained there until his retirement. He died August 13, 1975.

Wilbur Willis Swingle was born January 11, 1891, in Warrensburg, Missouri. He was an undergraduate at the University of Kansas where he took an AB in Zoology in 1915 amd an MA in 1916. Swingle was awarded a Ph.D. in Biology at Princeton University in 1920. He returned briefly to Kansas as an instructor in zoology before moving to Yale University as an assistant professor in biology. In 1926, Swingle moved west again and for three years was Professor of Zoology at the University of Iowa before being appointed professor of biology at Princeton University in 1929. In the early 1930s, Swingle was a regular summer visitor in the Biological Laboratory, Cold Spring Harbor, where he pursued his researches in endocrinology. He was appointed the Henry Fairfield Osborn Professor of Biology in 1956. Swingle remained at Princeton until his retirement when he became Professor Emeritus and continued research at the New Jersey Neuro-Psychiatric Institute in Skillman. He was awarded the Fred Koch Medal of the Endocrine Society in 1959. He died on May 20, 1975.

Addison T. 1855. *On the constitutional and local effects of disease of the suprarenal capsules*. Highley, London.

Brown-Séquard C.E. 1856. Récherches experimentales sur la physiologie et la pathologie des capsules surrénales. *Comp. Rendu.* **43:** 422–425.

Harris R.G. 1933. Introduction. *Cold Spring Harbor Symp. Quant. Biol.* **1:** v–vi.

Kendall E.C. 1937. A chemical and physiological investigation of the cortex. *Cold Spring Harbor Symp. Quant. Biol.* **5:** 299–326.

Reichstein T. and Shoppe C.W. 1945. The hormones of the adrenal cortex. *Vitam. Horm.* **1:** 345–413.

Rowntree L.G., Greene C.H., Swingle W.W., and Pfiffner J.J. 1930. The treatment of patients with Addison's disease with the "cortical hormone" of Swingle and Pfiffner. *Science* **72:** 482–483.

Swingle W.W. and Pfiffner J.J. 1930a. An aqueous extract of the suprarenal cortex which maintains the life of bilaterally adrenalectomized cats. *Science* **71:** 321–322.

———. 1930b. The revival of comatose adrenalectomized cats with an extract of the suprarenal cortex. *Science* **72:** 75–76.

Reprinted from SCIENCE, November 7, 1930, Vol. LXXII, No. 1871, pages 482–483.

THE TREATMENT OF PATIENTS WITH ADDISON'S DISEASE WITH THE "CORTICAL HORMONE" OF SWINGLE AND PFIFFNER

THE preparation of an aqueous extract of the suprarenal cortex which would maintain the life of bilaterally suprarenalectomized cats indefinitely was announced by Swingle and Pfiffner in a brief article published in SCIENCE of March 21, 1930. Subsequently they have reported that by the administration of this extract they were able not only to revive comatose animals, on the verge of death from suprarenal insufficiency, but also to restore them to a normal condition and to keep them in perfect health by daily injections.

The significance of such an announcement and the interest aroused by the possibility of using this extract in clinical medicine are obvious. An extensive experience in the use of the so-called Muirhead regimen in cases of Addison's disease has convinced us of the futility of ordinary therapeutic measures in combating the crises of acute suprarenal insufficiency which develop in the course of this disease and of the great need for a more active cortical preparation which can be administered either hypodermically or intravenously. This point was further emphasized by a patient with Addison's disease who was brought to the hospital in a state of complete collapse, May 31, 1930. The outlook seemed hopeless under ordinary conditions, but as a last resort a telegram was sent to Drs. Swingle and Pfiffner and they forwarded a supply of cortical extract by air mail. The patient, who was in a state of typical collapse, was restored to activity within two to three days. A summary of the clinical history in this case follows:

2

The patient was a farmer, aged thirty-nine years, and first came to the clinic in January, 1930. He had had pleurisy with effusion eleven years previously and symptoms of Addison's disease had been present for eight months. He was in a critical state when admitted; he was in collapse, the systolic blood pressure was 78 mm of mercury, and the blood urea 48 mg for each 100 cc. Treatment was given with solutions of sodium chloride and glucose intravenously, and the Muirhead regimen was instituted. The patient improved slowly; he was dismissed from the hospital thirty-nine days after admission.

Progress at home on the Muirhead regimen was satisfactory for a while, but the patient was brought back to the clinic in a state of collapse, May 31. Treatment with solutions of sodium chloride and glucose was instituted again, with only partial success. The extract of the suprarenal cortex sent by Drs. Swingle and Pfiffner arrived on the sixth day after the patient's admission to the hospital and treatment was begun with daily doses of 20 cc given subcutaneously. Within thirty-six hours a marked effect on appetite and strength was apparent. The patient, who had been so nauseated as to retain water with difficulty, now asked for wieners and sauerkraut and in lieu of the latter ate a double order of beefsteak with relish.

This extract produced considerable local irritation at the site of injection and because of the content of epinephrine could not be given intravenously in therapeutic doses. A further supply of the extract was not available at that time; therefore the patient was put back on the Muirhead regimen. He did well for a few weeks, but gradually failed and again went into collapse, from which the timely arrival of a fresh supply of extract sufficed to insure temporary recovery.

This cycle has been repeated three times in this case. The last time it was possible to use Swingle and Pfiffner's newest extract, which is free from epinephrine. This was given intravenously in divided doses in a quantity of 20 cc daily with a total dosage of 50 cc. Before its use the patient was excessively weak, bedridden, depressed, nauseated, losing weight and showed evidence of

3

failing circulation. Within forty-eight hours he had
taken a new lease on life, his appetite was excellent, his
strength was greatly improved and he appeared to be in
a state of perfect health. He gained 9 pounds in weight
in the next eight days and has been in good condition
since then.

Since that time it has been possible to observe the
effect of the preparation on three other patients suffering from Addison's disease. The condition of one
patient was not considered serious at the time of his
examination and he was kept on the treatment for
only four days. There were no spectacular changes
during this period and the small supply of extract
precluded its further trial. In the other two cases
the clinical condition of the patients and the results
obtained by treatment were similar in character to
those observed in the first case. Metabolism studies
were made in one case during the period of observation. The results will be reported later, but preliminary observations indicate disappearance of creati-
nuria and retention of nitrogen in consequence of the
administration of the suprarenal extract.

The results in these cases convince us of the apparent efficacy of this cortical extract. There was no
striking change in the blood pressure, but the disappearance of anorexia, increase of appetite to the
point of hunger, the gain in weight and the definite
feeling of increased strength and well-being were
striking. As long as the preparation was administered, the results were all that could be desired. However, our supply of the preparation has been limited,
so that we have not been able to observe the results
following consistent dosage and continued administration. The first preparation was not free from epinephrine and caused local irritation when given subcutaneously. The later supply, however, is almost free
from epinephrine; it is suitable for intravenous

4

administration and is almost non-irritating locally. As has been shown, the immediate results in a crisis were excellent. Addison's disease, however, is chronic, and it will be necessary for several years to elapse before a final appraisal can be made of the value of this new therapeutic agent in its treatment.

LEONARD G. ROWNTREE
CARL H. GREENE

DIVISION OF MEDICINE,
THE MAYO CLINIC,
ROCHESTER, MINNESOTA

THE suprarenal cortical extract used intravenously by Dr. Rowntree on patients with Addison's disease represents the modification of our original aqueous preparation mentioned in an earlier communication to this journal.[1] This extract, 1 cc of which represents 30 gm of fresh beef cortex, contains only 0.3 per cent. of solids. The epinephrine content as measured by blood pressure assay on dogs is at most between $1:1,000,000$ and $1:2,000,000$. The method of fractionation used is based on our observation that, by the proper use of permutit, epinephrine can be practically quantitatively separated from the cortical hormone. The 70 per cent. alcohol-soluble fraction obtained by our previously described method[2] is simply filtered in alcoholic solution through an adequate amount of permutit which removes the epinephrine. Much inert material including most of the contaminating pigment is also removed by this fractionation step.

W. W. SWINGLE
J. J. PFIFFNER

PRINCETON UNIVERSITY AND
BIOLOGICAL LABORATORY,
COLD SPRING HARBOR, L. I.

[1] SCIENCE, 72: 75–76, 1930.
[2] SCIENCE, 71: 321–322, 1930.

The Genetics of Taste and Smell

Blakeslee A.F. 1932. **Genetics of Sensory Thresholds: Taste for Phenylthiocarbamide.**
(Reprinted, with permission, from *Proc. Natl. Acad. Sci.* **18:** 120–130.)

Albert Blakeslee with a subject at the 1938 meeting of the American Association for the Advancement of Science.

I
N 1918, BLAKESLEE (1874–1954) AND HIS ASSIStant, B.T. Avery, were examining *Verbena* plants, classifying the colors of the flowers. Blakeslee noted that pale pink flowers were strongly fragrant but he was surprised to find, on pointing this out to Avery, that these flowers were without any scent for Avery. On the other hand, the red *Verbena* flowers were fragrant to Avery, but Blakeslee could not detect any fragrance at all. They carried out a more systematic survey of 46 individuals and found that for about two-thirds, the red flowers were fragrant while the pink flowers had no scent. The reverse was the case for the remaining one-third. Unfortunately, an early frost killed the plants before Blakeslee was able to carry out large-scale testing to determine the genetic basis of this olfactory discrimination (Blakeslee 1918).

There the matter rested for 13 years until 1931, when Arthur Fox, a synthetic chemist with Du Pont, reported that he and a co-worker differed in their responses to phenyl thiocarbamide (PTC) that Fox had synthesized (Fox 1931). His coworker complained of the extreme bitterness of PTC, but Fox insisted that it had no taste at all. In 1932, Fox reported how people responded to a large number of structural variants of PTC and showed that the bitter taste was associated with the presence of a C=S group (Fox 1932).

Blakeslee and the human geneticist Leonard Snyder at Ohio State University read Fox's brief 1931 report and recognized that the ability to taste PTC might be a Mendelian trait. Both wrote to Fox asking for samples of PTC to test it more thoroughly. Snyder published his study of 100 families in 1931 (Snyder 1931) and Blakeslee reported his more comprehensive results in 1932, in a paper published back-to-back with Fox's more detailed study (Blakeslee 1932; Fox 1932).

Snyder and Blakeslee in their first reports had concluded that tasting was inherited as a Mendelian recessive trait, with "taste" being dominant to "non-taste." In accord with this hypothesis, Blakeslee's new results showed that all the children of two non-tasters were themselves nontasters, whereas a taster child always had at least one taster parent. However, Blakeslee showed the situation to be more complex than this. Instead of using small crystals of PTC and scoring for taste or no-taste, he used varying dilutions of PTC and recorded the threshold at which individuals detected bitterness. Blakeslee took special precautions in administering the tests. For example, those who were about to be tested were not allowed to see the responses of those being tested, and he noted the possibility that people who already knew that PTC was bitter might have scored a lower threshold than those who did not.

Blakeslee found that tasters were not a single group but were distributed across the dilutions he used (1:5000; 1:20,000; 1:80,000; 1:320,000) with a cutoff for those who were non-tasters. Furthermore, the ability of children to taste varied with the abilities of their parents. Children of a non-taster and a 1:5000 taster were poorer tasters than the children of a non-taster and a 1:320,000 taster. However, Blakeslee found also that the non-tasters could detect bitterness if saturated solutions—cold or hot—were used in place of crystals, so he redefined his non-taster group as comprising individuals who could not taste PTC except at concentrations higher than 1:5000.

In an interesting extension of the survey, Blakeslee tested how PTC tasting correlated with the ability to detect other substances. Although there were differences in people's responses to different concentrations of such substances as quinine, hydrochloric acid, and saccharine that Blakeslee believed were genetic in origin, there were no general correlations between tasting particular substances.

Blakeslee's interest in the genetics of sensory perception continued for many years (Blakeslee and Salmon 1935a,b), and at several meetings he set up stations at which participants could test themselves. These occasions ranged from the Eugenics Conference in 1932 where 6377 participants were tested for their responses to PTC, to the International Flower Show in New York in 1935, where over 8400 attendees sampled a pair of flowers to determine whether or not they could detect an odor. The most remarkable occasion was a dinner at the American Association for the Advancement of Science in 1935, where the tasting tests were interspersed between the courses. One can only hope that those diners who found the taste of PTC to be nauseating enjoyed the rest of their dinner (Blakeslee 1935).

Modern linkage analysis has indicated that the locus associated with PTC tasting is on the long arm of human chromosome 7, but the gene itself has not yet been cloned. The pattern of inheritance for tasting PTC and a similar compound, 6-*n*-propylthiouracil (PROP) is not the simple autosomal dominant proposed by Blakeslee and Snyder. More recent analyses point to a multifactorial pattern with the taster group itself being composed of tasters and "supertasters."

There is much discussion about the dietary health consequences of aversion to the taste of PTC and PROP. The same bitter taste is characteristic of many vegetables and fruits that may reduce the risk of cancer. If PTC/PROP tasters avoid these foods, they may be increasing their risk of cancer. However, it is very difficult to do tasting experiments with real food where other factors—color, texture, social context—come into play. Still, there is evidence that preschool tasters reject raw broccoli whereas non-tasters like it, indicating that genetic factors may influence the acquisition of food preferences during childhood (Tepper 1998).

Clearly, Blakeslee was fascinated with what his studies on taste revealed of the ways human beings interact with their environment. Just as human beings are born with different sensory levels, he wrote, so "...different people live in different worlds...so far as their sensory reactions are concerned."

Biographical details on Blakeslee can be found on page 48.

Blakeslee A.F. 1918. Unlike reaction of different individuals to fragrance in verbena flowers. *Science* **48:** 298–299.

———. 1932. Genetics of sensory thresholds: Taste for phenyl-thio-carbamide. *Proc. Natl. Acad. Sci.* **18:** 120–130.

———. 1935. A dinner demonstration of threshold differences in taste and smell. *Science* **81:** 504–507.

Blakeslee A.F. and Salmon T.N. 1935a. Genetics of sensory thresholds: Variations within single individuals in taste sensitivity for PTC. *Proc. Natl. Acad. Sci.* **21:** 78–83.

———. 1935b. Genetics of sensory thresholds: Individual taste reactions for different substances. *Proc. Natl. Acad. Sci.* **21:** 84–90.

Fox A.L. 1931. Tasteblindness. *Science* **73:** 14.

———. 1932. The relationship between chemical constitution and taste. *Proc. Natl. Acad. Sci.* **18:** 115–119.

Snyder L.H. 1931. Inherited taste deficiency. *Science* **74:** 151–152.

Tepper B.J. 1998. Genetics of perception '98. 6-*n*-Propylthiouracil: A genetic marker for taste, with implications for food preference and dietary habits. *Am. J. Hum. Genet.* **63:** 1271–1276.

GENETICS OF SENSORY THRESHOLDS: TASTE FOR PHENYL THIO CARBAMIDE

By Albert F. Blakeslee

Carnegie Institution of Washington, Department of Genetics, Cold Spring Harbor, N. Y.

Read before the Academy Tuesday, November 17, 1931

Dr. A. L. Fox[1] first showed that many people cannot detect the bitter taste in crystals of phenyl thio carbamide. In an earlier publication, Salmon and the writer[2] showed that inability to taste the crystals appears to be inherited as a Mendelian recessive. (The same conclusion has been reached independently by L. H. Snyder.[3]) In addition to "non-tasters" of the crystals, we were able to classify the "tasters" of the crystals roughly according to their taste acuity by means of dilutions at which the bitter taste was first detected. This work on the genetics of taste thresholds has been extended, and the results on testing 103 families is presented in the present paper.

Table 1 summarizes the tests regarding "tasters" and "non-tasters." When both parents were non-tasters all the children have been non-tasters. When both parents were tasters, most of the children have been tasters but also a considerable percentage have been non-tasters. These relations would be expected if the difference between tasters and non-tasters were due to a pair of Mendelian factors. If all the tasters were heterozygous dominants one would expect a 3:1 ratio in the mating $T \times T$ and a 1:1 ratio in the back-cross $O \times T$. The deficiency in the non-taster class would be explainable by the frequent occurrence of parents homozygous for the factors for tasting. The critical mating for the hypothesis is that between two non-tasters ($O \times O$). If inability to taste the substance is due to a recessive factor, all the children should be non-tasters. Our tests from this mating furnish 22 children all of whom are non-tasters. Snyder's published results[3] add 17 more, making a total of 39. So far as the tests have gone one is warranted in making certain predictions regarding parents and offspring. A taster child will have at least one parent who is a taster and all the children of two non-taster parents will be non-tasters.

TABLE 1

HEREDITY OF TASTE DEFICIENCY FOR PHENYL THIO CARBAMIDE
(Percentages in parentheses)

TYPE OF MATING	PARENTS NO. OF MATINGS	CHILDREN NON-TASTERS (O)	TASTERS (T)	TOTAL
$O \times O$	10	22 (100)	0 (0)	22
$O \times T$	39	32 (43.2)	42 (56.7)	74
$T \times T$	54	22 (16.8)	109 (83.2)	131
Totals	103	76 (33.5)	151 (66.5)	227

In tables 1, 2 and 3, 6 persons were recorded as both parents and children.

The heredity differences between people in their taste perceptions of phenyl thio carbamide are not so simple as the preceding paragraph might lead one to suppose. Innate differences have been found to exist in regard to the weakest concentration (threshold) at which the substance could be detected, the apparent strength of the sensory reactions, the ability to detect differences in concentration and ability to distinguish the taste of the thio compound from that of other substances such as acids for example.

Table 2 gives the thresholds for the families tested. The technique has been already described. Solutions were made up in artesian well water from a stock solution of 1:5000. The stock solution kept its strength without sensible deterioration for a considerable length of time. The solutions in the small bottles from which the tests were made were frequently renewed since they tended to lose their strength, perhaps on account of soluble substances on the soda fountain straws used in the test. About 0.6 cc. was usually taken up in the straw pipettes.

TABLE 2

PHENYL THIO CARBAMIDE

206 PARENTS			TASTE THRESHOLDS OF 227 CHILDREN				
TYPE OF MATINGS	NO. OF MATINGS	O	A (1:5,000)	B (1:20,000)	C (1:80,000)	D (1:320,000)	TOTALS
$O \times O$	10	22					22
$O \times A$	2	4					4
$O \times B$	8	9	1	4			14
$O \times C$	25	18		10	19	4	51
$O \times D$	4	1			2	2	5
$A \times C$	1			3	1		4
$B \times B$	5	2	1	3	2	2	10
$B \times C$	12	8			10	9	27
$B \times D$	2	2			4	4	10
$C \times C$	17	5			19	16	40
$C \times D$	12	3			10	12	25
$D \times D$	5	2		1	5	7	15
Totals	103	76	2	21	72	56	227

To avoid the psychological influence of expectation, it was a rule not to allow those about to be tested to see the test given to others. A bottle of plain water was of value at times when the subject was in doubt of his reaction or when psychological rather than taste reactions were suspected. In order to make the tests more closely comparable, all the recorded tests except two[5] were made by the writer. The first bottle contained a 1:1,280,000 solution in which concentration none was able to detect a bitter taste. Each solution tested was four times as strong as that previously tested. The threshold was recorded as the first concentration at which the taste was distinct.

Frequently the subject would say that a certain concentration was not water but could not decide what the taste was, even when the four tastes sweet, bitter, sour and salty were suggested to him. At the next higher concentration the subject might at once call the taste bitter and sometimes say that he now knew it was a bitter taste that he had been doubtful about in the more dilute solution. In such cases the threshold was recorded for the stronger concentration at which the subject first recognized the taste was bitter. The subject was generally given also a grade stronger than his threshold if he did not object to the bitter taste. Those who had heard about the test and therefore expected a bitter taste may have scored a lower threshold than those who knew nothing about the chemical that was being given them. Tests of a few individuals at different times of day with a series of solutions in which each was twice as strong as the one previously tested indicated that environmental factors might have some slight influence upon the tasting ability. These differences were not great, however. Greater accuracy in determining thresholds and a smoother curve of distribution would have been possible if the factor 2, or 3 instead of 4, had been used in making up the solutions. This change, however, would have doubled the number of bottles needed for a test and, although it would have been better for a laboratory experiment, it might have been too inconvenient for use in survey work among non-scientific families where the technique employed must be simplified. A larger amount of fluid at each concentration would have had similar advantages and disadvantages.

Despite the shortcomings in method, some of which have been discussed, and the relatively small numbers for a classification of this kind, it is possible to draw some definite conclusions from the data assembled in table 2. All types of matings, for which there is more than a single family represented, have given some negative children. Homozygous cannot be separated from heterozygous tasters by their thresholds, nor by the use of di ortho tolyl thio carbamide, to which some "tasters" are insensitive. In matings with O's, the effect of the grades of parents upon offspring is most clearly marked. $O \times O$ as well as $O \times A$ give only O offspring. $O \times B$ give about half as many tasters as O's. $O \times C$ give about twice as many tasters as O's and $O \times D$ give about four times as many tasters as negative children. The O grade forms the bottom of a series not only in respect to taste acuity of the individuals but also in respect to the effect upon the proportion of negative children as well as the proportion of acute tasters in the offspring. The threshold A appears to behave like a O, but if a considerable number, instead of only 4 children, had been recorded from this mating, some tasters might have appeared among the offspring. As will be seen later, there are probably different grades even of O's and, considered from the standpoint of a continuous series from D to O, it would not be surprising that, if a very large number of $O \times O$ matings were studied, they might be found to yield a few taster offspring.

The acuteness of taste among the taster children also increases with increase of taste acuity in one of the parents. Thus from $O \times B$ matings, all the tasters are B or poorer tasters. From the mating $O \times C$, nearly a third of the tasters are B's, two-thirds C, and about an eighth in the most acute grade D.   From the mating $O \times D$ there are no B tasters and half of the tasters belong to the acute grade D.

When both parents are tasters, the effect of their thresholds upon the grades of their offspring is not so clear.   The mating $B \times B$, however, is seen to be set apart from the matings of higher grade by having a large proportion of poor tasters.   The matings of higher grade all have about the same proportion of acute (D) and medium (C) tasters in their offspring.   The proportion of negative offspring, however, is somewhat different.   If we disregard the single mating $A \times C$, and group together the matings $B \times B$, $B \times C$ and $B \times D$ we have 35 tasters to 12 negative children; a ratio of 3:1.   Taking the matings of higher grade ($C \times C$, $C \times D$ and $D \times D$) we have in the offspring 70 tasters to 10 non-tasters; a ratio of 7:1 or half as many non-tasters as from the group of lower grade parents.

In table 3 is given a summary of the threshold tests of table 2 classified under parents and children as well as according to male and female.   The data suggest that more of the children than of the parents are acute tasters (grade D).   This is true for both males and females.   There are factors which might tend to make children score lower than their parents for the same taste acuity.   Young children are not as familiar as their parents with bitter tastes.   As we have pointed out earlier, familiarity with a taste will probably lead to a lower recorded threshold for the same sensation.   To avoid as much as possible the difficulties of language it has been the rule not to include in the records children under 10 years of age.   Older children often have difficulty in describing the taste but when asked to compare it with something they have tasted before, country children are likely to say it tastes like dandelion stems which they have stuck in their mouths in making "curls" while city children are likely to compare the taste with some medicine they have taken.   The data in table 3 also suggest that there are more acute tasters among females and this difference appears in both parents and offspring.   The proportions of non-tasters among males and females are reversed in parents and children and it is questionable how much significance should be attributed to apparent differences between parents and offspring and between males and females in respect to taste acuity.   Age is not necessarily associated with poor taste since a subject 82 years old could detect bitterness in a 1:320,000 dilution.

The grand totals at the bottom of table 3 indicate that there is a curve of distribution with a discontinuity between the "non-tasters" (O) and those with a threshold at 1:5000.   A further word is desirable at this point in regard to the manner of making the tests.   A preliminary test had shown that there were few (12%) who had a threshold as high as 1:20,000.   The most concentrated solution, therefore, was taken at 4 times as strong, or 1:5000. Later a few persons (1%) were found who could not taste the bitterness until the 1:5000 solution. Our practice was to try the crystals if the subject could not taste the 1:5000 solution. The fact that a few could taste bitterness in the solutions even in high (D) dilutions but were unable to taste the crystals suggested trying saturated solutions of the compound on negative subjects. The majority of non-tasters thus retested could detect bitterness in a cold saturated solution while the rest could detect it in a hot saturated solution with crystals in suspension or better in a similar hot solution in weak alcohol. Dr. Fox has found phenyl thio carbamide to be 0.26% soluble in water at

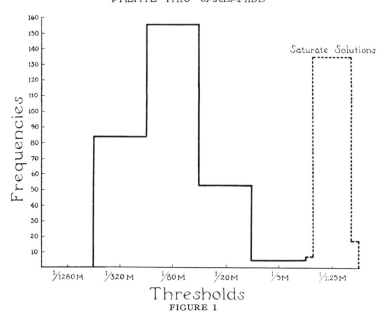

FIGURE 1

18°C. and 5.93% at 100°. In 95% alcohol it is 5.6% soluble at 16°C. and 68% in boiling alcohol. The fact that all who could not taste the crystals whom we have retested with saturated solutions have been found to be positive (25 cases) indicates that all persons can taste bitterness in the compound if only it can be gotten to the taste organs in a sufficiently concentrated condition. In a diagram (Fig. 1) we have indicated the "non-tasters," as previously defined, by a dotted line. Taste ability for the substance therefore forms a bimodal curve. The outer hump on the dotted line represents those who needed a hot saturated solution to sense the bitter taste. The inner hump represents a subject who tasted bitterness in the crystals when a large amount was taken and who had a threshold of 1:5000 after chewing gum for a couple of hours. Gum chewing did not change taste of other *O*'s, however. Some *O*'s had no bitter taste from an excess of the dry crystals while some got a feeble taste under these conditions.

The fact that in the case of phenyl thio carbamide an apparent deficiency in taste has been resolved into merely a difference in thresholds suggests that threshold differences may be responsible for other apparent deficiencies in taste and smell, such as the differences in peoples' reactions to odors of Verbenas earlier discovered by the writer.[4]

TABLE 3

TASTE ACUITY FOR PHENYL THIO CARBAMIDE

(Percentages in parentheses)

	O	*A* 1:5000	*B* 1:20,000	*C* 1:80,000	*D* 1:320,000	TOTALS
PARENTS						
Male	26	3	16	48	10	103
	(25.2)	(2.9)	(15.5)	(46.6)	(9.7)	
Female	33	0	16	36	18	103
	(32.0)	(0)	(15.5)	(35.0)	(17.5)	
Totals, parents	59	3	32	84	28	206
	(28.6)	(1.5)	(15.5)	(40.8)	(13.6)	
CHILDREN						
Male	44	1	9	34	19	107
	(41.1)	(0.9)	(8.4)	(31.8)	(17.8)	
Female	32	1	12	38	37	120
	(26.7)	(0.8)	(10.0)	(31.7)	(30.8)	
Totals, children	76	2	21	72	56	227
	(33.5)	(0.9)	(9.3)	(31.7)	(24.7)	
Grand totals	135	5	53	156	84	433
	(31.2)	(1.2)	(12.2)	(36.0)	(19.4)	

The *O*'s in the tables, it may now be said, represent those who have a threshold above 1:5000. The results would be about the same if the *O*'s were defined as those who cannot taste the crystals.[5] The factors responsible for the bimodality of the curve in figure 1 and for the fact that some who have low thresholds for bitterness in the solutions cannot taste the crystals are not entirely clear. Probably solubility of the compound in the saliva may have something to do with the phenomena. From the pH of salivas shown in table 4, it is obvious that acidity of saliva is not the controlling factor. These pH determinations were kindly made by Miss Satina by the use of brom thymol blue as an indicator. These pH's probably represent innate differences as well as possible effects of environmental factors.

The inquiry has frequently been made whether or not acuteness of taste for the thio compound indicates acuteness of taste for other substances. This was early seen not to be the case since those in a family who were said to be the first to detect sourness in milk were sometimes good and sometimes poor tasters for the thio compound. It seemed desirable, however, to compare acuity of taste for different bitter compounds as well as for representatives of the other three primary tastes, sweet, sour and saline. Accordingly, the tests shown in table 4 were made of 21 individuals all of whom except No. 17 were connected in some way with the Department

TABLE 4

THRESHOLDS IN TASTE ACUITY

SUBJECT NO.	P. THIO CARB.	BITTER QUININE	PICRIC ACID	SOUR HCl (38%)	SWEET SAC-CHARINE	SALT NaCl	ACIDITY OF SALIVA (pH)
1	0	1:5M	1:20M	1:800	1:20M	1:400	7.0
2	0	1:5M	1:20M	1:1600[a]	1:80M	1:400	6.8
3	0	1:20M	1:20M	1:200	1:20M	1:400	6.6
4	0	1:20M	1:20M	1:400	1:10M	1:400	7.1
5	0	1:20M	1:20M	1:1600[a]	1:20M	1:400	7.0
6	0	1:20M	1:80M	1:800	1:40M	1:400	6.8
7	0	1:20M	1:80M	1:800	1:40M	1:400	7.0
8	0	1:80M	1:80M	1:800	1:40M	1:400	6.8
9	0	1:80M	1:80M	1:1600	1:40M	1:400	6.2
10	1:20M	1:80M	1:20M	1:1600	1:40M	1:400	6.9
11	1:80M	1:5M	1:20M	1:400[b]	1:10M	1:200[c]	...
12	1:80M	1:5M	1:320M	1:400	1:40M	1:200[c]	6.9
13	1:80M	1:20M	1:80M	1:1600	1:40M	1:200	7.0
14	1:80M	1:20M	1:80M	1:1600	1:80M	1:400	6.7
15	1:80M	1:20M	1:320M	1:400	1:80M	1:200	6.9
16	1:80M	1:320M	1:80M	1:1600	1:80M	1:800[c]	7.1
17	1:80M[d]	1:320M[d]	1:80M	1:1600[d]	1:80M	...	...
18	1:80M	1:320M	1:320M	1:1600	1:40M	1:400	6.9
19	1:80M	1:320M	1:320M	1:1600	1:80M	1:400	6.5
20	1:320M	1:20M	1:80M	1:800	1:40M	1:400	6.8
21	1:320M	1:20M	1:80M	1:1600[b]	1:40M	1:400	6.7
Dilution factors	4	4	4	2	2	2	

[a] All solutions of HCl taste like alum, not sour. [b] This solution of HCl tastes astringent, stronger solutions taste sour. [c] 1:200 and 1:100 NaCl taste sour and astringent. [d] Solutions of quinine, phenyl thio carbamide and HCl taste alike. [e] This solution of NaCl tastes sour, next stronger solution tastes salty.

of Genetics. The tests were made with soda fountain straws and efforts were made to avoid influence of previous tastes by taking the tests of different substances at different times or by drinking water between tests when a longer separation in time was not convenient. The method employed is somewhat crude and subject to errors already discussed in tests of the thio compound. The table gives information, however, that appears not to have been assembled elsewhere. No one of the group is the poorest taster in respect to all the substances tested and no one is in the best grades for all the substances. Some, however, are relatively acute tasters for all substances tested and others are relatively poor for all. It was surprising further to learn that there was no close connection between ability to taste two different bitter substances. Some of these negative for the thio compound have relatively low thresholds (1:80,000) for quinine sulphate and for picric acid. Tests on a larger number might show that some who were negative for the thio compound were in the most acute grade for quinine and picric acid. The most striking difference in reaction to a bitter substance is shown by subject No. 12, who was in the poorest grade for quinine but in the most acute grade for picric acid. It had earlier been seen by the use of powdered quinine sulphate, as an attempted standard for the word bitter, that the reaction to quinine did not run parallel to that for the thio compound. Incidentally two cases were found who were apparently negative to powdered quinine sulphate. It would be interesting to test out the taste buds for different bitters.

For sour, sweet and salt test substances it was necessary to use a dilution of 2 instead of 4 to get a reasonable spread to the thresholds. Salt showed the least differences between the reactions (the highest threshold was only 4 times the concentration of the lowest) despite the differences between people in respect to the amount of salt they like in their food. The threshold at which a substance is detected does not of itself give information regarding likes or dislikes of the subject nor does it indicate the strength of sensation felt. Some were able to detect the thio compound even in the lowest concentration (1:320,000) but found little difference in sensation between this and the 1:5000 solution which is 64 times stronger. A few could not detect the compound until they were given the relatively strong solution of 1:20,000 but at this threshold made an extremely strong complaint of the bitter taste. There appears to be no close correlation between thresholds and emotional response.

The saccharine had a fleeting sweet taste at the thresholds and was difficult to record. The highest threshold was 8 times the strength of the lowest. The dry powder tastes both sweet and bitter to most subjects. For the sour hydrochloric acid the lowest threshold was 8 times as dilute as the highest. The sour sensation was easier to grade than that from saccharine and salt. Despite the recognized differences between members of a household in their ability to detect sourness in milk and also similar differences in professional milk testers, several dairy investigators who were consulted believed that no comparable tests of milk tasters had ever been made to discover how they differed in taste perceptions. The bitter substances show greater differences between the high and low thresholds than the other substances tested.

It is probable that the differences shown in table 4 represent innate heritable differences in the subjects tested, although this has been established only for the phenyl thio carbamide. By tabulating the thresholds of parents and offspring for other substances than the thio compound, it may be possible to demonstrate an hereditary basis for other taste perceptions. Certain weakly acid salts have been found tasteless to a person with a high threshold for HCl.

From the notes to table 4, it will be seen that there were some qualitative differences in reactions to the same substance, which need further study. In earlier tests, it was not infrequent to have the taste of the thio compound described by some other term than bitter, such as sour, sweet, astringent, like lemons, rhubarb, cranberries or vinegar. They have been recorded as tasters under the appropriate threshold. It has been possible as yet to retest carefully only a few of such cases. Some have been shown to be due to using the wrong term for the taste since these subjects could distinguish bitter from sour substances when they had these substances given them in comparable tests. There were several cases found, however, which were clearly due to what may be called *bitter-sour indiscrimination.* In the first (no. 17 of table 4) quinine sulphate, phenyl thio carbamide and hydrochloric acid were all described as puckery or astringent plus a taste like vinegar. Picric acid was said to have the same taste without the astringency. It will be noted that subject No. 17 detected the acid and the bitters in low concentrations although (except for the picric acid) he could not discriminate what most call bitter from what most call sour. A second subject described hydrochloric acid and the thio compound as lemony in taste but the quinine and picric acid as sour. A third, an assistant in a western department of zoölogy, was retested by a fellow assistant. She reports that quinine and hydrochloric acid are indistinguishable to her. When the solutions are strong she calls them both bitter, when they are dilute she calls them sour. Sour and bitter are terms to denote to her quantitative and not qualitative differences. Perhaps a majority of the bitter-sour indiscriminators would react like this subject and call the thio compound bitter rather than sour. It is probable therefore that a considerable number among those calling the substance bitter would be shown by tests to be examples of taste indiscrimination. There is also evidence of various grades of acuteness in taste discrimination.

A further investigation of taste indiscriminations, which appear to correspond to color blindness in vision, is deferred to a later study. Such indiscriminations in taste appear not to have been noted before. Parker in his interesting book on taste and smell[6] makes no mention of them. A number of physiologists and psychologists who have been consulted are also unaware of mention of such phenomena in the literature. That they have not been discovered before may be due to the fact that the emphasis has generally been placed upon the reactions of a single individual rather than upon the sensory differences between different individuals.

Differences in taste thresholds for a number of other substances than phenyl thio carbamide have been found by us and other investigators.[7] The same is true for odors. In the single case investigated we have found these differences in powers of sensory perceptions innate and hereditary. Evidence is thus given for the belief that humans are born with innate differences in respect to all their senses and that different people live in different worlds, therefore, so far as their sensory reactions are concerned.

[1] Fox, A. L., immediately preceding paper in this issue, These Proceedings. Cf. also note in *Science News Letter*, April 18, 1931.

[2] Blakeslee, A. F., and Salmon, M. R., *Eugenical News*, **16,** 105–108, (July, 1931).

[3] Snyder, L. H., *Science, N. S.*, **74,** 151–152, August 7 (1931). Dr. Snyder used para-ethoxy-phenyl-thio-carbamide, a slightly different substance from that which we used.

[4] Blakeslee, A. F., *Science, N. S.*, **48,** 298–299, September 20 (1918).

[5] One child from the mating $C \times C$ who could not taste the 1:5000 solution had a delayed weakly bitter taste from the crystals and 2 children from the mating $O \times C$ showed a similar reaction. One parent in the $O \times O$ mating, who at first reported both solutions and crystals to be tasteless, later, after others in his family had been tested and he had heard their discussions, said the crystals tasted bitter. Other than these mentioned none could detect bitter in the crystals who did not taste bitter in the 1:5000 solution. Retests with a large amount of the crystals have given a weak bitter taste to some recorded as negative. All the parents in table 2 were tested with solutions as well as all their children except 4 children from the mating $O \times O$ who were tested with crystals. Two of the latter were tested by some one else. Otherwise all the records in the table were obtained by the writer personally. To the $O \times O$ mating might have been added 4 negative children tested with both solutions and crystals, the parents of whom were tested with crystals by the writer's assistant. The records in the tables are not entirely comparable with those in our previous paper[2] in which the O's include those who tasted the solutions but found the crystals tasteless. A few errors in records in the earlier tables have been corrected.

[6] Parker, G. H., *Smell, Taste and Allied Senses in Vertebrates*, Lippincott & Co., 1922.

[7] Biester, Alice, Wood, Mildred W., and Wahlin, Cecile S., on sugars, *Am. J. Physiol.*, **73,** 387–396 (1925). Ward, J. C., and Munch, J. C., on strychnine, *J. Am. Pharm. Assoc.*, **19,** 1057–1060 (1930). Munch, J. C., on capsicums, *J. Am. Pharm. Assoc.*, **18,** 1236–1246 (1929).

Pigeons, Hormones, and Lactation

Riddle O., Bates R.W., and Dykshorn S.W. 1932. **A New Hormone of the Anterior Pituitary.**
(Reprinted from *Proc. Soc. Exp. Biol Med.* **29:** 1211–1212 [by permission of Blackwell Science, Inc.].)

Oscar Riddle with one of his pigeons in 1938.

AT THE SAME TIME THAT SWINGLE AND PFIFFNER were pursuing their cortical hormone in the Biological Laboratory (pages 91–94), Oscar Riddle (1877–1968) was hunting a hormone of his own in the Department of Genetics.

Davenport had invited Riddle (Corner 1974) to join what was then the Station for Experimental Evolution in 1912. Riddle's Ph.D. was on the genetic and biochemical basis of the light and dark bars on the wings of birds, specifically pigeons, in Charles O. Whitman's department of Zoology at the University of Chicago. This type of investigation might have been expected to appeal to Davenport but, apparently, the two men did not get along. This was due in part to Riddle's dedication to Whitman, who never accepted Mendelian genetics, in contrast to Davenport, who was a booster for Mendel. It is not clear why Davenport brought Riddle to Cold Spring Harbor. Perhaps Riddle's heterodox views on genetics developed later, but the situation cannot have been helped by the time Riddle devoted to preparing for publication Whitman's vast mass of data on evolution in pigeons (Riddle 1912). This was published in 1919 and must have occupied a good deal of Riddle's time during his first years at Cold Spring Harbor (Whitman 1919a,b,c).

Riddle tended to pursue research that was at variance with the rest of the Department. A major theme in Riddle's research, derived no doubt from his thesis work, was an emphasis on metabolic and physiological explanations for biological processes. For example, he proposed that sex determination depended on differences in basal metabolism (Riddle 1916). This brought him into conflict with the geneticists who had shown that sex depended on the inheritance of sex-determining genes, the metabolic differences between the sexes being an epiphenomenon. It is not surprising that Riddle was attracted to endocrinology and the hormonal control of metabolism, and he claimed to have found a hormone in the thymus gland—thymovidin—that controlled formation of eggs in birds (Riddle 1924). This was not confirmed and, as Riddle wrote later, "...my inexperience led me to an overconfidence and dogmatism which more mature years could only regret." Nevertheless, it is in endocrinology that Riddle made his major contribution. In a series of three papers, he reported the detection, isolation, and clinical application of a hormone controlling lactation. The second of these papers is reprinted here.

The control of lactation was an enigma until, in 1928, Stricker and Grueter found that extracts of the anterior pituitary induced lactation in rabbits, although they believed that prior ovulation was essential (Stricker and Grueter 1928). Corner, in work carried out in part during the summer of 1929 at the Biological Laboratory, independently showed the same but demonstrated that ovulation was not a prerequisite (Corner 1930). He failed to isolate the hormone responsible. According to Corner, Riddle read Corner's report and decided to exploit the extraordinary phenomenon of "crop milk" secreted by doves and pigeons to feed their young (squab). "Cropmilk" is made up of epithelial cells shed from the lining of the crop. The cells contain large amounts of protein and lipid so that the "milk" has a cheese-like quality. It is highly nutritious—made up almost entirely of protein and lipid—and squab body weight doubles during the two days they are fed cropmilk by both parents (Horseman and Buntin 1995).

It is not clear why Riddle chose to use the crop gland as an assay for a possible anterior pituitary hormone affecting lactation, for, as he pointed out, the mammary gland and the crop gland are not homologous. Nevertheless, he and Braucher began to assay extracts of the anterior pituitary and other hormones for their ability to stimulate crop gland enlargement. It is difficult to know exactly what they were testing, given the imperfect knowledge of what these extracts might contain. Their significant findings were that whereas extracts of anterior pituitary were active, pitocin (oxytocin) and pitressin (antidiuretic hormone) from the posterior pituitary were inactive. They were able to rule out one of the anterior pituitary hormones—luteinizing hormone purified from the urine of pregnant women was inactive.

In the following year, Riddle made extracts of sheep and beef anterior pituitary that were relatively free of contaminants. These extracts were effective in stimulating crop gland enlargement about sevenfold over untreated birds (Riddle et al. 1932). They named their new hormone "prolactin." Their most remarkable finding was that prolactin induced lactation in male and female guinea pigs and rabbits, demonstrating unequivocally that it was the hormone involved in mammalian lactation (Riddle et al. 1933). In the Department's 1933–1934 Report, Davenport was able to report that Riddle had provided prolactin to the Sloane Hospital of the Columbia University Medical Center for use in stimulating lactation in mothers. Production of prolactin was also being undertaken commercially through the clinical use of his extracts.

It is notable that Riddle's section of the Annual Report that had carried the title "Endocrine Studies in the Pigeon" was now called simply "Endocrine Studies" and had doubled in size from three pages in 1930 to seven pages in 1935. Although Riddle's section on endocrinology appeared anomalous among those on genes and chromosomes and mutations, it seems clear that his research was a significant contribution to the work of the Department. Working out the actions and interactions of the various pituitary hormones occupied Riddle for the rest of his scientific career. He retired from the Carnegie Institution in October, 1944, but not before completing two monographs summarizing his life's work on the metabolism and endocrinology of pigeons and doves (Riddle 1947; Riddle and associates 1947).

An unusual facet of Riddle's life was his interest in the high school teaching of biology. He had taught in high school before graduating, but his passionate belief in the power of education came from his childhood days. Then, he had experienced an epiphany of sorts when he learned of evolution and had felt intellectually liberated from the religious fundamentalism of his family. He did not resist the opportunity to speak his mind when, in 1936, he gave an address as a vice president of the American Association for the Advancement of Science in St. Louis. There he spoke of the suppression of the teaching of evolutionary biology in high schools by reactionary religious forces, in contrast to the rapid progress being made in universities and colleges. This was controversial stuff but led to Riddle chairing a committee to report on the state of biology teaching in high schools. Following his retirement, he continued to campaign vigorously for recognition of the importance of biology in contemporary society.

Oscar Riddle was born on September 27, 1877. He enrolled in Indiana University in 1896, but for two years, he spent some time teaching in Puerto Rico, graduating in 1902. Riddle went as a graduate student to the University of Chicago, completing his

Ph.D. in zoology in 1907. His supervisor was Charles Whitman who became "...nearer a father to me than anyone I have known." Riddle taught at the University of Chicago until 1910, when he left Chicago to spend a year in Europe. However, Whitman died and Riddle returned to help put Whitman's scientific affairs in order. Riddle moved to the Station for Experimental Evolution in 1912 and remained there until he retired in 1945. He remained active, especially in promoting the teaching of biology in high schools, until his death on November 29, 1968. He was elected to the American Philosophical Society in 1926; to the American Academy of Arts and Sciences in 1934; and to the National Academy of Sciences in 1939. He was president of the American Rationalist Society, 1959–1960.

Corner G.W. 1930. Hormonal control of lactation: Non-effect of corpus luteum; positive action of extracts of hypophysis. *Am. J. Physiol.* **95:** 43–55.

──────. 1974. Oscar Riddle. *Biogr. Mem. Natl. Acad. Sci.* **45:** 427–465.

Horseman N.D. and Buntin J.D. 1995. Regulation of pigeon cropmilk secretion and parental behaviors by prolactin. *Annu. Rev. Nutr.* **15:** 213–238.

Riddle O. 1912. A note on Professor Whitman's unpublished work. *University of Chicago Magazine* **4:** 208–211.

──────. 1916. Sex control and known correlation in pigeons. *Am. Nat.* **50:** 385–410.

──────. 1924. Studies on the physiology of reproduction in birds. XIX. A hitherto unknown function of the thymus. *Am. J. Physiol.* **68:** 557–580, p. 578.

──────. 1947. *Endocrines and Constitution in Doves and Pigeons.* Carnegie Institution of Washington. Publication No. 572, Washington, D.C.

Riddle O. and Braucher P.F. 1931. Studies on the physiology of reproduction in birds. XXX. Control of the special secretion of the crop-gland in pigeons by an anterior pituitary hormone. *Am. J. Physiol.* **97:** 617–625.

Riddle O. and associates. 1947. *Studies on carbohydrate and fat metabolism with especial reference to the pigeon.* Carnegie Institution of Washington. Publication No. 569, Washington, D.C.

Riddle O., Bates R.W., and Dykshorn S.W. 1932. A new hormone of the anterior pituitary. *Proc. Soc. Exp. Biol. Med.* **29:** 1211–1212.

──────. 1933. The preparation, identification and assay of prolactin—A hormone of the anterior pituitary. *Am. J. Physiol.* **97:** 617–625.

Stricker P. and Grueter R. 1928. Action du lobe anterieur de l'hypophyse sur la montee laiteuse. *C.R. Soc. Biol.* **99:** 1978–1980.

Whitman C.O. 1919a. *Posthumous works of Charles Otis Whitman, vol. I: Orthogenetic evolution in pigeons* (ed. O. Riddle). *Carnegie Institution of Washington Publication No. 257, Washington, D.C.*

──────. 1919b. *Posthumous works of Charles Otis Whitman, vol. II: Inheritance, fertility, and the dominance of sex and color in hybrids of wild species of pigeons* (ed. O. Riddle). *Carnegie Institution of Washington Publication No. 257, Washington, D.C.*

──────.1919c. *Posthumous works of Charles Otis Whitman, vol. III: The behavior of pigeons* (ed. H.A. Carr). *Carnegie Institution of Washington Publication No. 257, Washington, D.C.*

Reprinted from the PROCEEDINGS OF THE SOCIETY FOR EXPERIMENTAL BIOLOGY AND MEDICINE.
1932. xxix. pp. 1211-1212

6288

A New Hormone of the Anterior Pituitary.

OSCAR RIDDLE, ROBERT W. BATES AND SIMON W. DYKSHORN.

From the Carnegie Institution of Washington, Station for Experimental Evolution, Cold Spring Harbor, N. Y.

Riddle and Braucher[1] showed that the effective stimulus for the enlargement and functioning (formation of "crop-milk") of the crop-glands of pigeons is a substance derivable from the anterior hypophysis only; and the luteinizing substance derived from pregnant urine is not that substance. They were unable to decide "whether the principle activating the crop-gland is the growth, the sex maturity, or a third and now unknown anterior pituitary hormone." We have now shown (a) that the principle which evokes the crop-gland response is a third anterior pituitary hormone, and (b) have identified this same hormone, which we shall here call "Prolactin" as the hitherto undefined pituitary principle which is essential for lactation in mammals.

To prepare this hormone relatively free from the growth and gonad-stimulating principles frozen anterior pituitaries of beef or sheep were ground, defatted with acetone and alcohol and dried. This powder was extracted 3 times in aqueous medium at a pH of approximately 2.5. The acid extracts were precipitated isoelectrically. The isoelectric precipitate was redissolved and reprecipi-. tated 3 times to free it of maturity hormone and then dried with acetone. About 10% of the original weight of dried powder is thus obtained in an acid-soluble isoelectric-insoluble form. The addition (to suspensions) of 0.2% cresol to complete the destruction of the growth principle does not markedly affect the "Prolactin".

Assays of several of our preparations, and some of the growth

[1] Riddle, O., and Braucher, P. F., *Am. J. Physiol.*, 1931, **97**, 617.

PROCEEDINGS

TABLE I.

Intramuscular injections of anterior pituitary preparations into male ring doves showing that growth and gonad-stimulating principles do not activate the crop-gland; "prolactin", practically free of the "maturity" principle, does activate the gland.

Preparation	cc. per day	Dura- tion	Age of bird	Body wt.	Testes Treated animal	Average control	Single crop gland
		days	mo.	gm.	mg.	mg.	mg.
Growth hormone*	.2	5	3.1	139	28.9	10.5	99
No. 36	.4	5	3.3	142	61.7	13.9	74
	.8	5	3.3	128	168.6	13.9	128
Growth hormone No. 31[1]	.6	5	3.1	135	35.3	10.5	73
Growth hormone[2]	.4	7	2.6	155	26.1	7.0	105
No. 31 + Prolan	.4	7	3.0	145	31.2	9.5	159
Prolan[3]	.2	5	3.0	143	6.9	9.5	84
	.2	7	2.9	150	10.1	8.6	120
Antuitrin	.3	9	2.7	132	67.0	7.6	96
(anterior lobe)	.3	9	2.8	144	118.3	8.5	67
"Prolactin"	.5[4]	6	2.9	129	8.8	8.6	740
	.5	6	2.8	167	8.9	8.5	1050
	.3	8	2.6	131	9.0	7.0	390
	.4	5	2.1	164	3.9	5.9	880
	.2	5	2.4	155	7.5	6.8	960
	.5[4]	4	2.6	148	7.5	7.0	985
	.5	4	2.3	123	7.4	6.3	925
	.5	4	2.1	144	8.3	5.9	765

* Lee and Schaffer; 0.4 cc. daily per 100 gm. rat gave maximum growth.
[1] Lee and Schaffer; 0.15 cc. daily per 100 gm. rat gave maximum growth.
[2] Each 0.2 cc. daily.
[3] Antuitrin S (Parke, Davis and Co.), from pregnant urine; 100 R. U. per cc.
[4] Represents 10 mg. of acid-soluble isoelectric-insoluble fraction.

and gonad-stimulating extracts of others, have been made on immature common pigeons and ring doves of both sexes. Table I shows only a part of data from male ring doves. The crop-gland response is equally decisive in either species and sex; the gonad-stimulating response is more pronounced in males. Neither species has been adequately tested as to its suitability for the assay of the growth hormone; we therefore show here the absence of the "Prolactin" effect from growth hormone preparations of Drs. M. O. Lee and M. K. Schaffer, Boston, who kindly supplied very potent samples assayed on hypophysectomized rats. Antuitrin (fresh) gives good gonad-stimulation in immature birds, but is devoid of the crop-gland stimulant. Prolan added to growth hormone was also without effect.

Male and female mature guinea pigs and mature female rabbits have responded to daily or twice daily subcutaneous injections of "Prolactin" by milk secretion—the males after preliminary treatment with theelin and progestin. The gonad-stimulating principle,

A New Anterior Pituitary Hormone

freed of "Prolactin", fails in these tests; and the fresh growth principle is, by these and other tests, practically excluded.   The secretion
begins after 2-3 days in rabbits, 3-5 days in guinea pigs; the quantity is highly variable, and the term of secretion also variable.   A
resumption of milk secretion 5-7 days after cessation of treatment
occurs in some cases.

Quantitative Biology: Biophysics in the Biological Laboratory

Fricke H. and Curtis H.J. 1934. **Electric Impedance of Suspensions of Yeast Cells.** (Reprinted, with permission, from *Nature* **134:** 102–104 [©Macmillan Magazines Ltd.].)

Hugo Fricke at work in his biophysics laboratory in 1930.

THE PRIMARY GOAL OF REGINALD HARRIS'S PLANS for the Biological Laboratory was to create a research institute that would challenge the status of its arch rival, the Marine Biological Laboratory at Wood's Hole, and come to rank with the best institutes in North America. An essential step in this transformation was the conversion of the Biological Laboratory from a summer residence for scientists based elsewhere to a full-time research institute. Harris began with the appointment of Hugo Fricke (1906–1972), the first scientist to conduct year-round research at Cold Spring Harbor (Adams 1972; Hart 1972; Meikle 1988).

Fricke's background was in radiation physics, and he had been an assistant to Nobel laureates Niels Bohr in Copenhagen and Manne Siegbahn in Lund. In 1919, Fricke moved to the United States, first to Columbia University and then, in 1920, to Harvard University. He became interested in biophysics and pursued this new interest when he was appointed director of the Biophysical Laboratory at the Cleveland Clinic. There Fricke began two lines of research that occupied him throughout his life; one dealt with the effects of X-ray radiation and the other with the properties of the cell membrane. He continued this work after his move to the Biological Laboratory in 1928. In the following year, a laboratory purposely built for biophysics research was opened. Its cost of $12,000 was covered by Mrs. Walter James, and the building was named the Doctor Walter B. James Memorial Laboratory in honor of her husband who had been a member of the Board of Directors of the Long Island Biological Association for 26 years and was its President at the time of his death.

For the first half of this century, until the development of the electron microscope, studies of the cell membrane were largely the province of the biophysicists. They used a variety of techniques to probe the membrane and hoped that the data so obtained could be interpreted in terms of the physical structure of the membrane. For example, measurements of the permeability of the cell to many different molecules indicated that the membrane consisted of a layer of lipid molecules sandwiched between layers of protein molecules (Davson and Danielli 1943). Another popular approach used electrical measurements of cells where the capacitance and resistance of the cells were interpreted in terms of membrane properties.

While at Cleveland Clinic, Fricke had derived mathematical models for interpreting data from impedance measurements of cells that could be used, with appropriate assumptions, to calculate the thickness of a cell membrane (Fricke and Morse 1925). Suspensions of cells were placed in cylindrical chambers and the resistivity of the suspension and its parallel capacity were measured over a wide range of frequencies. The impedance (equivalent to the resistance measured using direct current) is made up of two components, one due to the cell membrane and the other due to the interior of the cell. At low frequencies, the former predominates. Fricke determined that the capacity of the red cell membrane was 0.81 $\mu F/cm^2$ and, making assumptions about the chemical nature of the membrane, he calculated that the thickness of the membrane was about 3.3 nM (Fricke and Morse 1925). Cole (1972) described this as "...an achievement of truly historic importance" and, as late as 1972, thought it "...probably still the best indication of the molecular dimension of the membrane in an intact, functional, living cell." After his move to Cold Spring Harbor, Fricke continued to refine these measurements and apply them to cells, such as yeast, that were more complex and required the derivation of more complex mathematical models. The paper reprinted here (Fricke and Curtis 1934b) is representative of his work at Cold Spring Harbor.

Fricke's work in radiation chemistry was of great significance (Fricke et al. 1938). In particular, he carried out a series of very thorough and quantitative experiments examing the products obtained on irradiating water with X rays. This was regarded as a necessary preliminary to understanding the radiation chemistry of cells and later became important in deisgning water-moderated reactors. The topic was also controversial but Fricke's carefully controlled experiments—particularly a scrupulous regard for the purity of the water used—came to be regarded as "...monumental contributions" (Adams 1972) and a "...brilliant series of researches" (Hart 1972).

Fricke's collaborator in his biological experiments was Howard Curtis, who joined the Biological Laboratory staff in 1932 (Zirkle 1973). They published eight papers together (see, e.g., Fricke and Curtis 1934a; 1935a,b) on impedance measurements of a variety of cell types as well as technical papers appearing in journals such as *Physical Review* and *J. Physical Chemistry*. Curtis later moved to join Kenneth Cole at Columbia University where they carried out fundamental electrophysiological studies of nerve conductance. These were influential, and Alan Hodgkin (who shared a Nobel Prize with Andrew Huxley for their analysis of nerve conductance) tells of working with Cole and Curtis at their summer laboratory at Woods Hole.

Cole was one of several biophysicists who came to the Biological Laboratory as part of its summer research program, to work in the Walter B. James Laboratory of Biophysics. This was especially true in the years between 1933 and 1940 when the Symposia were on biophysical topics. For example, in 1936, the Symposium on "Excitation Phenomena" led to an influx of biophysicists, some of whom stayed on to work in the Laboratory for some weeks. These included H.A. Abrahamson and L.S. Moyer (protein electrophoresis), H. Curtis and K. Cole (electrophysiology), and J.Z. Young (nerve conduction). Other notable biophysicists who carried out summer research included H. Davson and J.F. Danielli (permeability), H. Neurath (protein diffusion), W.J.V. Osterhout (membrane biophysics), N. Rashevsky (theoretical biology), and D. Wrinch (protein structure).

In 1934, quantitative biology at Cold Spring Harbor was strengthened further by the establishment of a laboratory for research in general physiology, to be directed by Eric Ponder of New York University and formerly of the University of Edinburgh. Ponder's research was centered on the red blood cell and he was no stranger to the Laboratory, having been collaborating with Fricke. He was the author of a standard monograph on the red blood cell and its properties (Ponder 1934), and he continued this research at the Biological Laboratory. Following Harris's untimely death in 1934, Ponder was appointed director. Through the 1930s biophysics flourished at the Laboratory, with between 10 and 18 individuals working on a wide variety of topics. However, these numbers mask the fact that the all-year biophysicists were very few in number, and only Fricke and Ponder were there for more than a few years. By 1940, Fricke and two full-time support staff had gone, leaving only Ponder. Demerec became Director the next year.

The state of biophysics research paralleled that of the Biological Laboratory as a whole—it was in serious decline. In 1939 the year-round staff totaled seven, of whom only three were scientists, compared with the corresponding figures of seventeen and nine 10 years earlier. When Demerec became director in 1941, full-time research in the Biological Laboratory had come to an end and the only research was done by summer visitors. But what visitors! Participants in the 1941 Symposium on *Genes and Chromosomes: Structures and Organization* who stayed on to do research included M. Delbruck, S.E. Luria, Bentley H. Glass, Alexander Hollaender, E.B. Lewis, Barbara McClintock, Alfred Mirsky, H. Muller, James Neel, Marcus Rhoades, and Sewall Wright—a group including five future Nobel laureates. This Symposium was the first in the classic series on genetics and the first "condensed" version, lasting just two weeks instead of five.

As Demerec wrote in the Annual Report, his appointment marked the beginning of the third epoch in the life of the Biological Laboratory, in which there was to be close cooperation between the Laboratory and the Department of Genetics. It marked also the beginning of a new epoch in research for the Laboratory, which was transformed into a center for genetic analysis using microorganisms.

Hugo Fricke, August 15, 1892–April 5, 1972, was born in Aarhus, Denmark. His Ph.D. was in physics. He worked with Niels Bohr at the University of Copenhagen and with Manne Siegbahn at the University of Lund before moving to the United States in 1919. Fricke established the Biophysical Laboratory at the Cleveland Clinic (1921–1928) and then the Walter B. James Laboratory of Biophysics at the Biological Laboratory. Fricke remained at the Biological Laboratory until 1955. In his retirement Fricke was a consultant to the Argonne National Laboratory and the Danish Atomic Energy Commission.

Howard James Curtis was born on December 11, 1906 in Lansing, Michigan. He received his Ph.D. from Yale University in 1932 and that same year came to the Biological Laboratory to work with Fricke. He joined Kenneth Cole at Columbia in 1935, and they carried out some of the fundamental studies on excitation and conduction in nerves. In 1943, Curtis moved to secret work on animal radiation at Oak Ridge, later the site of the Oak Ridge National Laboratory. In 1946, he returned to Columbia before going to Vanderbilt University as head of the department of physiology. Curtis returned to Long Island and was chairman of the Department of Biology at Brookhaven National Laboratory until his retirement in 1965. He died on September 13, 1972.

Adams G.E. 1972. Hugo Fricke, 1892–1972. *Int. J. Radiat. Biol.* **22:** 309–310.

Cole K.S. 1972. *Membranes, ions and impulses: A chapter in classical biophysics*, p. 17. University of California Press, Berkeley.

Davson H. and Danielli J.F. 1943. *The permeability of natural membranes.* Cambridge University Press, Cambridge, United Kingdom.

Fricke H. and Curtis H.J. 1934a. Specific resistance of the interior of the red blood corpuscle. *Nature* **133:** 651.

———. 1934b. Electric impedance of suspensions of yeast cells. *Nature* **134:** 102–104.

———. 1935a. Electric impedance of suspensions of leukocytes. *Nature* **135:** 436.

———. 1935b. Electric impedance of hemolyzed suspensions of leukocytes. *J. Gen. Physiol.* **18:** 821–836.

Fricke, H. and Morse 1925. The electric resistance and capacity of blood for frequencies between 800 and 41/2 million cycles. *J. Gen. Physiol.* **9:** 137–152.

Fricke H., Hart E.J., and Smith H.P. 1938. Chemical reactions of organic compounds with X-ray activated water. *J. Chem. Physics* **6:** 229–240.

Hart E.J. 1972. Hugo Fricke, 1892–1972. *Radiat. Res.* **52:** 642–646.

Meikle T. 1988. *A guide to the Hugo Fricke collection at the Cold Spring Harbor Laboratory library.* Cold Spring Harbor Laboratory Archives, Cold Spring Harbor, New York.

Ponder E. 1934. *The mammalian red cell.* Gebrüder Borntraefer, Berlin.

Zirkle R.E. 1973. Howard James Curtis, 1906–1972. *Int. J. Radiat. Biol.* **23:** 1–3.

(*Reprinted from* NATURE, *Vol.* 134, *page* 102, *July* 21, 1934.)

ELECTRIC IMPEDANCE OF
SUSPENSIONS OF YEAST CELLS

BY

DR. HUGO FRICKE

AND

HOWARD J. CURTIS

Dr. Walter B. James Laboratory for Biophysics,
Biological Laboratory, Cold Spring
Harbor, Long Island, N.Y.

THE electric impedance of suspensions of yeast cells, suspended in solutions of electrolytes, has been measured as a resistance, R, and a parallel capacitance, C, with a Wheatstone bridge. In Fig. 1 are shown C and R, as functions of frequency, for a 63 per cent suspension of yeast cells in a $0 \cdot 1$ per cent sodium chloride solution. The form of the curve for C is interesting, particularly when it is compared

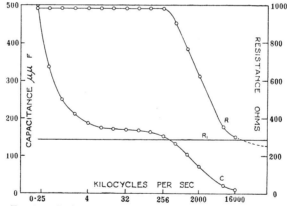

FIG. 1.   Resistance (R), capacitance (C) and resistance of suspending fluid (R_1) for a 63 per cent suspension of yeast in 1 per cent sodium chloride*.

with the curve usually obtained for tissues[1], for it seems that it provides evidence regarding characteristics of the cell surface which may be obscured in the case of tissues by reason of their lack of homogeneity. For suspensions of red blood corpuscles, curves[1] have been obtained similar to those for yeast, although the increase of C at low frequencies is of much smaller magnitude.

Up to a frequency of about 128,000 cycles per second, the conductance of the yeast cell is very low compared with that of the suspending fluid. Over this range of frequencies, the impedance is derived from the surface of the cells (as well as from the suspending fluid) which acts as a complex impedance with a large capacitative component and a small phase angle. The drop in C and R at 128,000 cycles is considered to represent the point at which the impedance of the cell surface has been lowered sufficiently to allow an appreciable part of the current to pass into the cells. The nearly constant value of C at frequencies between 16,000 and 128,000 cycles is interesting. Within this range, C is also independent of the suspending fluid, as shown in Fig. 2, which shows C and R for 63 per cent suspensions of yeast in different concentrations of sodium chloride.

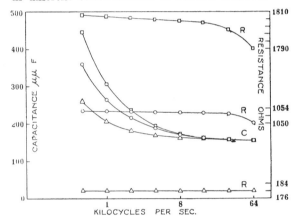

Fig. 2.   Resistance (R) and capacitance (C) for 63 per cent suspensions of yeast in different concentrations of sodium chloride ; R_1 is resistance of suspending fluid*.

□, 0·01 per cent sodium chloride, R_1 = 513 ohms.
○, 0·1 „   „   „   „ R_2 = 299 „
△, 1·0 „   „   „   „ R_3 = 53 „

* By the test of the electric conductance, the yeast cells are in equilibrium with 0·25 per cent sodium chloride. When the concentration of the suspending fluid is different from this, there is a slow change in the conductance of the suspending fluid.

In interpretation it may be assumed that, in this range of frequencies, the impedance at the surface of the yeast cell is derived from a poorly conducting membrane which acts as a static condenser. The increase of C and R at lower frequencies may be due to the polarisation of a slight conductance current through the membrane. Polarisation would be expected to occur if the permeability of the membrane were different for anions and for cations and the polarisation would be larger in the more dilute solutions.

The static capacitance per square centimetre of membrane can be calculated to be 0·6 μF at 16,000 cycles. Taking arbitrarily the dielectric constant of the membrane as 3, this capacitance would correspond to a thickness of 40×10^{-8} cm. This value is slightly larger than that found for the red blood corpuscle[2], although the difference scarcely exceeds the experimental error.

[1] H. Fricke, Cold Spring Harbor Symposia on Quantitative Biology, **1**, 117 : 1933.

[2] H. Fricke and S. Morse, *J. Gen. Physiol.*, **9**, 137 ; 1925.

Printed in Great Britain by FISHER, KNIGHT & CO., LTD., St. Albans

Broken Chromosomes

McClintock B. 1942. **The Fusion of Broken Ends of Chromosomes following Nuclear Fusion.**
(Reprinted, with permission, from *Proc. Natl. Acad. Sci.* **28:** 458–463.)

Barbara McClintock and Harriet Creighton at the 1956 Symposium on "Genetic Mechanisms: Structure and Function."

THIS PAPER IS ONE OF A SERIES BY BARBARA McClintock (1902–1992) describing the behavior of chromosomes with broken ends. These findings showed that the ends of chromosomes were special and presaged research on telomeres some 40 years later.

The primary goal of the United States Department of Agriculture's corn breeding program was to develop high-yielding strains of corn. However, the USDA also promoted basic genetic research on corn in the expectation that this knowledge would be useful to the breeder. Such genetic approaches moved into the ascendancy, so that by the start of the 1920s, the more traditional crop improvement strategies of selecting and cross-breeding high-yield varieties had been eclipsed by research based on Mendelian principles (Fitzgerald 1990).

In 1922, the USDA decided that studies on corn would be enhanced by the appointment of a cytologist whose work would complement that of the geneticists, and although the appointment was to be made by the USDA, the scientist would be attached to one of the land-grant institutions. Cornell, the pre-eminent center for maize genetics, was selected as the institution, and Lowell Randolph joined Rollins Emerson's group at Cornell. Three years later in 1925, a young graduate student, Barbara McClintock, was appointed as Randolph's assistant (Keller 1983; Fedoroff and Botstein 1992; Fedoroff 1995). Unfortunately, the relationship between McClintock and Randolph rapidly soured so that, although they published a joint paper in 1926 (Randolph and McClintock 1926), by 1928, McClintock was practically independent of him (Rhoades 1984, 1992).

McClintock's first important paper—derived from her thesis work—was a study of triploid maize plants (McClintock 1929a). She had modified John Belling's aceto-carmine stain (see page 47) and used squashed preparations of root tips rather than the tissue sections used by Randolph (McClintock 1929b). The preparations were of such quality that, within a few weeks, she was able to distinguish the ten chromosomes of maize—something Randolph had failed to do despite several years of effort (McClintock 1929c). McClintock continued to produce important results at a rapid rate, including a classic paper with Harriet Creighton that showed that genetic recombination was accompanied by a physical rearrangement of the chromosomes involved (Creighton and McClintock 1931).

When, in 1984, Marcus Rhoades listed seventeen important results in maize cytogenetics published between 1929 and 1935, no fewer than ten were by Barbara McClintock (Rhoades 1984). However, despite the fact that she was doing research that put her in the first rank of cytogeneticists, she was not given a job at Cornell. Instead, she received successive fellowships from the National Science Council, the Guggenheim Foundation, and the Rockefeller Foundation. In 1936, Lewis Stadler, who was establishing a center of maize genetics at the University of Missouri, created a post for McClintock. This, too, was only a stop-gap measure, and it was not until 1942 when Milislav Demerec invited her to join the Department of Genetics at Cold Spring Harbor that she found a stable position. The paper reprinted here is the first that McClintock published after her move to Cold Spring Harbor. She remained there until her death in 1992.

In 1931, McClintock published a study describing the various structural changes induced by X-irradiation on maize chromosomes (McClintock 1931). The common feature of these changes—inversions, translocations, and "ring chromosomes"—was that they involved breakage of chromosomes followed by the joining together of the broken ends derived from the same or different chromosomes. McClintock was particularly struck by the behavior of the broken ends of chromosomes, and these became the subjects of her research over the next 15 years. She observed that broken chromosome ends found each other and fused with remarkable efficiency, and that broken chromatids of a single chromosome could fuse (McClintock 1938a). This led to complications when the chromatids separated during cell division, giving rise to what McClintock called the "breakage-fusion-bridge" cycle that perpetuated the production of aberrant chromosomes (McClintock 1938b).

Chromosomes do not normally fuse end-to-end, suggesting that there must be some difference between a normal chromosome end and one created by breakage. She regarded the latter as being "unsaturated," presumably meaning that it required something to make it "saturated" and unable to fuse. McClintock also noticed that, at specific developmental stages, broken ends could "heal"—that is, no longer fuse with other broken ends—and once healed in this manner, the ends continued to behave like normal ends (McClintock 1939). In the paper reprinted here, she further explored and confirmed these findings by asking whether two chromosomes with broken ends had to be in "intimate contact" (within the same nucleus) when breakage occurred for them to fuse subsequently. She did this by producing plants that had received one broken chromosome 9 from the male gamete and a second broken chromosome 9 from the female gamete. These chromosomes carried markers that affected the pigmentation and surface of the resulting kernels, so that she was able easily to recognize plants that had received two broken chromosomes 9.

McClintock examined 18,243 kernels and found only 20 that had received a broken chromosome 9 from each parent of this type (she found more, but these did not have embryos). Plants were grown from these kernels. Ten were abnormal, and her cytogenetic analysis showed that in these, fusion of the chromosome 9 pairs had occurred. Of the morphologically normal plants, analysis showed that in one half, there was no sign that the chromosomes 9 had ever fused—instead they had healed. In the remainder, the broken end of one chromosome 9 had fused with the broken end of another chromosome, while the broken end of the remaining chromosome 9, not having another broken end with which to fuse, had healed. What McClintock showed in these experiments was the first and only clear evidence for healing of broken ends in the diploid, mitotically dividing cells of growing plants. She had previously shown that healing does not occur in the gametes or endosperm cells, a triploid cell type that does not develop into plants. Diploid plant cells, she showed in this paper, become competent for such healing beginning after zygote formation. Although not the stated objective of the paper, this important evidence for cell-type-specific control of chromosome healing has important implications for later work on telomere formation.

McClintock's observations on broken ends between 1932 and 1941 demonstrated that the intact ends of chromosomes differ in some way from ends made by breaking a chromosome (Blackburn

1992). Similar conclusions were reached at about the same time by H.J. Muller, who was examining the structural abnormalities induced in *Drosophila* chromosomes by X-rays (Muller 1938; Gall 1995). The equivalent molecular formulation came when it was pointed out that DNA polymerase could not replicate the very tip of the 3′ end of a length of DNA (Watson 1972). This would lead to the gradual loss of the ends of a DNA molecule, so some special structure or mechanism must have evolved to avoid the disastrous loss of genes that would occur as the ends of chromosomes became nibbled away. Molecular analysis has shown that both structure—telomeres (named by Muller)—and mechanism—the enzyme, telomerase—preserve the ends of chromosomes. Telomeres are extended tracts of repeated DNA (GGGTTAA in human chromosomes) that are maintained at the appropriate length by telomerase, thus protecting the genes (Blackburn and Greider 1995).

> *Barbara McClintock was born in Hartford, Connecticut, June 16, 1902, and went to Erasmus Hall High School in Brooklyn, New York. Her undergraduate and graduate work was done at Cornell University (B. S., 1923; M. A., 1925; Ph.D., 1927). She was an instructor in botany at Cornell University, 1927–1931, before holding fellowships from the National Research Council (1931–1933) and Guggenheim Foundation (1933–1934). McClintock was a research associate at Cornell University, 1934–1936, before moving as an assistant professor to the University of Missouri in 1936. She was appointed a staff member of the Department of Genetics, Carnegie Institution of Washington, Cold Spring Harbor in 1942, a post she held until 1967, when she was made a Distinguished Service Member of the Institution. McClintock was the third woman elected to the National Academy of Sciences (1944) and the first female President of the American Genetics Society (1945). She received many honorary degrees and prizes, most notably the Achievement Award, Association of University Women, 1947; the Merit Award, Botanical Society of America, 1957; Kimber Genetics Award, National Academy of Sciences, 1967; the National Medal of Science, 1970; the Wolf Prize in Medicine, 1981; the Albert Lasker Basic Medical Research Award, 1981; the MacArthur Prize Fellow Laureate, 1981; and the Nobel Prize for Physiology or Medicine, 1983. Barbara McClintock died on September 2, 1992.*

Blackburn E.H. 1992. Broken chromosomes and telomeres. In *The dynamic genome: Barbara McClintock's ideas in the century of genetics* (ed. N. Fedoroff and D. Botstein), pp. 381–388. Cold Spring Harbor Laboratory Press, Cold Spring Harbor.

Blackburn E.H. and Greider C.W., eds. 1995. *Telomeres.* Cold Spring Harbor Laboratory Press, Cold Spring Harbor, New York.

Creighton H.B. and McClintock B. 1931. A correlation of cytological and genetical crossing-over in *Zea mays. Proc. Natl. Acad. Sci.* **17:** 492–497.

Fedoroff N. 1995. Barbara McClintock. *Biogr. Mem. Natl. Acad. Sci.* **68:** 211–235.

Federoff N. and Botstein D., eds. 1992. *The dynamic genome: Barbara McClintock's ideas in the century of genetics.* Cold Spring Harbor Laboratory Press, Cold Spring Harbor, New York.

Fitzgerald D. 1990. *The business of breeding: Hybrid corn in Illinois, 1890–1940.* Cornell University Press, Ithaca, New York.

Gall J.G. 1995. Beginning of the end: Origins of the telomere concept. In *Telomeres* (ed. E.H. Blackburn and C. Greider), pp. 1–10. Cold Spring Harbor Laboratory Press, Cold Spring Harbor, New York.

Keller E.F. 1983. *A feeling for the organism.* W.H. Freeman, New York.

McClintock B. 1929a. A cytological and genetical study of triploid maize. *Genetics* **14:** 180–222.

———. 1929b. A method for making aceto-carmin smears permanent. *Stain Technol.* **4:** 53–56.

———. 1929c. Chromosome morphology in *Zea mays. Science* **69:** 629.

———. 1931. Cytological observations of the deficiencies involving known genes, translocations and an inversion in *Zea mays. Miss. Agric. Exp. Stn. Res. Bull.* **163:** 1–30.

———. 1938a. The fusion of broken ends of sister half-chromatids following chromatid breakage at meiotic anaphases. *Miss. Agric. Exp. Stn. Res. Bull.* **290:** 1–48.

—————. 1938b. The production of homozygous deficient tissues with mutant characteristics by means of the aberrant mitotic behavior of ring-shaped chromosomes. *Genetics* **23:** 315–376.

—————. 1939. The behavior in successive nuclear divisions of a chromosome broken at meiosis. *Proc. Natl. Acad. Sci.* **25:** 405–416.

Muller H.J. 1938. The remaking of chromosomes. *Collecting Net* **8:** 182–198. Reprinted in *Studies in genetics: The selected papers of H.H. Muller,* 1962. pp. 384–408. Indiana University Press, Bloomington, Indiana.

Randolph L.F. and McClintock B. 1926. Polyploidy in *Zea mays. Am. Nat.* **60:** 99–102.

Rhoades M.M. 1984. The early years of maize genetics. *Annu. Rev. Genet.* **18:** 1–29.

—————. 1992. The early years of maize genetics. In *The dynamic genome: Barbara McClintock's ideas in the century of genetics* (ed. N. Fedoroff and D. Botstein), pp. 45–69. Cold Spring Harbor Laboratory Press, Cold Spring Harbor, New York.

Watson J.D. 1972. Origin of concatemeric T7 DNA. *Nat. New Biol.* **239:** 197–201.

Reprinted from the Proceedings of the NATIONAL ACADEMY OF SCIENCES, Vol. 28, No. 11, pp. 458–463. November, 1942

THE FUSION OF BROKEN ENDS OF CHROMOSOMES FOLLOWING NUCLEAR FUSION

BY BARBARA MCCLINTOCK

DEPARTMENT OF GENETICS, CARNEGIE INSTITUTION OF WASHINGTON, COLD SPRING HARBOR

Communicated September 22, 1942

When, through radiation or other causes, chromosomes are broken within a single nucleus, 2-by-2 fusions may occur between the broken ends. These fusions may lead to rearrangements of parts of the chromatin complement, giving rise to various chromosomal aberrations which are detected as reciprocal translocations, inversions, deficiencies, etc. Since, in the well-investigated cases, the breakages occurred within a single nucleus, the conditions that lead to fusions of broken ends could not easily be ascertained. The following questions have been asked: (1) Must two or more chromosomes be in intimate contact at the time of breakage in order that fusions may occur? (2) If no intimate contact is necessary at the time of breakage, are the broken ends "unsaturated," that is, capable of fusion with any other unsaturated broken end? (3) If question (2) can be answered in the affirmative, what forces are involved which lead to the contact and subsequent fusions of the two unsaturated broken ends? Likewise, (4) how long will these broken ends remain unsaturated, i.e., capable of fusion?

Questions (1) and (2) could be answered if the following conditions were present: Assume that fusion occurs between two nuclei each of which possesses one chromosome, one end of which has been broken. Each nucleus will then have a single broken end. When these nuclei fuse and their chromosomes intermix within a single nucleus, the chromosome with a broken end contributed by one nucleus could fuse with the chromosome with a broken end contributed by the second nucleus. The chromosome fusion should occur between these two broken ends. This experiment may easily be conducted in maize. The two nuclei that fuse can be the male and the female gametes, respectively. The method of obtaining gametes having a chromosome with a single "unsaturated" broken end has been reported previously.[1] This method may be briefly summarized. Plants were obtained which possessed one normal chromosome 9 and one chromosome 9 with a duplication of the short arm. This duplicated arm extended beyond the normal short arm, the serial order of parts within the duplicated segment being the reverse of that of the normal short arm. When the duplicated segment is involved in crossing-over, a dicentric chromatid may be produced which is the equivalent of two chromosomes 9 attached at the ends of their short arms. Breakage of this dicentric chromatid during a meiotic anaphase results in the entry into a spore nucleus of a chromatid with a single broken end.   Fusion then occurs at the position of breakage between the two sister halves of this broken chromatid, forming a new dicentric chromatid. This, in turn, produces a bridge configuration in the first gametophytic division, as the two centromeres of the dicentric chromatid pass to opposite poles in the anaphase figure. Again, a broken chromatid enters each telophase nucleus.   In each nucleus, fusion again occurs between the two sister halves of this broken chromatid at the position of this latter breakage.   This *chromatid* type of breakage-fusion-bridge cycle continues in successive gametophytic mitoses.   Therefore, all the nuclei of the fully developed male or female gametophyte will possess one chromosome with a single broken end. Following fertilization, two nuclei from the female gametophyte and one from the male gametophyte fuse to form the primary endosperm nucleus.   If the nuclei of one of the gametophytes possesses such a chromosome carrying dominant genes in the arm with the broken end, and if the other gametophyte possesses a normal, non-broken chromosome carrying the recessive alleles, variegation for these genes will be apparent in the fully developed endosperm.   This is because the chromosomes with the broken ends continue the breakage-fusion-bridge cycle in successive nuclear divisions during the development of the endosperm.   Following various non-median breakages of the bridge configurations, the dominant genes may be deleted from some nuclei and duplicated in the sister nuclei.   Continued repetition of this type of breakage during the development of the endosperm produces a conspicuous variegation pattern.   The behavior of the chromosome with the broken end in the sporophytic tissues of these kernels is entirely different.   The chromatid type of breakage-fusion-bridge cycle ceases when the zygote is formed.   The newly broken end "heals."   Following this healing, no further fusions occur.   The broken end is as stable in its subsequent behavior as any normal chromosome end.

There is no reason to suspect that the condition of those chromosomes in the gamete nuclei that participate in endosperm fusions differ from those participating in zygotic fusions.   However, in one case, the broken end remains unsaturated, and in the other case the broken end heals.   It is reasonable to believe, therefore, that the healing process occurs subsequent to zygotic fusions and not before.   One may tentatively assume that a chromosome end, broken in the pre-gametic division, is unsaturated at the time of zygotic fusion.   If each gamete contributes a chromosome with an unsaturated broken end, one could expect fusion to occur between these two broken ends.   The reasoning behind this expectation is based on the behavior of ring-shaped chromosomes in sporophytic tissues.[2] Ring-shaped chromosomes are frequently broken during mitotic anaphases. Following such breakage, the telophase nuclei receive a chromosome both ends of which are broken.   Fusion may then occur between these unsaturated broken ends, reëstablishing the ring-shape.   This is a chromosome fusion, not a chromatid fusion.   This behavior has suggested that fusion of unsaturated broken ends in sporophytic tissues may be chromosomal in contrast to gametophytic and endosperm tissues where chromatid fusions may occur.   The experiment to be described furnishes proof of the correctness of these assumptions.

To test whether the broken ends are unsaturated in the gamete nuclei, two such ends were introduced into the zygote nucleus, one contributed by the male gamete and one by the female gamete.   Detection of kernels whose zygote nuclei had received such chromosomes was accomplished by introducing contrasting endosperm markers carried by the chromosomes with the broken ends (i, aleurone color, and wx, waxy starch, located in the short arm of one parental chromosome 9 and the alleles I, inhibitor of aleurone color, and Wx, normal starch, carried by the other parental chromosome 9).   The endosperms of those kernels that receive a broken chromosome 9 from each parent should show a very distinctive type of variegation.   All three broken chromosomes 9 would undergo the break-fusion-bridge cycle.   This would lead to variegation for I-i and Wx-wx, variegation for depth of color in the i regions due to multiplication of the number of i genes, and scarred and pitted regions in both the I and i sectors due to the presence of cells which are homozygous deficient for segments of the short arm of chromosome 9.[3]

Plants heterozygous for the duplication chromosome 9 and homozygous for i wx were crossed by plants heterozygous for the duplication and homozygous for I and Wx.   Three types of kernels, with respect to endosperm characters, should be produced; (1) I Wx, non-variegated kernels following fusion of a nucleus carrying a non-broken I Wx chromosome with nuclei carrying either a non-broken or a broken i wx chromosome; (2) kernels variegated for I-i and Wx-wx following fusion of a nucleus carrying a broken I Wx chromosome with nuclei carrying a non-broken i wx chromosome; (3) kernels resulting from the fusion of a nucleus carrying a broken I Wx chromosome with nuclei carrying a broken i wx chromosome.   As stated above, the endosperm character of this latter type of kernel could be

Vol. 28, 1942   *GENETICS: B. McCLINTOCK*   461

anticipated. When observations were made of the kernels resulting from this cross, these latter kernels were very conspicuous. From a total of 18,243 kernels examined, 20 possessing an embryo were of this latter type. More of this type were present but they were germless. These 20 kernels were germinated to determine what had happened to the two broken chromosomes 9 which had entered the zygote. If both broken ends had healed without fusion, normal-appearing plants would be expected to arise from these kernels. If the two broken ends had fused, a dicentric chromosome would have been produced. It would be composed of the chromosome 9 contributed by the male gamete and the chromosome 9 contributed by the female gamete, with their short arms fused end-to-end. When this dicentric chromosome divided and when the two centromeres of each chromatid passed to opposite poles at a mitotic anaphase, two contiguous bridges should be formed. Following breakage of these two bridges, two chromosomes, each with a freshly broken end, should enter each sister telophase nucleus. As stated previously, one could expect fusions to occur between the broken ends of these two chromosomes in each sister telophase nucleus. This would reëstablish the dicentric condition, for again the two chromosomes 9 would be joined to form one chromosome with two centromeres. Repeated anaphase bridge configurations should be expected in subsequent divisions following this *chromosomal* type of breakage-fusion-bridge cycle. The cells of a plant possessing such a dicentric chromosome undergoing this behavior should be composed of various types of homozygous and heterozygous duplications and deficiencies of the short arm of chromosome 9, following repeated non-median breaks in the anaphase bridges. Consequently, these plants should be conspicuously modified in appearance, because of the variation in degree of duplication or deficiency in the many nuclei of the plant.

In the seedling stage, 10 of the 20 plants arising from the kernels classified as having received a broken chromosome 9 from each parent were obviously of the type expected if a dicentric chromosome were present. The presence of the dicentric chromosome was confirmed by examination of the division figures in the young roots, where nearly half of the observed anaphase figures showed two contiguous bridges derived from a dicentric chromosome. Nine of the remaining plants were normal in appearance, and one plant was normal in morphological growth but pale yellow[4] in color and died in the seedling stage. No bridges were observed in the roots of these latter 10 plants.

Due to aberrant growth and death of many cells, 5 of the plants with a dicentric chromosome died in the seedling stage. The remaining 5 plants continued to grow. In all 5 plants, as growth continued, sectors of tissue developed which showed no aberrant growth patterns. These sectors were quite normal in appearance. Gradually, these recovered sectors gained the ascendency in growth until most of the plant was normal in appearance.   In one plant, 3 normal side shoots developed from the base of a very aberrant main shoot which was obviously dying.   Root tips were taken at various times from all of the 5 plants that had survived the seedling stage.   In all cases, vigorous growth of some side branches of the root system was noted. Examination of division figures in these roots no longer showed any dicentric bridges.   The examined cells possessed the normal chromosome number of 20, instead of the 18 monocentric chromosome plus 1 dicentric chromosome previously observed in the younger roots.   Sporocytes for examination of the chromosomes at pachytene were obtained from two of the three recovered shoots of one plant and from the recovered main shoots of the four other plants.   In all cases, 10 bivalent chromosomes were present, one of which was a bivalent chromosome 9.   The two chromosomes 9 were not fused at the ends of their short arms.   In most cases, the composition of the short arm of each member of the bivalent was greatly modified, although in each tassel sample the two chromosomes 9 maintained their respective morphologies in all examined cells.   Of the 12 chromosomes 9 examined from these six samples, no two were alike. The composition of the short arms of the chromosomes 9 in the two re-covered branches from one plant originally possessing a dicentric chromo-some was entirely different.   In each case, it was apparent that the cells of the examined part of the tassel had originated from one individual cell whose cell ancestors had previously been undergoing the chromosomal type of breakage-fusion-bridge cycle involving the original dicentric chromosome 9.   This could be determined in each case by the comparative morphologies of the short arms of the two members of the bivalent.   In several cases it was possible to determine the minimum number of fusions, bridges and breaks that must have occurred before "healing" of the two broken ends had taken place in the nucleus of the particular cell that gave rise to the recovered sector.   The factors involved in the process of healing of two such broken ends within a single nucleus are still undetermined.

The pachytene chromosomes were examined in the surviving 9 of the 10 plants which were normal in morphological growth from the earliest seed-ling stage and which showed no bridges in the earliest roots.   This exami-nation showed that 4 of these plants had received a broken chromosome 9 from each parent.   However, morphological analysis of the short arms of the two chromosomes 9 in each case gave no indication that fusions had ever occurred between their broken ends.   In one plant, one parent had contributed a broken chromosome 9, but it was not possible to determine whether the other parent had contributed a broken chromosome 9.   In the remaining 4 plants, each parent had contributed a broken chromosome 9. However, the broken end of one chromosome 9 had fused with a broken end of a chromosome other than that of the chromosome 9 introduced by the second gamete. In each case, the broken end of the second chromosome 9 had no unsaturated end with which it could unite. As expected, this single broken end thereafter healed.

Conclusions.—The experiments outlined allow some specific answers to be given to the questions presented in the first paragraph of this paper. Question (1) may be answered in the negative. Two chromosomes do not have to be in contact at the time of breakage in order that fusions may occur between their broken ends. This was shown by the fusion that occurred in the zygote or in an early embryonic nucleus between a broken end of chromosome 9 contributed by the male gamete and a broken end of the chromosome 9 contributed by the female gamete. Question (2) may be answered in the affirmative. This was shown by the fusion of these two chromosomes, which produced a dicentric chromosome, and the subsequent behavior of this dicentric chromosome which, for some time, followed the chromosomal type of breakage-fusion-bridge cycle. Question (3) cannot be answered directly from the present observations. Nevertheless, the observations imply that some force exists which accounts for the fusion of unsaturated broken ends of chromosomes. Question (4) likewise cannot be answered directly. Nevertheless, it is certain that the unsaturated state does not persist indefinitely. An unsaturated broken end will become saturated or healed and incapable of fusions when only one such broken end is present in sporophytic tissues;[1] or, as shown in this report, two such broken ends may heal without fusions even when these two ends are present in the same nucleus.

[1] McClintock, B., *Genetics*, **26**, 234–282 (1941).

[2] McClintock, B., *Ibid.*, **23**, 315–376 (1938).

[3] McClintock, B. (unpublished).

[4] This mutant type appears when a plant is homozygous deficient for a small terminal segment of the short arm of chromosome 9.

Penicillin: The Department of Genetics at War

Demerec M. 1948. **Production of Penicillin.** (Reprinted from United States Patent Office, #2,445,748, filed September 16, 1946.)

Demerec M. 1945. **Development of a High-yielding Strain of *Penicillium*.** (Reprinted, with permission, from *Carnegie Inst. Wash. Year Book* **44:** 117–119.)

Milislav Demerec in 1944.

ON JULY 2, 1941, HOWARD FLOREY AND NORMAN Heatley arrived in the United States. It was wartime, and they had flown in planes with blacked-out windows from Bristol via Lisbon, the Azores, and Bermuda, landing in New York on Independence Day weekend. Florey's Oxford team of biochemists and clinicians had shown, just weeks before in a clinical trial of only six patients, that penicillin was a lifesaving antibiotic. However, in wartime Britain, with Germany bombing London and other major industrial cities, there was little likelihood of large-scale production of penicillin getting under way with the necessary urgency. Florey, instead, pinned his hopes on America (Hobby 1985).

Florey and Heatley went first to stay with Joseph Fulton, professor of biochemistry at Yale University. The Floreys and Fultons were old friends, and the Fultons were looking after the Floreys' two children who had been evacuated from Britain. Fulton introduced them to the eminent Yale embryologist, Ross G. Harrison, who was chairman of the National Research Council—an important position in a country soon to be at war. The key contact was initiated by Harrison, who introduced Florey and Heatley to Charles Thom of the U.S. Department of Agriculture. (In an odd twist of fate, it was Thom who had made the original identification of Fleming's mold as *Penicillium notatum*.) The Northern Regional Research Laboratory (NRRL) in Peoria had the necessary equipment and, most importantly, the NRRL routinely tested corn steep liquor as the culture medium for fermentation. *Penicillium* grew so well on corn steep liquor that the yield of penicillin increased tenfold from 1 to 2 units to 20 to 24 units per milliliter of culture medium.

Further increases came through the selection of high-yield strains of *Penicillium*, in particular, a strain of *Penicillium chrysogenum*. NRRL 1951 was isolated from a moldy cantaloupe from a market in Peoria and yielded 60 milligrams of penicillin per liter. A mutant of this—NRRL 1951.B25—

produced more than double this amount in submerged culture, and it may have been this result that initiated a systematic search for useful mutants. It was here that the Department of Genetics made its great contribution to the war effort, as a consequence of research begun by Demerec many years before.

Beginning in the early 1930s, Demerec had pursued his research on mutations by inducing them in *Drosophila* using X-rays and chemicals (Demerec 1933). In 1939, he and Alexander Hollaender began a project to study the effects of ultraviolet radiation on *Drosophila*, but in 1942 they turned to the mold *Neurospora crassa*. *Neurospora* was the organism specially favored for biochemical genetics—Edward Tatum and George Beadle had based their Nobel Prize-winning "one gene, one enzyme" hypothesis on a genetic analysis of *Neurospora* biochemical mutants. Demerec was also studying the origins of penicillin resistance and perhaps this, together with his experience of inducing mutations in a mold, led him to inducing mutants of penicillin, using X-rays and UV irradiation. They began these experiments in September of 1943, but, apparently, it was not funded until May 1944, when a contract from the War Production Board provided money for equipment and extra assistants.

The project got under way with small-scale experiments using strain NRRL-832 before turning to the high producing strain NRRL-1951. These experiments determined that treatment with 75,000 Roentgens was a suitable compromise between inducing mutations and killing the mold spores (Demerec 1944). The spores were irradiated at the Memorial Hospital in New York and spread out on agar plates back in the Department of Genetics. As soon as they had germinated, individual spores were inoculated into test tubes and cultures were grown in shaking incubators. Samples were tested for penicillin production, and those with high yields were saved and the rest were discarded. It must have been hard work and rather depressing. They tested some 5000 mutants and found 504 that were worthy of further analysis (Demerec 1945).

The Department did not have the facilities needed to test mold growth and penicillin yield under large-scale production conditions, so the promising cultures were tested elsewhere. They were shipped first to the University of Minnesota where larger incubators were available, and mutants passing that test went to the University of Wisconsin where they were tested in 80-gallon fermentation tanks. The end result was strain X-1612, which produced 300 milligrams of penicillin per liter, twice as much as the starting strain. A contemporary assessment of X-1612 was given by Robert Coghill who had directed the penicillin efforts at NRRL: "...X-1612 is now very widely used by the penicillin industry and is producing what formerly would have been considered fantastic yields" (Coghill and Koch 1945).

This work was not written up for formal publication, and the description here is taken from Demerec's Annual Reports. Demerec did file a patent on X-1612, an early example of biotechnology entrepreneurship by Cold Spring Harbor. However, neither Demerec nor the Carnegie Institution of Washington benefited—the patent rights were assigned to "the United States of America, as represented by the Administrator, Civilian Production Administration."

This was not the only work on penicillin carried on at Cold Spring Harbor during the war. Vernon Bryson, better known later as a phage geneticist, joined the Biological Laboratory in 1943. Bryson worked on a project studying aerosol delivery of penicillin under grants from the Josiah Macy, Jr. Foundation and the Chemical Warfare Service (Bryson et al. 1944). It was thought that the high potency of penicillin made it likely that aerosol inhalation would be an efficient means of delivery. Bryson found that penicillin activity was not diminished by making it into an aerosol with a mean particle size of 0.54 μm. Using rabbits, he showed that the aerosol penetrated into the lung and that penicillin could be found in the blood. However, the efficiency was very low compared with intravenous injection, and, although the paper ended on an optimistic note, it is unlikely that this approach was pursued.

See pages 78–79 for biographical information on Demerec.

Bryson V., Sansome E., and Laskin S. 1944. Aerosolization of penicillin solutions. *Science* **100:** 33–35.

Coghill R.D. and Koch R.S. 1945. Penicillin: A wartime accomplishment. *Chem. Eng. News* **23:** 2310–2316.

Demerec M. 1933. The effect of X-ray dosage on sterility and number of lethals in *Drosophila melanogaster.* *Proc. Natl. Acad. Sci.* **19:** 1015–1020.

———. 1944. X-ray experiments with Penicillium. *Carnegie Inst. Wash. Year Book* **43:** 112–113.

———. 1945. Development of a high-yielding strain of Penicillium. *Carnegie Inst. Wash. Year Book* **44:** 117–119.

———. 1948. Production of penicillin. United States Patent Office, #2,445,748, filed September 16, 1946.

Hobby G.L. 1985. *Penicillin: Meeting the challenge.* Yale University Press, New Haven, Connecticut.

Patented July 27, 1948

2,445,748

UNITED STATES PATENT OFFICE

2,445,748

PRODUCTION OF PENICILLIN

Milislav Demerec, Cold Spring Harbor, N. Y., assignor to the United States of America as represented by the Administrator, Civilian Production Administration

No Drawing. Application September 16, 1946, Serial No. 697,380

2 Claims. (Cl. 195—36)

This invention relates to the production of antibiotic substances, and more particularly to the production of penicillin.

It is an object of this invention to produce penicillin in extremely high yields. Another object is to produce mutations of molds of the genus mycetes capable of yielding extremely large amounts of antibiotic substances. Other objects will appear hereinafter as the ensuing description proceeds.

These objects are accomplished in accordance with this invention wherein an antibiotic-yielding organism of the genus mycetes is subjected to controlled dosage with ultra-short-wave radiation at an intensity insufficient to kill such an organism but amply sufficient to produce chromosome rupture whereby a mutant is produced, and thereafter such a mutant is cultured in a suitable propagation medium whereby a high yield of antibiotic substance is secured. Suitable mold organisms of the genus mycetes for use in accordance with this invention include *Penicillium notatum, Penicillium chrysogenum,* Actinomyces, and similar mold organisms of the genus mycetes. Preferably, there is employed an already high-yielding strain of *Penicillium chrysogenum.*

Suitable short wave irradiation for the production of mutants includes ultra-violet irradiation, cosmic irradiation, atomic fission irradiation, and preferably X-ray irradiation. For each of the above types of irradiation a dosage is chosen such that chromosome rupture is accomplished without killing the organism. In the case of X-ray irradiation a suitable dosage has been found to be between 10,000 and 150,000 roentgen units, and preferably there is employed a dosage of between 50,000 and 100,000 roentgen units. Under this dosage, the mutation is brought about in the spore form of the organism principally. In irradiating a culture containing numerous spores with the selected dosage of irradiation, it is usually found that the viability of many of the organisms is detrimentally affected. However, numerous of the remaining organisms are found, upon subsequent culture, to be mutants of the parent strain. Obviously, for commercial use of penicillin in antibiotic substances only the mutant or mutants producing the highest yields are propagated.

After irradiation, the mutant spores are spread on the surface of a suitable potato-dextrose-agar culture medium and allowed to germinate. Upon germination, the germinated organisms are isolated into the standard liquid lactose-corn steep liquor culture medium, made in accordance with the formula of the Northern Regional Research Laboratory at Peoria, Illinois, as described in the Journal of the Elisha Mitchell Scientific Society, vol. 61, page 78, for August, 1945. Thereupon, the cultures are agitated for a period of from 2 to 12 days. The resulting growth of the mold is treated in any suitable way for the recovery of its antibiotic content. For example, it may be extracted with amyl acetate, then treated with a small amount of activated carbon to purify the extract, and then further extracted with aqueous acetone. After separation from the activated carbon and the amyl acetate, the aqueous acetone solution is evaporated to yield an aqueous solution of penicillin or other antibiotic substance. Thereupon, by careful evaporation crystalline penicillin can be secured as a residue.

The following illustrative example shows how the invention may be carried out, but it is not limited thereto:

A seven-day-old culture of *Penicillium chrysogenum* (strain NRRL 1951.B25), grown in a test tube having the approximate dimensions, 1 x 7.5 cm., on standard potato-dextrose-agar medium, was subjected to X-ray radiation of 75,000 roentgen units at a rate between 2,000 and 3,000 units per minute. The X-ray irradiation was carried out with a usual d-therapy type of equipment commonly used for cancer treatment. After irradiation, spores were collected from the culture and spread on the surface of a potato-dextrose-agar culture medium contained in a Petri dish where they were allowed to germinate. Immediately after germination, the spores were isolated in separate test tubes containing 2 cc. each of liquid lactose-corn-steep medium made in accordance with the formula of the Northern Regional Research Laboratory at Peoria, Illinois. Thereupon, the test tubes were placed in a shaker machine having a four-inch horizontal stroke and operated at 250 strokes per minute. Each test tube had an approximate inside diameter of 10 mm. and a length when stoppered of 110 mm. The tubes were stoppered and the shaking machine operated for five days, after which the contents of each tube were diluted with 100 volumes of distilled water and assayed in duplicate for penicillin content by the Oxford cup method, using *Staphylococcus aureus* (NRRL strain B313). The tubes which showed a high yield of penicillin were saved for bulk fermentation tests and the low-yielding cultures were discarded. The following table shows the diameter of the inhibited region in millimeters for each Oxford

2,445,748

3

cup test and the asterisk indicates the high-yielding cultures which were retained:

Culture No. Plate	1	2	3	4	5	6
1	15	13	20	24	7	21
2	15	12	20	22	14	21
	7	8	9	10	11	12
3	24	18	21	20	21	25
4	*24	19	22	24	22	*24
	13	14	15	16	17	18
5	19	15	16	15	16	17
6	(¹)	21	14	15	18	16
	19	20	21	22	23	24
7	15	15	26	25	24	18
8	14	15	*24	*25	(¹)	18
	25	26	27	28	29	30
9	22	20	15	(¹)	12	7
10	24	16	15	21	11	17
	31	32	33	34	35	36
11	20	17	13	23	24	21
12	24	20	12	17	17	20
	37	38	39	40	41	42
13	20	20	21	14	13	20
14	21	20	21	13	13	20
	43	44	45	46	47	48
15	24	21	20	25	18	25
16	20	20	19	*28	18	*23

¹ Leaked.

Each of the above high-yielding strains were then subjected to bulk fermentation on standard lactose-corn-steep-liquor medium and one of them was found to consistently rate 369 on the Oxford penicillin scale. By comparison, previous strains had yielded only 169 and 98, respectively.

It will be seen from the foregoing description that there has been provided a highly desirable technique for the production of mutants of antibiotic molds, as well as a process for producing penicillin in yields many times those previously secured.

4

Since many apparently differing embodiments of this invention will occur to one skilled in the art, various changes can be made without departing from the spirit and scope of this invention.

What is claimed is:

1. A process for the production of penicillin which comprises inoculating a growth medium with a fungus mutant produced by X-ray irradiation of *Penicillium chrysogenum* with 50,000 to 100,000 roentgen units, then agitating and aerating said medium at a temperature between 20° and 25° C. for about 4 to 12 days, and recovering penicillin from the resulting fungus growth.

2. The process of claim 1 wherein the *Penicillium chrysogenum* is exposed to about 75,000 roentgen units of X-ray irradiation.

MILISLAV DEMEREC.

REFERENCES CITED

The following references are of record in the file of this patent:

UNITED STATES PATENTS

Number	Name	Date
2,107,830	Liebesny	Feb. 8, 1938

OTHER REFERENCES

Jahiel, Science, September 29, 1944, 195/P., page 298.

Bonner et al., Penicillin Research Progress Report #3, OPRD Contract #169, Department of Biology, Stanford University, April 25, 1944, pages 1 to 3.

Summary of Discussion at the Penicillin Technical Meeting, Hotel Astor, New York, January 21–22, 1944, pages 1 and 11.

Nadson et al., Comptes Rendus, Acad. des Sciences, T. 186, 1928, pages 1566 to 1568.

Development of a High-Yielding Strain of Penicillium

Beginning in September 1943, Mrs. Sansome and M. Demerec, in collaboration with Dr. A. Hollaender, of the National Institute of Health, Bethesda, Maryland, started experiments to produce, by means of X-ray and ultraviolet irradiations, strains of *Penicillium* that would give high yields of penicillin. Experiments were conducted on a small scale until May 1944, when a contract with the War Production Board became effective and funds were made available for additional equipment and special assistants to carry on routine tests. At that time Dr. H. E. Warmke joined the group. The work was continued at the Department until November 1944. Since early in 1944, similar work, also under contract with the War Production Board, had been going on at the laboratories of Stanford University, the University of Minnesota, and the University of Wisconsin.

Penicillin may be obtained from a culture medium when *Penicillium* is grown on its surface, or when it is submerged and aerated by shaking or by bubbling air through the medium. At the time we joined the project it was known that high surface yielders may not be high yielders in submerged cultures, and vice versa. Manufacturing experience had indicated also that submerged culturing is more efficient and economical than surface culturing. Therefore, the aim of the project was to develop high-yielding strains with submerged culturing. Since a considerable amount of equipment is necessary for complete tests of the yielding capacity of submerged strains, it was decided to divide the work so that the irradiation and the preliminary rough screening tests to isolate possible high yielders would be carried out at our laboratory; further tests for yielding capacity would be made at the University of Minnesota laboratory, where large shaking machines were already available; and the final tests would be carried on at the University of Wisconsin in 80-gallon tanks.

For efficient planning of experiments, it was essential to acquire certain fundamental knowledge about the reaction of *Penicillium* to X-rays. It is known that the frequency of mutations is proportional to the dosage, and also that the rate of killing of the treated spores increases with the dosage. The most efficient dosage for our experiments, therefore, was that which would produce a sufficiently high frequency of mutations and at the same time leave enough survivors.

In order to establish this dosage, the mutation rate and killing rate were determined on spores treated with 25,000, 50,000, 75,000, and 100,000 r-units. Spores were treated both dry and in saline suspension. Detailed results of these experiments are published in last year's report (Year Book No. 43, p. 113). On the basis of these results, a dosage of 75,000 r-units was selected for our experiments.

Tests were also conducted to find out whether the mutation rate or the germination rate of treated spores deteriorates when they are stored in a refrigerator. Since it was found that deterioration does not occur, it was possible to treat large batches of spores and to store them for subsequent use.

The majority of cultures obtained by irradiation of a high-yielding strain may be expected to have the same high-yielding capacity as the original strain. Consequently, the quick assay methods designed to screen out low yielders are not applicable to these cultures. What is needed is a quick assay method which will pick out the exceptional high yielders. In a search for such a quick screening method it was thought worth while to investigate the possibility that the pellets formed in shaker flasks originate from single spores, in which case the penicillin-producing capacity of single pellets could be tested directly. An experiment designed to test this possibility revealed that pellets are formed from a mixture of mycelia originating from several spores.

After experimentation with various techniques, a standard procedure for making tests was developed. Spores of *Penicillium chrysogenum* were X-rayed on agar slants with 75,000 r-units. The irradiation was given by Mr. L. D. Marinelli at the Memorial Hospital in New York, at an intensity of 2420 r per minute. Treated spores were spread on the surface of potato-dextrose-agar plates; and immediately after germination they were isolated into test tubes containing 2 cc. of liquid culture medium. These were put into the shaker machine, which had a 4-inch horizontal stroke and operated at 250 strokes per minute. The tubes we used had an inside diameter of 10 mm. and were 110 mm. long. The size of tubes and the amount of nutrient in each tube is determined by the properties of the shaker. After 5 days of continuous shaking, a sample of the medium taken from each tube was diluted 100 times and assayed for penicillin content by the cup method, using *Staphylococcus aureus* (NRRL strain B313). Tubes showing a high yield were saved and the fungus growing in them was cultured; the remaining tubes were discarded. In this way about 90 per cent of the cultures were eliminated as low or average yielders, and 10 per cent were saved as possible high yielders and were shipped to the Division of Plant Pathology, University of Minnesota, St. Paul, for further tests. All together, 504 selected strains were sent to Minnesota. One among these was the strain now known as X-1612, which yields

DEPARTMENT OF GENETICS   119

about twice as much penicillin as the strain
1951.B25 from which it originated. This
new high-yielding strain is now used in
production of penicillin.

The Unstable Genome—Movable Genes

McClintock B. 1951. **Chromosome Organization and Genic Expression.** (Reprinted from *Cold Spring Harbor Symp. Quant. Biol.* **16:** 13–47.)

Tracy Sonneborn and Barbara McClintock relax at the 1951 Symposium on "Genes and Mutations."

THE WORK REPORTED IN THIS PAPER—A SUMMARY of McClintock's 10 years' research on peculiar genetic phenomena in maize—brought about a revolution in the way that geneticists thought about genes and the stability of the genome. McClintock demonstrated that her findings could be accounted for by assuming that small genetic elements could move from place to place in the maize genome (Fedoroff 1992, 1995). This radical explanation was not easily accepted, and it was not until 34 years after publication of this paper that she was awarded the Nobel Prize for Physiology or Medicine.

Barbara McClintock had continued to pursue her research on the behavior of broken chromosomes in maize (page 125), but increasingly from the late 1930s, her interests changed from morphological studies of chromosomes to the genetic consequences of the chromsomal changes she was finding. In the previous paper (page 129), McClintock used plants that produced gametes with a broken chromosome 9 because of recombination between duplicated segments at meiosis. By noting an unusual pattern of kernel pigmentation, McClintock realized that in some families breakages were occurring at the same site on chromosome 9 that she called *Ds* (for "dissociation," her word for these breakages). Further genetic analysis showed that another locus, *Ac* (for "activator"), was needed for *Ds* to make breaks and that, apparently, *Ds* was able to move and cause breaks at another locus.

McClintock tracked the consequences of these breaks by using chromosomal markers in the region of the breaks that led to detectable changes in the kernels. For example, the *C* locus controls production of anthocyanin, a pigment that leads to a deep purple color of the aleurone (the colored outer layer of the endosperm), and the *wx* locus affects starch production; wild-type endosperm stains blue with iodine whereas the mutant is stained red.

A consequence of these chromosomal breakages was that they initiated a breakage-fusion-bridge cycle that continued to generate chromosomal abnormalities, in particular, small deletions in the short arm of chromosome 9. In 1946, McClintock observed a single remarkable kernel, the patterning of which was completely unexpected, on an ear of corn derived from crossing plants homozygous for *C* (colored aleurone) with plants homozygous for *c*, the recessive colorless allele of

C (McClintock 1948; reprinted in McClintock 1987). Approximately one-half of the kernels were dark purple from expression of *C*, and one-half were purple with colorless spots of varying size, depending on when breakage at the *Ds* locus had occurred, leading to the loss of *C*. If *C* was lost early, then the cell could give rise to a large spot. However, there was a single kernel out of 4000 resulting from this cross, in which there were small purple spots on a colorless background. How could this be? If the colorless background had been caused by chromosome breakage and loss of *C*, how was it possible for the recessive *c* allele to be converted in *C*? McClintock realized that *C* had been mutated by *Ds* moving into that locus. Its function was restored when *Ds* moved again. And, just as in *Ds*-induced breakage, *Ac* had to be present. McClintock found other examples of mutable loci, for example, at the *wx* locus (McClintock 1950), and later other examples of movable elements such as *Suppressor-mutator* (*Spm*) (McClintock 1953). The paper from the 1951 Symposium reprinted here was her first public presentation of her conclusions.

Quite apart from her conclusion that genetic elements could move from one place in the genome to another and affect expression of a gene, her discussion, especially in the section "Mutable loci and the concept of the gene," was not calculated to appeal to the majority of her contemporaries. She began by pointing out that she had deliberately avoided using the word "gene," and, although not denying the existence of such units, she believed that the lack of knowledge of how these units operated in the chromosomes was such that "...no truly adequate concept of the gene can be developed until more has been discovered about the function of the various nuclear components." That McClintock invoked the name of Richard Goldschmidt, who was acerbic in his opinions that the idea of the gene as a discrete entity was wrong, cannot have helped (Goldschmidt 1958). McClintock herself wrote that "The response to it [the presentation reprinted here] was puzzlement and, in some instances, hostility" (McClintock 1987). However, some participants in the meeting have different recollections of McClintock's presentation and her reception (Comfort 1999).

McClintock emphasized also the importance of *control* of gene expression: "The knowledge gained from the study of mutable loci focuses attention on the components in the nucleus that function to control the action of the genes in the course of development" (Comfort 1999). It was for this reason that McClintock was enthusiastic about the work of Jacob and Monod (1961) that dealt with control of gene expression. She wrote a detailed commentary drawing parallels between the regulator, operator, and structural genes system proposed by Jacob and Monod for bacteria and the *Spm* system that she had studied (McClintock 1961). Her enthusiasm is evident in the fact that her commentary appeared within just a few months of the classic Jacob and Monod paper. It is significant in the light of the supposed difficulty of her papers that both her commentary and the Jacob and Monod paper were included in a selection of papers on molecular genetics published as early as 1965 (Taylor 1965).

Despite the continuing skepticism of her peers for her emphasis on control and despite her largely confining her publications to the Annual Report of the Carnegie Institution of Washington (McClintock 1987), beginning in the early 1960s there was increasing evidence that transposable elements were at work in organisms other than maize (Shapiro 1995). For example, Green (1967) examined unstable loci in *Drosophila* (see pp. 81–86); the bacteriophage Mu was found to insert its chromosome into the *E. coli* chromosome (Taylor 1963); insertion sequences, plasmids, and transposons were found in bacteria (Shapiro 1995); and an insertional "cassette model" was proposed to account for mating-type interconversion in yeast (Hicks et al. 1977). By May 1976, there was more than enough material for the first meeting devoted to movable elements (Bukhari et al. 1977) and, fittingly, it was held at Cold Spring Harbor Laboratory.

McClintock had emphasized that elements like *Ac* and *Dc* affect the expression of genes, and she believed that such controlling elements were part of the normal cellular and developmental mechanism for regulating gene expression. As she put it in the discussion to a paper she gave in 1955, these elements "...modulate the activities of the primary genic components and in very precise ways... . They could be viewed as 'genes' controlling nuclear processes in contrast to the other class of 'genes' whose activities result in products that reach the cytoplasm" (McClintock 1956). However, these ideas were rejected (Comfort 1999), and the lasting legacy of McClintock's work is that, as the title of the book published to mark her 90th birthday has it, we have come to take it for granted that genomes are *dynamic* (Fedoroff and Botstein 1992). Continuing studies have demonstrated the ubiquity of movable genetic elements and have shown that the genomes of higher organisms are littered with their debris, perhaps contributing to evolution by generating genetic diversity (Britten 1997). Some 30% of the human genome is made up of such DNA fragments and there is good evidence that some of these are active—moving around in our genomes (Kazazian 1998).

Biographical details of Barbara McClintock can be found on page 127.

Britten R.J. 1997. Mobile elements inserted in the distant past have taken on important functions. *Gene* **205:** 177–182.

Bukhari A.I., Shapiro J.A., and Adhya S.L., eds. 1977. *DNA: Insertion elements, plasmids and episomes.* Cold Spring Harbor Laboratory, Cold Spring Harbor, New York.

Comfort N.C. 1999. "The real point is control": The reception of Barbara McClintock's controlling elements. *J. Hist. Biol.* **32:** 133–162.

Fedoroff N.V. 1992. Maize transposable elements: A story in four parts. In *The dynamic genome: Barbara McClintock's ideas in the century of genetics* (ed. N. Fedoroff and D. Botstein), pp. 389–415. Cold Spring Harbor Laboratory Press, Cold Spring Harbor, New York.

———. 1995. Barbara McClintock. *Biogr. Mem. Natl. Acad. Sci.* **68:** 211–235.

Fedoroff N. and Botstein D., eds. 1992. *The dynamic genome: Barbara McClintock's ideas in the century of genetics.* Cold Spring Harbor Laboratory Press, Cold Spring Harbor, New York.

Goldschmidt R. 1958. *Theoretical genetics.* University of California Press, Berkeley, California.

Green M.M. 1967. The genetics of a mutable gene at the white locus of *Drosophila melanogaster. Genetics* **56:** 4674–4682.

Hicks J.B., Strathern J.N., and Herskowitz I. 1977. The cassette model of mating-type. In *DNA: Insertion elements, plasmids and episomes* (ed. A.I. Bukhari et al.), pp. 457–462. Cold Spring Harbor Laboratory, Cold Spring Harbor, New York.

Jacob F. and Monod J. 1961. Genetic regulatory mechanisms in the synthesis of proteins. *J. Mol. Biol.* **3:** 318–356.

Kazazian H.H., Jr. 1998. Mobile elements and disease. *Curr. Opin. Genet. Dev.* **8:** 343–350.

McClintock B. 1948. Mutable loci in maize. Annual Report of the Director of the Department of Genetics. *Carnegie Inst. Wash. Year Book* **47:** 155–169.

———. 1950. The origin and behavior of mutable loci in maize. *Proc. Natl. Acad. Sci.* **36:** 344–355.

———. 1953. Induction of instability at selected loci in maize. *Genetics* **38:** 579–599.

———. 1956. Intranuclear systems controlling gene action and mutation. *Brookhaven Symp. Biol.* **8:** 58–74.

———. 1961. Some parallels between gene control systems in maize and bacteria. *Am. Nat.* **95:** 265–277.

———. 1987. Introduction. In *The discovery and characterization of transposable elements: The collected papers of Barbara McClintock.* Garland Publishing, New York.

Shapiro J. 1995. The discovery and significance of mobile genetic elements. In *Mobile genetic elements* (ed. D. Sherratt). Oxford University Press, Oxford, United Kingdom.

Taylor A.L. 1963. Bacteriophage-induced mutation in *E. coli. Proc. Natl. Acad. Sci.* **50:** 1043–1051.

Taylor J.H., ed. 1965. *Selected papers on molecular genetics.* Academic Press, New York.

CHROMOSOME ORGANIZATION AND GENIC EXPRESSION

BARBARA McCLINTOCK

Department of Genetics, Carnegie Institution of Washington, Cold Spring Harbor, N. Y.

During the past six years, a study of the behavior of a number of newly arisen mutable loci in maize has been undertaken. This study has provided a unique opportunity to examine the mutation process at a number of different loci in the chromosomes. For some of these loci, several independent inceptions of instability have occurred during the progress of this study. The types of mutation that appear, and the types of instability expression, need not be the same at any one locus. In fact, comparisons of the behavior of these different mutable conditions at a particular locus have shown striking diversity, not only with regard to the changes in phenotypic expression that result from mutations at the locus, but also with regard to the manner in which mutability is controlled. Knowledge of the genetic constitutions, with respect to mutable loci already present in the plants in which new mutable loci have arisen, and the subsequent behavior of the newly arisen mutable loci, have provided evidence that allows an interpretation of their mode of origin and also their mode of operation. As a consequence of this study, some rather unorthodox conclusions have been drawn regarding the mechanisms responsible for mutations arising at these loci. The same mechanisms may well be responsible for the origins of many of the observed mutations in plants and animals.

Instability of various loci—whether referred to by the terms mutable loci, mutable genes, or variegation, position effect, etc.—has been known for many years, and many such cases have received considerable study. The conditions associated with the more obvious position-effect phenomena in *Drosophila* are well known. Those associated with instability of phenotypic expressions in other organisms have been less well understood. It is because of the distinctive advantages that the maize plant offers for such a study that it has been possible to obtain precise evidence concerning some of the events associated with the origin and behavior of mutable loci. The first of these advantages relates to the ease of observing the chromosomes, and thus determining the nature of some of the changes that occur in them. The presence of a triploid endosperm in

the kernel provides a second advantage. This endosperm, with its outer aleurone layer that can develop pigments, and the underlying tissues that may develop starches of several types, or sugars, or carotenoid pigments, permits the detection of differences in phenotypic expression of various types. Some of these may be quantitatively measured. Thirdly, there are a number of different loci known in which heritable alterations have given rise to changes in the expression of these several endosperm components. The mutations at some of these loci affect characters of both the endosperm and the plant tissues. This applies particularly to those mutations that affect the development of the anthocyanin pigments. In the studies to be described, the presence in the short arm of chromosome 9 of four marked loci that affect endosperm characters has been of particular importance for analyzing the events occurring at mutable loci. The necessity of having such markers will become evident in the discussion. For this study, the accumulated knowledge of the behavior of newly broken ends of chromosomes in maize has been of particular importance. Its significance for interpreting the origin of mutable loci will be indicated in the sections that follow.

THE CHROMATID AND CHROMOSOME TYPES OF BREAKAGE-FUSION-BRIDGE CYCLE

The diagrams of Figure 1 illustrate the mode of origin of newly broken ends of chromosomes at a meiotic mitosis and the subsequent behavior of these ends in successive mitotic cycles. A chromosome with a newly broken end entering a telophase nucleus in the gametophytic or endosperm tissues will give rise in the next anaphase to a chromatid bridge configuration (McClintock, 1941). The bridge is produced because fusion occurs between sister chromatids at the position of previous anaphase breakage. This sequence of anaphase breaks and sister-chromatid fusions will continue in successive mitoses. It has therefore been designated the chromatid type of breakage-fusion-bridge cycle. This cycle is illustrated in A of Figure 1. In the sporophytic tissues, however, this cycle usually does not occur. The broken end entering a telophase

14   *BARBARA McCLINTOCK*

A

THE CHROMATID TYPE OF BREAKAGE-FUSION-BRIDGE CYCLE

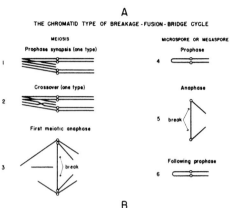

B

THE CHROMOSOME TYPE OF BREAKAGE-FUSION-BRIDGE CYCLE: SAME AS A FROM
MEIOSIS TO FORMATION OF EGG AND SPERM NUCLEI

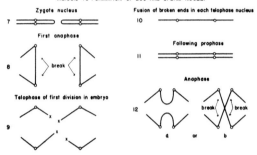

FIG. 1. Diagrams illustrating the origin of a newly broken end of a chromosome at the meiotic anaphase and its subsequent behavior. A. The chromatid type of breakage-fusion-bridge cycle. 1. One type of synaptic configuration at the first meiotic prophase between homologous arms of a pair of chromosomes, one member of which carries a duplication of this arm in the inverted order. 2. The production of a dicentric chromatid as the consequence of a crossover. It is composed of two complete chromatids of this chromosome. 3. Anaphase *I*. Bridge configuration produced by separation of centromeres of the dicentric chromatid. A break in the bridge occurs at some position between the two centromeres. 4. Fusion of sister chromatids at the position of the previous anaphase break is exhibited in prophase of the microspore or megaspore nucleus. 5. Separation of sister **centromeres** at anaphase in the microspore or megaspore produces a bridge configuration. This bridge is broken at some position between the two centromeres. 6. Fusion of sister chromatids occurs at the position of the preceding anaphase break. Separation of sister centromeres at the next anaphase again produces a bridge which is broken at some position between the two centromeres. This cycle continues in successive mitoses during the development of the gametophyte and the endosperm.

B. The chromosome type of breakage-fusion-bridge cycle. It may be initiated in the sporophyte if each gamete contributes a chromosome which has been broken in the anaphase of the division preceding gamete formation. The zygote nucleus will then contain two such chromosomes. In the prophase of the first division of the zygote, 7, each of these is composed of two sister chromatids fused at the position of the previous anaphase break. In the first anaphase of the zygotic

nucleus heals, and its subsequent behavior resembles that of a normal, nonbroken end of a chromosome. (Note: The chromatid type of breakage-fusion-bridge cycle can continue throughout the development of the sporophytic tissues under certain conditions. These conditions are usually not present in the genetic stocks of maize.) If, however, a chromosome with a newly broken end is introduced into the zygote by each gamete nucleus, the broken ends of the two chromosomes are capable of fusion (McClintock, 1942). This establishes a dicentric chromosome. A different type of breakage-fusion-bridge cycle is thereby initiated. In the telophase nuclei, the fusions now occur between the broken ends of chromosomes rather than between the broken ends of sister chromatids, as described above. This sequence of events has been called the chromosome type of breakage-fusion-bridge cycle, and is illustrated in B of Figure 1. A study of the consequences of these cycles has revealed that they may initiate breakage events in chromosomes of the complement other than those undergoing the cycle. This complication has been of significance, for it appears that these unanticipated alterations of the chromosomes may be responsible primarily for the origin of mutable loci and of other types of heritable change.

UNEXPECTED CHROMOSOMAL ABERRATIONS
INDUCED BY THE BREAKAGE-FUSION-
BRIDGE CYCLES

In the course of an experiment designed to induce small internal deficiencies within the short arm of chromosome 9, a number of plants were obtained that had undergone the chromosome type of breakage-fusion-bridge cycle in their early developmental period. The short arm of each chromosome 9 was involved in this cycle. It is

division, 8, these two chromosomes give rise to bridge configurations as the centromeres of the sister chromatids pass to opposite poles. Breaks occur in each bridge at some position between the centromeres. In the telophase nuclei, two chromosomes, each with a newly broken end, are present as diagrammed in 9. The crosses mark the broken ends of each chromosome. Fusion of broken ends of *chromosomes* occurs in each telophase nucleus, 10, establishing a dicentric chromosome. In the next prophase, 11, each sister chromatid is dicentric. At the subsequent anaphase, several types of configurations may result from separation of the sister centromeres, two of which are shown in 12. Separations as shown in *b* of 12 give rise to anaphase bridge configurations. Breaks occur in each bridge at some position between the centromeres. The subsequent behavior of the broken ends, from telophase to telophase, is the same as that given in diagrams 9 to 12.

known that the cycle will often cease suddenly in certain cells and that these cells are then capable of developing sexually functional branches of the plant. In order to determine the nature of the chromosome changes produced by this cycle, the sporocytes of many of these plants were examined at the pachytene of meiosis. The expected types of altered constitution of the short arm of chromosomes 9 were found. In addition, other quite unexpected types of chromosome aberration appeared in a number of the plants. These alterations had been produced in the early developmental periods when the breakage-fusion-bridge cycles were occurring. With a few exceptions, the chromosome parts in which alterations had been initiated were the knobs and the centromeres, or the nucleolus organizer of chromosome 6. In the majority of cases, either the knob or the centromere of one of the chromosomes 9 that had been undergoing the breakage-fusion-bridge cycle was involved in the structural rearrangement. Nonrandomness was apparent with regard to the other chromosome involved in the aberration. For example, four cases were found in which the centromere of chromosome 9 had fused with the centromere of another chromosome—chromosome 2 in three of the four cases. Chromosome 8 was also very frequently involved in these structural changes.

The breakage-fusion-bridge cycle was obviously responsible for the induction of these alterations in the knobs, centromeres and the nucleolus organizer. That alterations in such elements were occurring without obvious direct participation of the knob or the centromere of the chromosome 9 undergoing the breakage-fusion-bridge cycle has also been indicated. This was made evident by the presence in one plant of an inversion involving the nucleolus organizer and the centromere region in chromosome 6, by an inversion in chromosome 5 in another plant involving the centromere and the knob regions, and by an inversion in chromosome 7 in a third plant involving the centromere region and the knob region in the long arm of this chromosome. In addition, some of the plants examined showed the presence of a ring chromosome that was not composed of segments of chromosome 9, so far as could be determined. It now must be emphasized that it was in the self-pollinated progeny of plants that had undergone the chromosome type of breakage-fusion-bridge cycle in their early developmental period that the initial burst of newly arisen mutable loci appeared. It might be suspected that this burst was a reflection of the mechanism that

had produced the alterations mentioned above. If so, the origin of mutable loci would be associated with change in these particular elements of the chromosome complement. It was some time, however, before sufficient evidence had accumulated to allow deductions to be drawn regarding this presumptive relationship. A description of the origin and behavior of some of the representative types of mutable loci should be given before this topic is again considered.

RECOGNITION OF THE RELATION OF MUTATION TO THE MITOTIC CYCLE

Interest in these mutable loci, appearing unexpectedly and in large numbers in the self-pollinated progeny of plants that had undergone the chromosome type of breakage-fusion-bridge cycle in their early developmental periods, was aroused when it was realized that in each case some factor was present which controlled the time or the frequency of mutations. This factor could be altered as a consequence of some event associated with the mitotic process. This was made evident by the appearance of sectors of tissue, derived from sister cells, that exhibited obvious differences in time of mutations, mutation frequency, or both. In many cases, it was also apparent that the mutations themselves arose as a consequence of some event associated with the mitotic cycle. This basic behavior pattern was exhibited by all the various newly arisen mutable loci. It directed attention to the mitotic mechanism as the responsible agent. It was concluded, therefore, that further investigation of these mutable loci might produce some evidence leading to an appreciation of the nature of the responsible mitotic events.

During six years of study of a number of newly arisen mutable loci, some well-established facts have accumulated concerning the processes associated with the origin of mutable loci and their subsequent behavior. Observation of consistent behavior in many mutable loci, where the cytological events associated with a change in phenotype could be determined, and comparison of the behavior of these loci with others in which cytological determinations could not readily be made, have provided an assemblage of interrelated facts upon which the conclusions to be stated later are based.

THE ORIGIN OF *Ds* AND ITS BEHAVIOR

The first evidence of the type of chromosomal event that is associated with the expression of mutability came with the discovery of a locus in the short arm of chromosome 9 at which chromosome breaks were occurring. This was observed in the self-pollinated progeny of one of the plants that had undergone the chromosome type of breakage-fusion-bridge cycle in early development. When first seen, the "mutability" was expressed by the time and frequency of the breaks that occurred at this locus in some cells during the development of a tissue. Also, some change could occur in somatic cells that affected the time and frequency; and this latter event likewise was associated with the mitotic process. The behavior pattern resembled in considerable detail the patterns exhibited by the mutable loci. In this case, however, a mechanism associated with chromosome fusion and subsequent breakage was responsible for the behavior observed. The mutations from recessive to dominant exhibited by the mutable loci would not alone have suggested a chromosome-breakage mechanism as being responsible. Because of this similarity of the patterns of behavior, it was suspected that the basic mechanism responsible for mutations at mutable loci could be one associated with some form of structural alteration at the locus showing the mutation phenomenon. This conclusion was consistent with the very first observations of the behavior of mutable loci. These observations had indicated that the events at mutable loci leading to mutations and also other events controlling their time and frequency of occurrence were associated with alterations that were in some manner produced during the course of a mitotic cycle.

Intensive study of this locus in chromosome 9 at which structural alterations occur at regulated rates and at regulated times in development has been rewarding. A "break" in the chromosome at this locus was the event first recognized. The factor responsible was therefore given the symbol *Ds*, for "Dissociation." The nature of the breakage event was later determined. It arises from dicentric and acentric chromatid formations. The acentric fragment is composed of the two sister chromatids, from the *Ds* locus to the end of the short arm. The complementary dicentric component includes the sister segments from the locus to the centromere plus the long arms of the two sister chromatids. This is the type of recognizable event found to occur most frequently at *Ds*. Other recognizable aberrations, however, may sometimes arise. One of them is the formation of an internal deficiency in the short arm of chromosome 9. Such deficiencies include the

regions adjacent to *Ds*, and vary in extent from minute to quite large. Translocations between this chromosome and another chromosome of the complement may arise, with one of the points of breakage at the *Ds* locus. Duplications, or inversions, of segments within chromosome 9 may also be produced, one of the breakage points being at *Ds*.

It was realized early in this study of *Ds* that changes could occur at the locus leading to marked alterations in frequency of the detectable breakage events. The original isolate was showing high frequencies of formation of dicentric chromatids and the associated acentric fragments. Changes arose at the locus, however, as a consequence of some event occurring in a somatic cell. These changes resulted in the appearance, in subsequent cell and plant generations, of lowered frequencies of these events. Such changes in the behavior pattern of *Ds* were called "changes in state"; and the *Ds* with the altered state behaved in inheritance as an allele of the original isolate of *Ds*. A subsequent change could occur, which again was recognized by an altered frequency of detectable breakage events, and which behaved in inheritance as an allele of the initial state, of the derived state, or of other unrelated derived states. By selecting altered states of *Ds*, a series of alleles of the original *Ds* has been isolated. The changes in state of *Ds*, and those occurring at other mutable loci, are of considerable significance in understanding the nature of the events responsible for the patterns of behavior of all mutable loci. A discussion of this significance will be postponed until the behavior of some other mutable loci have been considered. The meaning of the term will then be readily apparent.

TRANSPOSITION OF *Ds*

An important aspect of this study, with regard to the origin of mutable loci and nature of their mutation process, is related to transposition of *Ds* from one location in the chromosome complement to another. The discovery of such transpositions occurred in the course of studies aimed at determining the exact location of *Ds* in chromosome 9. These tests involved linkage relationships. A sequence of six marked loci along the chromosome arm were used, and the linkage studies clearly established the location of *Ds* as shown in Figure 2. This genetically determined location fitted the position of breaks in the chromosome observed in some of the sporo-

CHROMOSOME ORGANIZATION AND GENIC EXPRESSION   17

cytes of plants having *Ds* in either one or both chromosomes 9. Such chromosome breaks are illustrated in the photographs of microsporocytes at pachytene given in Figures 4 to 8. This was the location of *Ds* when it was first discovered, and has been called the standard location.

In the course of studies of the inheritance behavior of *Ds*, an occasional kernel appeared which showed that *Ds*-type activity—that is, chromosome breakage—was occurring at a new position in the short arm of chromosome 9. Attempts were made to germinate such kernels when they were found. If a plant arose from one, a study was then commenced to determine the new location of the *Ds*-type activity. Over 20 cases of the sudden appearance of *Ds*-type activity in new locations in the short arm of chromosome 9, and several cases of its sudden appearance in other chromosomes of the complement, have been investigated. Within the short arm of chromosome 9, such activity has appeared at various positions. All the isolates studied have shown sharply defined locations of the *Ds*-type activity. In these cases, the cytological determination of breakage position and the genetic determination of location were in agreement. New positions of *Ds*-type activity have appeared between all of the marked loci shown in Figure 2. For example, in four independently arisen cases, the new position of *Ds* has been located between *I* and *Sh*. In two of these, it is to the right of *I*, at or close to the same position in each case—approximately one-fifth the crossover distance between *I* and *Sh*. In the other two it is to the left of *Sh*, with a very low percentage of crossing over between *Ds* and *Sh* in each case.

The mode of detecting new locations of *Ds*-type activity has been selective, in that those arising in the short arm of chromosome 9 are immediately revealed on many of the ears coming from test crosses. *Ds*-type activity has suddenly appeared, however, in other chromosomes of the complement. Only when appropriate genetic markers are present can it be detected readily; and in most tests, such markers have not been present.

Several questions must now be asked. How do new positions of *Ds* activity arise? And what conditions are responsible for their occurrence? The methods used in seeking answers to these questions may be described. In some cases, it could be established that the appearance of *Ds* activity at a new location was associated with its disappearance at the known former location. It has been emphasized that the mechanism underlying *Ds* events is one that can give rise to translocations, deficiencies, inversions, ring-chromosomes, etc., as well as the more frequently occurring dicentric chromatid formations with reciprocal formation of acentric fragments. It has also been stated that in each such case one breakage point is at the known location of *Ds*. The appearance of *Ds* at new locations is probably associated with such a break-inducing mechanism. This was indicated by extensive analysis of the constitutions of two independent duplications of segments of the short arm of chromosome 9 when a new location of *Ds* activity was also present in this arm. In both cases, only one of the many tested gametes of one of the parent plants carried the particular chromosome aberration with the new location of *Ds*. It was detected in two single aberrant kernels on separate ears coming from similar types of crosses made in two different years. The female parent carried two morphologically normal chromosomes 9, each with the markers *C*, *sh*, *bz*, and *wx*. No *Ds* (or *Ac*, see below) was present in these plants. The male parent (one *Ac* present) carried two morphologically normal chromosomes 9. The markers *I*, *Sh*, *Bz*, *Wx*, and *Ds* (at its standard location) were present in one chromosome 9. The homologous chromosome carried *C*, *sh*, *bz*, *wx*, but no

FIG. 2. Diagram showing the approximate locations of the genetic markers in the short arm of chromosome 9 that have been used in this study. In symbolization, dominance is indicated by a capital letter or capitalization of the first letter. Recessiveness is indicated by lower-case letters. The symbols refer to the following plant or endosperm characters: *Yg*, normal chlorophyll; *yg*, yellow-green chlorophyll color in early period of development of the plant. *Sh*, normal endosperm; *sh*, shrunken endosperm. *I*, *C*, and *c* form an allelic series associated with pigment development in the aleurone layer of the endosperm. *I*, inhibitor of aleurone color formation, dominant to *C*. *C*, aleurone color, dominant to *c*, colorless aleurone. The *Bz* factor is associated with development of aleurone and plant color. When homozygous, the recessive, *bz*, (bronze), gives rise to an altered anthocyanin color in the aleurone and plant tissues, from a dark red or purple to a bronze shade. When *Wx* is present, the starch in the pollen and endosperm stains blue with iodine solutions, due to the presence of amylose starch; when only the recessive *wx* (waxy) is present, no amylose starch is formed and with iodine solutions, the starch stains a reddish-brown color. The position of *Ds*, indicated in the diagram, is the standard location (see text).

18   *BARBARA McCLINTOCK*

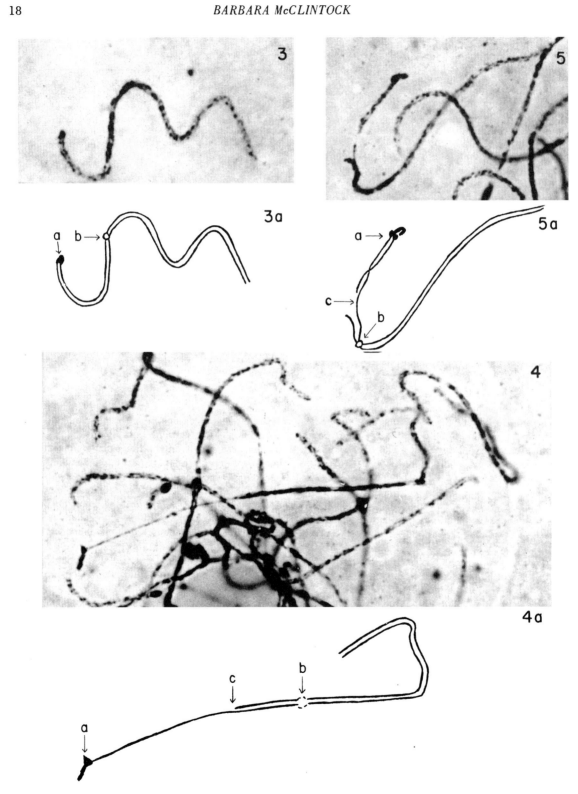

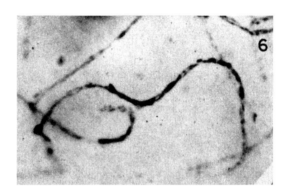

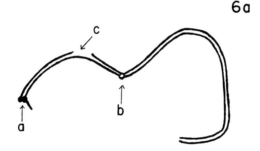

FIG. 3. Photograph of a normal bivalent chromosome 9 at pachytene of meiosis. In the accompanying diagram, 3a, the knob terminating the short arm is indicated by the arrow, *a*. The centromere is indicated by the arrow *b*. Mag. approximately 1800x. Fusion of homologous centromeres appears to occur at pachytene. Consequently, in the diagrams accompanying Figures 3 to 8, this region is indicated as single rather than double.

FIGS. 4 to 7 and accompanying diagrams, 4a to 7a. Illustrations of the position of breaks at the *Ds* locus as seen at pachytene of meiosis in plants having *Ds* at its standard location in one chromosome 9 and no *Ds* in the homologue. The two homologues are distinguishable. At the end of the short arm of the chromosome 9 having no *Ds*, a segment of deep-staining chromatin extends beyond the knob. The short arm of the chromosome 9 carrying *Ds* terminates in a knob. Magnifications approximately 1800x. In Figure 4, a break at *Ds* occurred in a premeiotic mitosis. The acentric fragment, from *Ds* to the end of the arm, was lost to the nucleus. Consequently, this segment is missing in the bivalent. The homologous segment in the chromosome 9 having no *Ds* is therefore univalent. In making the preparation, this segment was considerably stretched. In the accompanying diagram, arrow *a* points to the knob and the small deep staining segment extending beyond the knob. Arrow *b* points to the centromere region, not clearly shown in the photograph. Arrow *c* points to the position of the break in the chromosome 9 that carried *Ds*. Figures 5 and 6 show the appearance of the bivalent chromosome 9 when a break in the *Ds* carrying chromosome occurred at the meiotic prophase and when the free segment, from *Ds* to the end of the arm, paired with its homologous segment in the chromosome 9 having no *Ds*. In the accompanying diagrams, arrow *a* points to the knobs, arrow *b* points to the centromeres and arrow *c* to the position of the *Ds* break in one of the homologues. Figure 7 is similar to Figures 5 and 6 except that the free fragment, from the position of *Ds* to the end of the arm, did not pair with its homologous segment in the chromosome 9 having no *Ds*. In the accompanying diagram, arrow a^1 points to the knob and the deep-staining chromatin extending beyond the knob in the chromosome having no *Ds*. Arrow a^2 points to the knob of the unpaired acentric fragment. Arrow *b* points to the position of the centromeres, not observable in the photograph. Arrow c^1 points to the broken end of the centric segment, and arrow c^2 points to the broken end of the acentric segment. (For Fig. 7, see next page.)

20   *BARBARA McCLINTOCK*

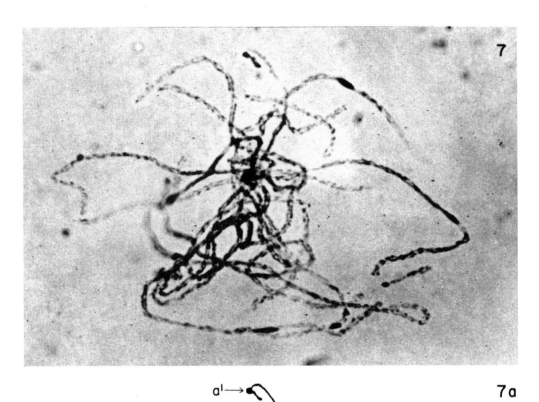

7

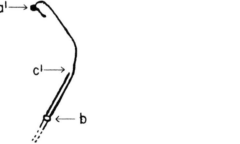

7a

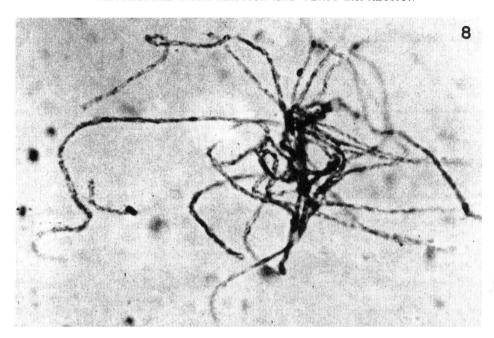

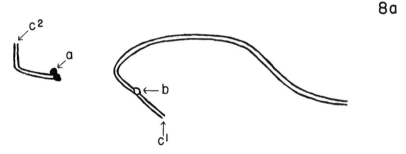

FIG. 8. The chromosome 9 bivalent at pachytene in a plant having Ds at its standard location in each chromosome 9. A Ds break occurred at the meiotic prophase in each chromosome. Consequently, the bivalent is composed of two free segments: one short acentric segment, from Ds to the end of the arm, and a long centric segment, from the position of Ds to the end of the long arm. In the accompanying diagram, arrow a points to the large knob terminating the short arm of each member of the acentric bivalent segment, and arrow b points to the centromeres of the centric bivalent segment. Arrow c^1 points to the broken ends of the centric bivalent segment and arrow c^2 points to the broken ends of the acentric bivalent segment.

Ds. The constitutions of the aberrant chromosomes 9, which were present in an individual pollen grain in each case, are shown in Figure 9. Each has a duplication of a segment of the short arm of chromosome 9. A study of these constitutions reveals that, in each case, the new position of Ds activity coincided with the position of one of the breaks that produced the duplication. Also, the position of the second break coincided with the previously known location of Ds in the morphologically normal Ds-carrying chromosome 9 of the male parent plant. It seems clear from the analysis of both these cases that the breaks were Ds initiated, and also that both breaks involved sister chromatids at the same location. Two other cases of newly arisen duplications associated with new positions of Ds activity have received study. These have proved to be similar in their modes of origin to the two cases diagrammed. Although these cases suggest that the new positions of Ds activity may arise from contacts with the Ds that is present in the chromosome complement, they do not constitute evidence that Ds is composed of a material substance and that the new positions arise from insertion of this material into new locations.

22   *BARBARA McCLINTOCK*

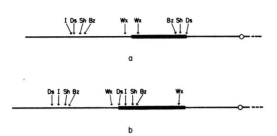

FIG. 9. Diagrams illustrating the constitution of two chromosomes 9, each of which carries a duplicated segment (heavy line). A transposition of *Ds* accompanied the formation of the duplication in each case. For details, see text.

Ac BEHAVIOR AND INHERITANCE: GENERAL CONSIDERATIONS

Before continuing discussion of the sudden appearance of *Ds* at new locations, it is necessary to consider another heritable factor called Activator (*Ac*). *Ac* is a dominant factor that must be present if any breakage events are to occur at *Ds*. If *Ac* is absent from the nucleus, no detectable events whatsoever occur at *Ds*, wherever it may be located; and no new positions of *Ds* activity appear. Because of this, and because new positions of *Ds* arise only in a few cells, during the period of development of a tissue when breakage events are occurring at *Ds*, it may be concluded that these new positions are one of the consequences of the mechanism that produces chromosome breakage events at *Ds*.

The location of *Ds* in chromosome 9 was first suggested by the altered phenotype of sectors of tissue derived from cells in which a break had occurred at *Ds*. Such sectors will appear if the factorial constitutions of the homologues differ. The chromosome arm carrying *Ds* must also carry dominant markers, and the homo-

logue must carry the recessive alleles. If one chromosome 9, delivered by the male parent, carries *I*, *Sh*, *Bz*, *Wx*, and *Ds* at its standard location, and if the homologue delivered by the female parent carries the recessive alleles *C*, *sh*, *bz*, and *wx*, and has no *Ds*, the events at *Ds* that form a dicentric chromatid and an acentric fragment will lead to elimination of the acentric fragment during a mitotic division. This fragment will form a pycnotic body in the cytoplasm, which subsequently disappears. All the dominant markers carried in this acentric fragment will be removed from the nuclei after such an event has occurred. The sector of tissue derived from these cells will exhibit the collective phenotype of the recessive alleles carried by the homologous chromosome 9 having no *Ds*.

If pollen of plants having one *Ac* factor and carrying *I*, *Sh*, *Bz*, *Wx*, and *Ds* (standard location) is placed upon silks of plants carrying *C*, *sh*, *bz*, and *wx* in their chromosomes 9, but having no *Ds* or *Ac*, two types of kernels will appear, those that received an *Ac* factor and those without *Ac*. The latter kernels will be colorless and nonshrunken, and will show the *Wx* phenotype. No variegation for the characters exhibited by the recessive alleles will appear, as shown in Figure 10. If *Ac* is present in the endosperm, however, the described *Ds* events will occur in some cells during the development of the kernel. This will result in elimination of the segment in the short arm of chromosome 9 carrying the dominant factors. All the cells arising from one in which such an event has occurred will exhibit the *C*, *sh*, *bz*, and *wx* phenotypes. Consequently, these kernels will be variegated. The photographs of kernels in Figures 11 to 13 will illustrate the nature of this variegation.

FIGS. 10 to 15. Photographs of kernels illustrating the effects produced by breaks at *Ds*, and the relation of the presence or absence of *Ac* and the doses of *Ac* on the time of occurrence of *Ds* breaks. The kernels arose from the cross of a plant (♀) carrying *C* and *bz* and having no *Ds* in each chromosome 9, by plants (♂) having *I*, *Bz*, and *Ds* at its standard location in chromosome 9. In Figure 10, no *Ac* is present; the kernel is completely colorless due to the inhibition of aleurone color when *I* is present. In Figures 11 to 13, one *Ac* is present. Breaks at *Ds* occur early in development with consequent loss of *I* and *Bz* to the nuclei. The cells without *I* and *Bz* give rise to sectors showing the *C bz* phenotype. It should be noted that *Bz* substance diffuses through several cell layers, from the *I Bz* areas into the *C bz* areas, producing a *C Bz* phenotypic expression in these genotypically *C bz* cells. Thus, all large areas of the *C bz* genotype are rimmed with the dark aleurone color of the *C Bz* phenotype. Small *C bz* areas may be mostly *C Bz* in phenotype because of this diffusion, and very small *C bz* areas are totally *C Bz* in phenotype. These relationships are clearly expressed in the photographs. In Figures 11 to 13, both large and small *C bz* sectors are present as well as areas in which no *C bz* sectors appear. This irregularity arises from alterations that occur to *Ac* during the development of the kernels. These give rise to sectors with no *Ac* or with altered doses of *Ac* (see text). Figure 14 shows the pattern that may be produced by *Ds* breaks when two *Ac* factors are introduced into the primary endosperm nucleus. There are many small sectors of the *C bz* genotype, all produced by relatively late occurring *Ds* breaks. This results in a heavily speckled variegation pattern. Figure 15 shows the pattern that may appear when three *Ac* factors are present. Only relatively few *C bz* specks, produced by late occurring *Ds* breaks, are present in this kernel.

CHROMOSOME ORGANIZATION AND GENIC EXPRESSION 23

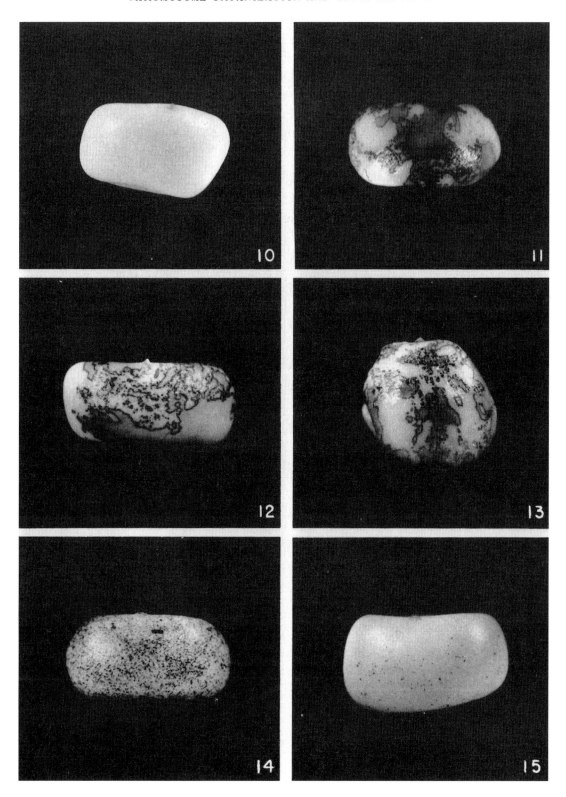

24   *BARBARA McCLINTOCK*

In the early studies, it soon became apparent
that the time of occurrence of *Ds* breakage events
in the development of a tissue depends upon the
dose of *Ac* present. The higher the dose of *Ac*,
the later in development will such events at *Ds*
occur. Because the endosperm is triploid—the
female parent contributing two haploid nuclei
derived from the female gametophyte, and the
male contributing one haploid nucleus—single
to triple doses of *Ac* may readily be obtained
after appropriate crosses. The time of response
of *Ds* to double and triple doses of *Ac* is shown
in Figures 14 and 15.

Initial studies of the inheritance behavior of
Ac showed that it follows the mendelian laws
known to apply to single genetic units. An illus-
tration of this inheritance behavior is given in
Figure 16. The ear in this figure was derived
from a cross in which the female parent carried
the recessive factor *c* in each chromosome 9, but
had no *Ds* or *Ac* factors, and the male parent
carried in each chromosome 9 the factors *C* and
Ds (standard location). It also carried a single
Ac factor. The expected 1-to-1 ratio of colored,
nonvariegated kernels (no *Ac* present) to kernels
showing sectors of the *c* phenotype (*Ac* present)
was apparent on this ear. Efforts to determine
the location of *Ac* in the chromosome complement
were commenced, but were soon abandoned when
it was realized that *Ac* need not remain at any
one location in the complement. It can appear
at new locations and in different chromosomes.
Because of this highly unexpected behavior of a
genetic factor, extensive studies were made to
determine the mode of inheritance of *Ac* and to
learn how these new positions arise. Much has
been learned from them about *Ac* behavior and
inheritance. The modes of investigating alterations of *Ac* will be considered later. Some of
the facts concerning its behavior, however, may
be stated here in summary form, by describing
the results obtained from several types of
experiments.

If plants having one *Ac* factor are self-pollinated,
the expected mendelian ratios may appear in the
F_2 populations. These are: one with two *Ac*, to
two with one *Ac*, to one with no *Ac*. The ratios
obtained in one such test were 61 : 145 : 68, which
is close to the statistical expectancy. When,
however, these F_2 plants having two *Ac* factors
are crossed by plants having no *Ac*, the expectation would be that all the progeny would have
one *Ac* factor. Usually, most of the plants do
have one *Ac* factor; but sometimes there are
plants displaying other conditions with respect

FIG. 16. Photograph of an ear produced when pollen
from plants having *C* and *Ds* (standard location) in
each chromosome 9 and carrying one *Ac* factor is
placed on silks of plants carrying *c* in each chromosome 9 and having no *Ds* or *Ac* factors. Approximately
half of the male gametes have no *Ac* factor. The
kernels arising from the functioning of such gametes
are fully colored; no variegation appears. The other
half of the male gametes carry *Ac*. The kernels arising
from the functioning of these gametes are variegated.
They show a number of sectors with the *c* phenotype.
Note the approximate 1 to 1 ratio of fully colored kernels to variegated kernels.

to *Ac*. The following unexpected types have appeared: plants having no *Ac* factor; others having
two nonlinked *Ac* factors; others having two *Ac*
factors that appear to be linked; still others
having an *Ac* factor that acts as a single unit in
inheritance but gives the same dosage action as
would two doses of the *Ac* factor in the parent
plant that contributed *Ac*. The dosage action of
Ac may be altered in other ways, moreover, so
that a single *Ac* factor, as determined genetically,
may exert an action either less than that of the
Ac factor contributed by the *Ac*-carrying parent
plant or falling between one and two doses of
this factor.

In the early studies of *Ac* inheritance, the *Ac*
factor was found not to be linked with the genetic
markers in the short arm of chromosome 9. In
one series of tests of *Ac* inheritance, where a
number of *Ac*-carrying plants were derived from
a cross between a plant having no *Ac* and a plant

having one *Ac*, a single aberrant plant was present in the F₁. Tests of this plant showed that it possessed a single *Ac* factor, which was obviously linked to the markers in the chromosome 9 delivered by the *Ac*-carrying parent plant. In the sister plants, and in the parent plant that contributed *Ac*, no such linkage was evident. Studies were then conducted to examine the linkage relationships of this *Ac* with the marked loci. The very same types of tests for linkage of *Ac* with the genetic markers in chromosome 9 were also conducted with the sister plants. A summary of

tance to the left of *I*. Studies were then undertaken to investigate not only the linkage behavior of *Ac* but also the types of events that alter *Ac* in these two sharply delimited locations. It was determined that in the majority of the *Ac*-carrying gametes produced by plants having an *Ac* at either of these two stated positions, no change in location or action of *Ac* occurred. Exchanges of *Ac* from one homologue to another took place as a consequence of crossing over, with consistent frequencies in each case. In a few of these gametes, however, the above-described types of

TABLE 1

Comparisons of *Ac* inheritance: (1) when *Ac* was not linked to markers in chromosome 9, and (II) when *Ac* was·linked to markers in chromosome 9, in crosses of

$$(I) \quad \text{♂} \frac{I \ Sh \ Bz \ Wx \ Ds}{C \ Sh \ Bz \ wx \ Ds} \ 1 \ Ac, \text{ nonlinked}$$

♀ *C sh bz wx/C sh bz wx*, no *Ds*, no *Ac* by (I) and

$$(II) \quad \text{♂} \frac{I \ Sh \ Bz \ Wx \ Ds \ Ac}{C \ Sh \ Bz \ wx \ Ds \ ac}$$

Chromosome-9 Constitution of ♂ Gamete	I		II	
	Variegated* *Ac* present	Nonvariegated* No *Ac*	Variegated* *Ac* present	Nonvariegated* No *Ac*
I Wx	268	255	928	246
C wx	248	242	164	893
I wx	88	91	62	387
C Wx	84	83	344	100

*Presence of *Ac* detected by sectors of the *C sh bz wx* phenotype in kernels on ears obtained from cross I or cross II. In the absence of *Ac*, no *Ds* events occur; the kernels are therefore nonvariegated.

one set of these comparative tests of *Ac* inheritance is presented in Table 1. In part I of this table, no linkage of *Ac* with *I* or *Wx* is evident; in part II, however, linkage is obvious. The data place *Ac* to the right of *Wx*. Approximately 20 per cent crossing-over appears to have occurred between *Wx* and *Ac*. Actually, this figure is only approximate, for a few kernels in the *Ac-wx* class do not carry *Ac* in chromosome 9. They do not belong in the crossover class. The *Ac* in these kernels had been transposed from chromosome 9 to another chromosome. Also, a few kernels in the *Wx*-no *Ac* class have not lost *Ac* because of crossing over but rather because it was removed from its location in chromosome 9.

The described case of sudden appearance of *Ac* in chromosome 9 is not the only one that has been found and similarly studied. Seven independent cases have so far been identified. In two of these, *Ac* appeared in the short arm of this chromosome: in the first case, a short distance to the left of *Wx*; and in the second case, a short dis-

aberrant *Ac* conditions were present, as follows: (1) *Ac* was no longer present in chromosome 9, but was carried instead by another chromosome. (2) Two *Ac* factors were present, one at the given location in chromosome 9, and one carried by another chromosome of the complement. (3) The *Ac* factor was unchanged in its location but showed an altered action; in a single dose it could be equivalent to the double dose before the alteration occurred, or could show an increased but not doubled dosage action, or a decreased dosage action.

The behavior of *Ac* was also studied in plants in which the *Ac* factor was present at an allelic position in each chromosome 9. Again, it could be determined that in a few gametes produced by such plants the above-described aberrant conditions with respect to *Ac* position and action were present. In addition, a few gametes were formed that had no *Ac* factor at all.

From what has been said about both *Ds* and *Ac*, it is apparent that with respect to inheritance behavior they are much alike. The same questions concerning mode of appearance in new locations in the chromosome complement apply to *Ac* as to *Ds*. With this relationship in mind, we may now return to further considerations of *Ds*. It should be emphasized again, however, that the described events occurring at *Ds*, wherever it may be located in the chromosome complement, depend upon the presence of an *Ac* factor in the nucleus, regardless of where this latter factor is located in the chromosome complement. New positions of *Ds* activity arise only when *Ac* is also present in the nucleus, and, again, regardless of where *Ac* may be located. In addition, any one altered state of *Ac*—for example, an altered dosage action—affects *Ds* wherever it may be located, and in exactly the same manner for every *Ds*, regardless of its state. In other words, it is the state and the dose of *Ac* that control just when and where *Ds* events will occur, and it is the state of a particular *Ds* that controls the relative frequency of any one type of event that occurs at *Ds*.

The Origin and Behavior of c^{m-1}

In the discussion of the appearance of *Ds* at new locations, the question was raised whether or not this involves the transposition of a material substance from one location to another. The question applies equally to *Ac*. If no material substance is transposed, a serious problem is presented regarding the basic action of any known genetic unit or factor that has been assigned to a particular locus in a particular chromosome. *Ac* clearly produces an obvious, measurable, phenotypic response, wherever it may be located. It shows dosage action, mendelian inheritance, and linkage behavior of the expected type, in any location—with the exception, already mentioned, of a few transpositions, changes in state, and losses of *Ac*. It might be considered that *Ds* and *Ac* represent forms of altered chromosome organization producing somewhat similar effects in each case, much like the Minutes in *Drosophila*. The evidence now to be presented, however, makes that assumption unlikely. This evidence considers the origin and the behavior of many different mutable loci. To begin this part of the discussion, we may consider the origin of mutable c^{m-1}, the first-detected mutable *c* locus that arose in a chromosome carrying a normal-behaving *C* locus (*C*, aleurone color; *c*, recessive allele, colorless aleurone).

The presence of an alteration at the known locus of *C*, which produced c^{m-1}, was detected probably within a few nuclear generations after it occurred. It was present only in one kernel among approximately 4,000 examined that had come from a cross of a single plant, used as a male, to 12 genetically similar female plants. All the other kernels on these ears gave the expected types of phenotypic expression. The male parent carried in both chromosomes 9 the genetic markers *Yg*, *C*, *Sh*, *wx*, and *Ds* (standard location). It also had one *Ac* factor, not linked to these markers. The female parents carried the stable recessive *yg*, *c*, *sh*, and *wx*. No *Ds* was present in their chromosomes 9, and also no *Ac* factor was present in these plants. The types of kernels to be expected from such a cross, and their relative frequencies, are the same as those shown in Figure 16. Approximately half the kernels should show the *C Sh wx* phenotype, with no variegation for the recessive characters since no *Ac* factor is present. The other half should carry *Ac* and thus be variegated. Sectors showing the *c sh* phenotype should be present. In these crosses, the expected classes of kernels appeared with the exception of one kernel. Instead of showing colorless areas in a colored background (resulting from losses of *C* following breakage events at *Ds*), this exceptional kernel showed a colorless background in which colored areas were present. The plant derived from this kernel was tested in various ways in order to determine the reason for this unexpected type of variegation. The early tests indicated, and subsequent tests proved, that mutations were occurring from the recessive *c*, to the dominant *C*, and that the mutable condition had arisen in one of the chromosomes 9 contributed by the *Ac*-carrying male parent plant.

Ds-type activity was also present in the chromosome carrying the new mutable *c* locus. The location of this activity was no longer to the right of *wx*, as would be expected since this was its location in both chromosomes 9 of the male parent plant. The new location was inseparable from that of the mutable *c*. All the recognizable breakage events associated with *Ds*-type activity now happened at this new location. In addition, mutations to *C* occurred. It was soon discovered that the mutations to *C* would appear only if *Ac* were also present in the nucleus, and that the time of occurrence of these mutations was controlled by the state and dose of *Ac* in precisely the same way that the state and dose of *Ac* controls *Ds* breakage events wherever *Ds* may be located. If *Ac* were absent, neither mutations to *C* nor *Ds*-type breakage events would occur. Thus when *Ac* is absent, the behavior of c^{m-1}

is equivalent to the previously known recessive *c*. If, however, *Ac* is again introduced into the nucleus by appropriate crosses, the potential mutability of this recessive is realized, for then mutations to *C* occur. The previously known recessive *c*, used for many years in genetic studies, is unaffected by the presence of *Ac*; and it remains stable, nonmutable, when present in nuclei in which *c^{m-1}* is also present and undergoing mutations.

In considering the mode of origin of *c^{m-1}* and its behavior, the following points may be reviewed: (1) the appearance of a new recessive that is mutable; (2) its derivation from a normal dominant *C*, which is nonmutable; (3) its appearance in a single gamete of a plant carrying *Ds* and *Ac*; (4) *Ac* control of the new mutable condition in exactly the same manner as *Ds* is *Ac*-controlled; (5) *Ds*-type chromosome breakage events also occurring at this mutable locus; and (6) disappearance of *Ds* from its former location in the same chromosome that carries the new mutable locus. This series of coincidences is striking enough in itself to command consideration of the possibility that this mutable recessive originated by transposition of *Ds* to the locus of the normal *C* factor. It is immediately apparent that, if this is true, the transposition of *Ds* from its former location to the new location created a condition that affects the formation of pigment in the aleurone layer; for no pigment is formed until some event occurs at this locus, and only when *Ac* is present. Previous tests have shown that the same *c* phenotypic expression can also arise if the tissues of the endosperm are homozygous deficient for the segment of chromatin carrying *C* (McClintock, unpublished). This might suggest that the presence of *Ds* has inhibited the normal action of the chromatin materials at the *C* locus. A final and most significant argument for the origin of *c^{m-1}* by a transposition of *Ds* to the *C* locus is derived from the fact that a mutation to *C* is associated with the loss of any further recognizable *Ds* events at the immediate location of *C*. It is apparent, therefore, that the mutation-producing event is associated with one involving *Ds* at this locus. All the evidence is consistent with the assumption that *c^{m-1}* arose by transposition of *Ds* to the locus of *C*, thereby inhibiting its action, and that removal of *Ds* is associated with removal of this inhibitory effect. The restored activity at the *C* locus is permanent, and its subsequent behavior resembles that present before *Ds* activity appeared at the locus. *Ac* no longer has any effect on its action and

behavior, just as it had no effect before *Ds* appeared at the *C* locus to give rise to *c^{m-1}*.

ADDITIONAL *Ds*-INITIATED MUTABLE LOCI

Simple coincidence rather than a relationship with *Ds* might still be claimed for the origin and behavior of *c^{m-1}* if it were the only case of such origin and behavior. Two other cases, similar to *c^{m-1}* and involving another marked locus, have appeared independently in the *Ds*- and *Ac*-carrying plants. Both involve the locus in chromosome 9 associated with the bronze phenotype (see Fig. 2). These two independent cases have been designated *bz^{m-1}* and *bz^{m-2}*. The description of the types of events occurring at *c^{m-1}* may be applied also to *bz^{m-1}* and *bz^{m-2}*. *Ds*-type breakage events occur at the mutable locus, as well as mutations from *bz* to an apparently full *Bz* expression. Both the *Ds*-type breakage events and the mutations to *Bz* are *Ac*-controlled; for if *Ac* is absent, neither will occur. In these cases also, the time when mutations to *Bz* or chromosome breaks occur is under the control of the state and the dose of *Ac*. Again, as with *C*, previous investigations had shown that a homozygous deficiency of the segment of chromosome including the *Bz* locus will reproduce the known recessive phenotype, *bz* (McClintock, unpublished).

That the presence of *Ds* close to a marked locus in a chromosome may result in frequent changes in the phenotypic expression of the marker has been indicated. Two independent cases of transposition of *Ds* from its standard location to a position near and to the left of *Sh* have been studied (see Fig. 2). In these two similar cases, less than one-half of one per cent crossing over occurs between *Ds* and *Sh*. In both cases, however, many gametes are produced that carry a "spontaneous" mutation from *Sh* to *sh*. These mutations occur only in those chromosomes carrying *Ds* immediately to the left of *Sh*, and only in plants that also have *Ac*. If *Ac* is absent, no such mutations appear. *Ac*-controlled events, therefore, in each of these two cases where *Ds* is near and to the left of *Sh*, are responsible for this high frequency of mutation to *sh*. When such a mutation occurs, *Ds* is not always lost to the chromatid; it is sometimes still present between *C* and *Bz*. Some of these "mutations" may prove to be newly arisen mutable *sh* loci, but the tests for this mutability have not been concluded.

Knowledge gained from the cases reviewed has led to the conclusion that the appearance of *Ds* at or close to the locus of a known genetic factor can give rise to frequent changes in the action of the factor. The initial change is to an action resembling that of the known recessive allele. In the cases of c^{m-1}, bz^{m-1}, and bz^{m-2}, a subsequent alteration produces a return to apparently full dominant expression of the factor. This common type of mutational expression in these cases is of considerable significance. Its importance will become evident in the discussion of the behavior of other newly arisen mutable loci.

ORIGIN AND BEHAVIOR OF ONE CLASS OF AUTONOMOUS MUTABLE LOCI

The behavior of some other types of mutable loci may now be considered for the additional and important knowledge they have contributed to an understanding of the basic processes involved. One of them is called mutable luteus. The luteus character is distinguished by a yellowish chlorophyll expression. This mutable luteus first appeared in the progeny derived from self-pollination of one of the original plants that had undergone the breakage-fusion-bridge cycle in early development. It resulted from some alteration at a normal locus, but the position of the locus in the chromosome complement was not known. This mutable locus is characterized, first, by its autonomous behavior. It required no recognizable, separate activator factor in order to undergo the mutation phenomenon. The mutations are registered in sectors of a plant as changes in the amount of chlorophyll that is produced. Alleles arise from germinal mutations. They are characterized by various quantitative grades of chlorophyll expression. These alleles, in turn, need not be stable; some of them may mutate to give higher or lower levels of chlorophyll expression. Even an allele apparently producing the full dominant expression may be unstable for it may mutate to or towards the lowest expression, which is luteus.

In studying one aspect of the behavior of this mutable luteus locus, a number of sister plants in one culture were all self-pollinated. On a resulting ear of one, and only one, of these plants, the presence of a new mutable locus was revealed. The mutability, registered in some of the kernels on this ear, involved a factor associated with the formation of pigment in the aleurone layer. Colorless kernels were present in which mutations to color occurred. None of the ears produced by the sister plants showed the presence of such a mutable factor; all kernels on these ears had the full aleurone color. Further study of the plants

derived from the variegated kernels on the aberrant ear showed that a mutable condition had arisen at a previously known locus, the A_2 locus in chromosome 5 (A_2, aleurone and plant color; a_2, recessive allele colorless aleurone, altered plant color). Before the appearance of the mutable condition in this one plant, the A_2 locus had given the normal dominant expression in both parent plants. It had shown no indication whatsoever of any instability. One parent had contributed mutable luteus to this culture. The mutable luteus locus, however, was not linked with the A_2 locus. It should be emphasized that this newly arisen mutable a_2 behaved in many respects like mutable luteus. It was autonomous; and quantitative alleles were produced, some of which, in turn, were mutable.

Another mutable locus arose in a culture derived from one having mutable luteus. It first appeared as a single aberrant kernel on one ear. This ear was produced from a cross in which a plant carrying mutable luteus was used as a female parent. The male parent was homozygous for the stable recessive a_1. (A_1, aleurone and plant color, located in chromosome 3; a_1, recessive allele, colorless aleurone and altered anthocyanin pigments in the plant.) This single aberrant kernel exhibited variegation. Sectors of colored aleurone appeared in an otherwise colorless kernel. Tests were initiated with the plant derived from this kernel, and continued with the subsequent progeny. From these tests it was learned that the aberrant kernel carried a newly arisen mutable a_1 locus, designated a_1^{m-1}, whose general behavior resembled that shown by mutable luteus. It was autonomous; it produced a series of alleles showing various grades of quantitative expression of anthocyanin, in both the aleurone and the plant; and of these alleles, in turn, some were unstable, mutating to give higher or lower levels of quantitative expression of the anthocyanin pigments in both aleurone and plant. Other important aspects of the mutation phenomenon at this locus will be considered later.

In the discussion of the Ac-controlled mutable loci, c^{m-1}, bz^{m-1} and bz^{m-2} that arose in Ds-Ac carrying plants, it was emphasized that the types of mutational response were similar. Here also, the mutational expressions of the two mutable loci that have arisen in the mutable luteus stocks are much alike, and they resemble that shown by mutable luteus itself. One further example of related mutable loci will be given. It also shows the similarity of behavior of the newly arisen mutable locus to the one already present in the plant. The direction this discussion is taking may now be apparent. It is towards the conclusion that the type of mutation occurring at a locus is a function of the type of chromatin material that is present at the locus or is transposed to it, and does not involve changes in the components of the genes themselves. Rather, it is this chromatin that functions to control how the genic material may operate in the nuclear system. With this in mind, a third example of related origins and behaviors of mutable loci may now be considered.

ORIGIN AND BEHAVIOR OF c^{m-2} AND wx^{m-1}: TWO RELATED MUTABLE LOCI

The progeny derived from self-pollination of another one of the original plants that had undergone the breakage-fusion-bridge cycle in early development was grown, and a number of these plants again self-pollinated. On a resulting ear of one of these plants, a new mutable locus was recognized. The factor involved was again associated with the production of pigment in the aleurone layer. Some of the kernels on this ear showed colored areas in a colorless background. Beginning with the plants derived from these kernels, a study was made of the condition responsible for the variegation. This proved to be due to another new mutable locus, and involved the previously discussed C locus in chromosome 9. This locus in the parent plant and in the sister plants of the culture, gave the normal dominant C expression. The new mutable condition was designated c^{m-2}, because it was the second case that appeared in this study. The types of mutation that arise from events at c^{m-2} are strikingly different from those shown by c^{m-1}. A series of alleles, as expressed by quantitative grades of pigment formation associated with the production of at least two different precursor-type diffusible substances, is produced by mutations at c^{m-2}. The intermediate alleles are not always stable, for some of them, in turn, can mutate to alleles showing higher or lower grades of color expression.

In the course of the study of c^{m-2}, a number of crosses were made, using pollen of plants homozygous for c^{m-2} and carrying a normal dominant Wx factor in each chromosome 9, on silks of plants carrying a stable recessive c and a stable recessive wx in both chromosomes 9 (see Fig. 2). A single aberrant kernel appeared on one of the ears resulting from this type of cross. It showed mutations to C and of the c^{m-2}

type, and, in addition, mutations from the wx to and towards the Wx phenotype. A plant was obtained from the kernel and a study commenced to determine the nature of this instability expression. It proved to be a new mutable wx, and was designated wx^{m-1}. The tests showed that it had arisen in the male parent plant, which carried c^{m-2} and normal dominant Wx in each chromosome 9. It was present, however, in only one of the many tested male gametes of this plant. The pattern of mutational behavior of wx^{m-1} strikingly resembled that shown by c^{m-2}. A series of quantitative alleles was produced by mutations of wx^{m-1} as registered by the amount of amylose starch produced. These alleles, in turn, could mutate to give greater or lesser amounts of amylose starch. Another endosperm character also was affected by some of the mutations of wx^{m-1}. This was expressed by an altered growth of the endosperm tissue, and accompanied some but not all of the mutations to the intermediate alleles, appearing particularly often in association with a mutation to one of the lower alleles. This accompanying mutation behaved as a dominant or a semidominant.

It is of particular significance, in comparison of the behavior of the series of mutable loci c^{m-1}, bz^{m-1}, and bz^{m-2} with that of the series c^{m-2} and wx^{m-1}, that the members of both series are controlled by the very same Ac factor—wherever it may be located—and in precisely the same manner, with respect to time and place of occurrence of mutations. When Ac is absent, no mutations occur in either series. In the latter series, mutations of the intermediate alleles also occur, but only when Ac is present. Because of this, it has been possible to isolate a series of quantitative alleles of C, and also a series of quantitative alleles of Wx, that are stable. Stability is maintained when Ac is removed. The percentages of amylose starch in the endosperm, produced by a number of the alleles arising from mutations at wx^{m-1} and freed of Ac, have been determined in terms of single, double, and triple doses of the particular allele. (Note: The writer is grateful to Dr. G. F. Sprague and to Dr. B. Brimhall, of Iowa State College, and also to Dr. C. O. Beckmann, of Columbia University, and Dr. R. Sager, of the Rockefeller Institute for Medical Research, for their chemical analyses of the amylose content produced by some of these mutants.) The preliminary tests suggest that alleles, falling into an almost continuous series with respect to the quantity of amylose starch they produce, may be obtained. It should be mentioned here that, as with C and Bz, previous investigation had shown that a tissue homozygous-deficient for the Wx locus will give the known recessive wx expression (McClintock, unpublished).

In order to compare the action of Ac on the members of these two series of Ac controlled mutable loci, crosses were made combining several of them in a single plant so that they might be present together in the nuclei of a tissue. By this means, it was possible to determine that the mutations at these various mutable loci arise as a function of the state and dose of Ac, irrespective of which mutable locus is involved or how many such loci from the same series or from the two different series are present in the nuclei of an individual plant.

In further comparison of the behavior of the different Ac-controlled mutable loci, one very significant correlation may now be given. It is known that all these loci show one other common characteristic. At all of them, some chromosome-break-inducing events occur, but only when Ac is also present in the nucleus and only at those times in the development of a tissue where mutations leading to changes in expression of the respective phenotypic character are also occurring. Again, it has been determined that such breaks may occur at the locus of Ac itself. The conclusion that the mutation-producing events in these two series of related mutable loci, and also at Ac itself, are associated with such a chromosome-break-inducing mechanism is difficult to avoid. This relationship will be explored after consideration has been given to a comparison of the types of mutability that may arise at any one known locus in a chromosome.

The descriptions of mutable loci given so far in this discussion have shown that the type of mutability and the mode of its control are not alike for all. Nevertheless, there appear to be classes of mutable loci, the members of which show similar types of changes in phenotypic expression—that is, of mutations—and similar types of control of these mutations. It is now necessary to indicate the extent to which various types of mutability expression may arise at any one particular locus. For this purpose the Wx, the C, and the A_1 loci will be chosen as examples.

COMPARISONS OF TYPES OF MUTABILITY ARISING AT ANY ONE LOCUS

a. The Wx locus

Six independent mutable conditions are known for the Wx locus. Five of them have arisen during the present study. One, wx^{m-1}, has been considered above. It is Ac-controlled, and produces quantitative alleles that may be unstable when Ac is present but are stable when Ac is absent from the nucleus. Mutations at this locus also give rise to an endosperm-growth-altering factor that is dominant in expression. The second mutable wx, wx^{m-2}, arose from a previously stable recessive wx carried in genetic stocks for many years. This mutable condition first appeared in a chromosome 9 in which a complex chromosomal rearrangement was present. Its mutations are expressed by different quantitative grades of the Wx phenotype. In the endosperm tissues, the sector produced by a cell in which a mutation has occurred is always markedly distorted in growth-type. The third mutable condition at this locus, wx^{m-3}, originated in a plant carrying a normal dominant Wx and also several mutable loci. It is autonomous, in that no separate activator factor is required for mutations to be expressed. It almost always mutates to give the full Wx phenotypic expression. The derived mutant giving the dominant expression is also mutable, for it produces mutants giving the full recessive expression, that is, wx. This recessive, in turn, may mutate again to give the full dominant expression. No altered growth conditions in the endosperm tissue accompany any of these mutations. The fourth case was recognized by a sudden change in the behavior of a previously normal Wx locus. It shows mutations producing various grades of quantitative expression between the full dominant and the full recessive. It is autonomous, and no alterations in growth conditions accompany the mutations. Another case, somewhat similar to the last, has recently arisen. It produces alleles giving various lowered expressions of the Wx phenotype, and appears to be autonomous although the information on its behavior is too incomplete to allow a full description. The sixth case is one that has been investigated by Sager (1951). It is autonomous, and gives quantitative grades of expression in the endosperm; but the germinal mutations that have been studied all give full or nearly full Wx expression, and the mutants are stable. No altered growth conditions appear to accompany these mutations.

Genetic analyses have indicated that all these various mutational changes occur at this one locus in chromosome 9, and yet all show a different kind of mutational behavior. It is evident that each arose in association with a particular type of alteration at the locus, and that different mutation-controlling mechanisms can be involved.

This is especially well illustrated by a comparison of the types of mutations produced by wx^{m-1} and wx^{m-3} and of their controlling mechanisms.

b. The C locus

The contrasts in the kinds of phenotypic expression produced by mutations at c^{m-1} and c^{m-2} have been discussed above. Although several other independent expressions of mutability at this C locus have also arisen, the study of them is too incomplete to allow detailed comparisons to be made. With respect to this locus in chromosome 9 it is necessary to mention, however, that in the cultures having mutable loci a heritable factor carried by a chromosome other than 9 has appeared on several occasions. The presence of such a factor results in the production of pigment in the aleurone when the endosperm is homozygous for the well-investigated stable recessive c, used for many years in genetic investigations. The pattern of pigment formation differs markedly from that produced by mutations at c^{m-1} or c^{m-2}. It resembles that associated with the factor Bh (Blotch), previously studied by R. A. Emerson (1921) and Rhoades (1945b). To complete the discussion of the series at this locus, it may be mentioned that a mutable condition has also arisen involving the expression of I, an allele of C. Changes in the degree of inhibitory action of I occur as a consequence of such mutations.

c. The A_1 locus

A study of changes at the locus of A_1 has contributed some very important information regarding the origin and behavior of mutable loci. For a number of years, a type of control of mutability of the recessive, a_1, has been known. A dominant factor, called Dotted (Dt), provokes mutability at the a_1 locus (Rhoades, 1936, 1938, 1941, 1945a). In many respects, Dt is comparable to Ac. It is an activator, for it produces mutations at a_1, just as Ac produces mutations at c^{m-1}, c^{m-2}, bz^{m-1}, bz^{m-2}, and wx^{m-1}. Moreover, when Dt is absent no mutations occur at a_1; the a_1 locus then gives a stable recessive phenotype. In the presence of Dt, mutations occur at a_1 to give mainly the higher alleles of the A_1 phenotypic expression. The time of occurrence of visible mutations at a_1 is usually late in the development of the plant or the endosperm tissues; and they occur in only some of the cells. This results in the presence of dots of the A_1 phenotype. The Dt factor has been located by Rhoades in the knob region terminating the short arm of chromosome 9. Dt and Ac appear not to be the same

activator, as plants and endosperm tissues that are homozygous for the recessive a_1 have not shown the dotted-type mutations to A_1 in the presence of Ac.

In the early period of this investigation of newly arisen mutable loci, the unexpected appearance of modifications in the knobs, and in the other chromosome elements previously mentioned, of plants that had undergone the breakage-fusion-bridge cycle in their early development, suggested that disturbances in these elements might have been responsible for the initial burst of mutable loci, including the origin of Ac itself. It was suspected, therefore, that this cycle might induce alterations in the heterochromatic elements that could initiate a Dt factor as they may have originated the Ac factor. Once initiated, this factor would activate a_1 to undergo alterations in somatic cells leading to A_1-type expression. The most direct way to induce changes in the heterochromatic elements was considered to be the breakage-fusion-bridge cycle itself. By subjecting tissues to this cycle during their developmental periods, and then examining the matured tissue, this hypothesis could be tested. A preliminary experiment, designed to test for production of mutations of a_1 to A_1 by the breakage-fusion-bridge cycle as an inductor, was performed in 1946. The experiment was repeated in 1950 on a much larger scale. Because this experiment was of particular significance in revealing the mode of origin of mutable conditions, and because it provided evidence about the relation of chromosome organization to genic expression and its control, the details will be given.

The silks of plants homozygous for a_1 and carrying no Dt factor received pollen from plants of similar constitution with respect to a_1 and Dt. The pollen parents carried one chromosome 9 with a duplicated segment of the short arm. The homologous chromosome 9 was deficient for a terminal segment of the short arm. Newly broken ends of chromosome 9 were produced in some meiotic cells, as diagrammed in Figure 1. Pollen grains of these plants carried either: (1) a deficient chromosome 9, which did not function in pollen-tube growth, (2) a chromosome 9 with a full duplication of the short arm—that is, the homologous chromosome 9, or (3) a chromosome 9 with a newly broken end. Among this last type, duplications or deficiencies of the short arm were present. Those carrying an extensive deficiency were nonfunctional but those carrying a relatively short duplication were better able to compete in functioning than those carrying the full duplication of the short arm. Thus the

32   *BARBARA McCLINTOCK*

majority of the functioning pollen grains of these plants carried a newly broken end of the short arm of chromosome 9 in their nuclei. These chromosomes had undergone the breakage-fusion-bridge cycle since the meiotic anaphase and continued to do so after being incorporated into the primary endosperm nucleus. Either before fertilization or during the development of the kernel, the breakage-fusion-bridge cycle might produce alterations in heterochromatic elements and some of them might include an alteration that would recreate the condition associated with Dt action. Mutations at the a_1 locus to give the A_1 phenotype could subsequently appear in the descendant cells. If this should occur early in development of the endosperm, a sector with dots of the A_1 phenotype would be produced. Examination of 95 ears resulting from the preliminary test conducted in 1946 revealed A_1 dots on 15 kernels from 14 different ears. In five of these kernels, more than one A_1 dot was present, and in one restricted region of the kernel in each case. The number of kernels with mutations to A_1 in this trial experiment was lower than anticipated, and the experiment was not expanded the following year. Later, as the probable relation between the origins of mutability and the alterations induced in the knobs or other chromosome elements by the breakage-fusion-bridge cycle became more clearly apparent, the same experiment was conducted in 1950 on a much larger scale. The results were rewarding, for now many kernels were obtained that had one or more A_1 dots. One hundred and twenty such kernels appeared in this second trial, and 24 of them had more than one A_1 spot. One of these kernels had 84 A_1 spots, distributed rather evenly over the kernel. In the other 23, the spots were not distributed at random over the aleurone layer but were restricted to well-defined sectors. In none of these kernels did any large areas of the A_1 phenotype appear. In all cases, the time of mutation, the pattern of mutation, and the type of mutation were much like those produced when the known Dt factor is present in endosperms homozygous for a_1. It was obvious that in each case the initial alteration had occurred in the ancestor cell that produced the dotted sector. This initial event was responsible for mutations that occurred at a_1 in some cells during the subsequent development of the endosperm. The observed mutations at a_1, therefore, were not produced directly by the breakage-fusion-bridge cycle but arose secondarily, as a consequence of an event that altered some particular component in the nucleus. It

was the alteration of this component that was responsible for the subsequent mutations at the a_1 locus. And this initial alteration was one that imitated the effect produced when the known Dt factor is present. It is difficult to avoid the conclusion that a new Dt-like factor has been produced in each such case, and that it was created by some event associated with the breakage-fusion-bridge cycle. Unfortunately, the plant grown from the one kernel having 84 dots distributed over the whole aleurone layer did not show any mutations to A_1, nor did mutations appear in the kernels when this plant was crossed to plants homozygous for a_1. The effective alteration probably was present in only one of the two sperms carried in the pollen grain. Because the break in chromosome 9 that initiated the breakage-fusion-bridge cycle was produced at the meiotic anaphase, the event giving rise to the Dt-like factor would have had to occur in the subsequent microspore division in order to be incorporated in the two sperm nuclei. An even larger experiment of this same type must be conducted if such a case is to be obtained. It should be mentioned in this connection that the size of the sectors within which A_1 spots appeared graded from large to small, the smaller sectors being most frequent. Also, about three-fifths of the kernels showing mutations to A_1 had only one A_1 spot. These frequencies are to be expected if the creation of a Dt-like factor is a consequence of an event, associated with the breakage-fusion-bridge cycle, that has a probability of occurrence in a limited number of mitotic cycles. In order to indicate why the dotted pattern of mutations to A_1 is to be anticipated, rather than any other, it is necessary to review the origin of the previously discovered $Dt - a_1$ mutable condition.

The $Dt - a_1$ mutable condition first appeared on one ear after self-pollination of a plant belonging to the commercial variety known as Black Mexican Sweet Corn. This variety is homozygous for A_1. The recessive a_1 in this case represented a new mutation from A_1, and was associated with the appearance of Dt. The original a_1 mutant, known for many years and used in genetic studies, had originally been found to be present in a commercial variety of maize. Both a_1 mutants responded in much the same manner when Dt was present in the nucleus. In both cases, the dotted mutation pattern was produced in the presence of Dt. The states of the two a_1 mutants thus appeared to be alike. This suggests that the older a_1 mutant may have been produced by a mechanism similar to that responsible for the origin of the newer a_1 mutant. A Dt factor may have arisen at the same time but subsequently been lost from the commercial variety during its propagation, leaving an apparently stable a_1 mutant. The change at C that produced c^{m-1} would have behaved quite comparably had Ac been absent from the nuclei in the initial gamete carrying c^{m-1}, or had it been removed by crossing before the change at this locus had been detected. If the mutation had been discovered several generations after its origin, and if Ac had been removed by a previous cross, it would have appeared to be a newly arisen, stable, recessive c. Only after an incidental cross to a plant carrying Ac would its potential mutability have been revealed. It is possible, therefore, that many known recessives may prove to be potentially mutable.

The essential similarity of the $Dt - a_1$ system to the $Ac - c^{m-1}$, etc. system is also expressed in the changes in state of a_1 that may occur in the presence of Dt. Such changes in state of a_1 have recently been described by Nuffer (1951). They are recognized individually by marked departures in frequency of visible mutations, in types of mutation and in time of occurrence of these mutations. The types of different phenotypic expression produced by mutations at altered states of a_1 are much the same as those produced by a mutable a_1 locus that has appeared in the Cold Spring Harbor cultures. This new mutable a_1 locus, called a_1^{m-1}, differs from the mutable a_1 studied by Nuffer in that it is autonomous and does not require Dt for mutability to be expressed. In this respect, mutability at the A_1 locus behaves like that at the C and the Wx loci, for both autonomous and activator-controlled mutable conditions may arise.

The origin of a_1^{m-1}, in a culture carrying mutable luteus, was described previously. It is autonomous, and produces a series of quantitative alleles, many of which are unstable in that they may mutate to give higher or lower levels of quantitative expression. Difference in degree of quantitative expression is only one of the consequences of mutations occurring at a_1^{m-1}, however. The diversity of phenotypic changes arising from these mutations is so great that an adequate analysis of all the observed types is a large task. They are distinguished not only by quantitative but also by qualitative differences in the anthocyanin pigments formed. Diversity is shown in other respects. For example, some of the mutations giving pale aleurone color are

related to changes involving the rate of a particular reaction responsible for pigment formation. Others appear to be related to the absolute amount of pigment that may be produced, regardless of a time factor. This becomes evident when comparisons are made of pigment-forming capacities in plants arising from kernels carrying different mutants of a_1^{m-1}, each producing a pale color in the aleurone of the kernel. In some cases, such plants are pale in their expression of anthocyanin color throughout their lives. Others are pale in anthocyanin color up to the time of anthesis, when growth of the plant terminates; but in the six or seven following weeks, as the plants mature their ears, the anthocyanin color gradually deepens, becoming intensely dark by the time the ears are mature. The kernels derived from both these two types of plants, however, may be equally pale. The fact that pigment forms late in the development of the kernel, and dehydration of the tissues occurs shortly thereafter, may explain this similarity in color of the kernels.

Other types of mutation occur at a_1^{m-1}. Some produce sectors of deep color that are rimmed by areas in which the color gradually fades off to colorless, as if a diffusible substance associated with pigment formation had been produced in excess in the mutant sector. The area of diffusion may be very extensive for some mutations, and only slight for others, whereas still other mutations give rise to no such diffusion areas at all. Some of the mutations that result in strong A_1 pigmentation are associated with failure of development or degeneration of some of the aleurone cells within the mutated sector.

Besides the mutational changes at a_1^{m-1} that affect the type and amount of pigment formation, a number of other changes occur which affect the subsequent behavior of a_1^{m-1}. These alterations are termed changes in state, since they affect not only the times when pigment-forming mutations will occur at the locus in future plant and endosperm cells but also the kinds and frequencies of such mutations, and their distributions and their sequences in the development of the tissues.

Any interpretations that attempt to explain the primary action of a specific locus in a chromosome, and how this action may be changed, must take into consideration the facts just enumerated concerning the behavior of this a_1^{m-1} locus. It is not reasonable to regard such changes in expression and action as being produced by changes in a single gene—that is, according to the usually accepted concepts of the gene that have been developed. The evidence suggests, rather, that the observed changes result either from alterations at a locus that has many individual components or from alterations at the locus affecting its relationship to other loci in the chromosome complement. If the latter is true, a combination of loci functions as an organized unit in the production of pigmentation. If such functional organizations exist within the nucleus —and it is reasonable to assume they do—then the large numbers of alleles known to arise at certain loci need not express altered genic action at the identified locus. Rather, any one alteration may affect the action of the organized nuclear unit as a whole. The mode of functioning of various other loci concerned may thereby be modified. In other words, the numerous different phenotypic expressions attributable to changes at one locus need not be related, in each case, to changes in the genic components at the locus, but rather to changes in the mechanism of association and interaction of a number of individual chromosome components with which the factor or factors at the locus are associated. According to this view, it is organized nuclear systems that function as units at any one time in development. In this connection it may be repeated that at $a_1{}^{m-1}$, and also at other mutable loci, many of the alterations observed represent changes in the potential for patterns of genic action during development (changes in state). Thus a pattern-controlling mechanism is being altered. If particular nuclear components are formed into organized functional nuclear units, the evidence would suggest that this may happen only at prescribed times in the development of an organism. In this event it may readily be seen that changes in pattern-controlling mechanisms would serve as a primary source of potential variability of genic expression without requiring any changes in the genes themselves.

A few more pertinent facts about the A_1 locus may now be mentioned. Mutability has arisen at this locus independently on a number of different occasions. Several cases have recently appeared in the stocks having Ds and Ac. Analysis of these cases has not proceeded to a stage where a complete description of their behavior may be given. Both Laughnan (1950) and Rhoades (1950) have found new cases of instability at this locus. Several have appeared in plants derived from kernels that had been aged for some time (Rhoades, 1950). Such aging is known to give rise to chromosomal aberrations as well as mutations; the observed instability may be an expression of one such structural change. Laughnan (1949, 1951) has shown that mutations at the A_1 locus may be associated with the mechanism of crossing over, suggesting again that mutations arise from structural changes at the locus. The crossover studies may elucidate the nature of some of these changes. Not only the cases described in this report but also others have produced evidence converging in support of a hypothesis that mutations originate from structural alterations in chromosome elements. The evidence derived from a study of progressive changes in state of c^{m-1} has shown the close relation between a structural change at a locus and one that so often has been called a "gene mutation." This study will now be described.

SIGNIFICANCE OF CHANGES IN STATE OF A MUTABLE LOCUS: SELECTED EXAMPLES

The foregoing review of the very different types of phenotypic expression that may be produced as a consequence of mutations arising at any one locus, and of the relation between the origin of a mutable condition and the type of mutations expressed, clearly indicates the necessity for caution in attempting to interpret the mutation process as one associated with a "change in a gene." With respect to events occurring at Ac, c^{m-1}, c^{m-2}, bz^{m-1}, bz^{m-2}, and wx^{m-1}, it has been established that a mechanism capable of producing chromosomal breaks at the locus is associated with the mutation-producing process. It could be argued from the evidence so far presented that these cases fall into a special category, and that what they may indicate regarding the mechanism of mutation at these mutable loci may not be used to interpret the mutation process in general. Knowledge gained from a study of the changes in state of c^{m-1} has shown, however, that no line may be drawn between those events at a locus that produce detectable chromosomal alterations and those that give rise to mutations but produce no readily detectable chromosomal alterations.

The origin of c^{m-1} by a transposition of Ds to the normal C locus has been discussed above. The state of Ds, when first transposed to the C locus, was one that produced many detectable chromosome breaks at this locus and few mutations to C. When plants having this state of c^{m-1} were crossed to plants that were homozygous for the stable recessive c, the majority of the resulting kernels showed this relationship. Some of these kernels, however, had sectors with higher rates of mutation to C and lower rates of detectable

CHROMOSOME ORGANIZATION AND GENIC EXPRESSION   35

chromosome breaks. In some sectors, no chromosome breaks were evident; but often a high frequency of mutation to C had occurred. There were also a few kernels on these ears that had this pattern throughout the kernel. When found, such kernels were selected from the ears (as well as others that showed changed mutational patterns). The plants grown from them were again crossed to plants homozygous for the stable recessive c. The kernels on the resulting ears now showed the types and frequencies of the different detectable events at c^{m-1} that had been observed in the kernel from which the plant had arisen. It was possible to determine in this manner that a heritable change had occurred at c^{m-1}. It was this change which was responsible for the altered frequencies of expression of the detectable events that subsequently occurred at this locus. That the altered response, in each isolated case, arose from a change at c^{m-1} and was not produced by a change at Ac, was determined by testing the responses of these c^{m-1} isolates with different isolates of Ac, each having a known type of action. Ac controlled, in each test, the time and place of the event occurring at c^{m-1}, but not its type. Because, in each case, the change in behavior of c^{m-1} was heritable, it must have arisen by an event that produced an alteration at this locus. It is this heritable altered condition that has been termed an altered state of the locus.

The various altered states of c^{m-1}, as previously mentioned, arise only when Ac is also present in the nuclei, and only at the times in development when mutations or chromosome breaks may occur at c^{m-1}. For our purposes, the most instructive of the changed states are those giving reciprocal frequencies of chromosome breaks and mutations to C. A series of isolates, each showing a particular relation between these two events at c^{m-1}, has been studied. The isolates ranged from those showing no mutations to C or only a very occasional one, but having a very high frequency of detectable chromosome breaks, to those showing a high frequency of mutations to C and no detectable breakage events or only an occasional one. The states of c^{m-1} that give high frequencies of chromosome breaks are unstable, for other altered states may be produced as a consequence of events at the locus, but only, as emphasized above, when Ac is also present in the nucleus. A particular state of c^{m-1} remains constant if maintained in plants having no Ac. The state of c^{m-1} giving no detectable chromosome breaks, and a correspond-

ingly high frequency of mutations to C, is very stable with respect to the absence of breakage events. Had the state of c^{m-1} been of this type when it first arose, there would have been no opportunity to discover that the chromosome-break-producing mechanism and the mutation-producing mechanism were related. If chromosome breaks are not exhibited by a mutable locus, therefore, it cannot be argued that because of this the basic mechanism producing the mutations must be different from that known to operate at c^{m-1} and at the other Ac-controlled mutable loci. The evidence obtained from this study of the origin and subsequent behavior of altered states of c^{m-1} argues, rather, for similarities if not identities in the basic mechanism.

Another type of change in state of c^{m-1} should be mentioned, although it occurs infrequently. It is detected by a much altered expression of the mutation at this locus that affects aleurone color. The frequency of origin of such states of c^{m-1} is so very low as to suggest that they represent entirely new modifications at this locus, comparable to the original inception of a mutable condition at any locus. They may well represent just such new inceptions, for these are to be expected in view of the fact that very different types of mutational behavior are exhibited at this same locus by c^{m-1} and c^{m-2}, both of which are Ac-controlled.

All the Ac-controlled mutable loci exhibit changes in state. These are characterized by changed relative frequencies of the different recognizable types of mutations that occur. In other words, as we have already seen, the types of mutation produced in Ac-controlled mutable loci are related to the state of the mutable locus itself. The time and place during development of occurrence of mutations, on the other hand, are controlled by the state and dose of Ac; and therefore alterations in them are related to changes in state of Ac rather than to changes in state of the mutable locus. The changes in state of Ac have been described. The autonomous mutable loci also undergo changes in state, as described previously for the a_1^{m-1} locus. In this group, however, the controller of the time and place of occurrence of mutations is a component of the locus itself. Consequently, the changes of state that arise at these loci are reflected in changes in the control of time and place of mutations as well as in the type or types of mutation that may occur, and also in the time and place of occurrence of each such mutation if several types are produced. Thus there is a much more diverse group of altered states associated with changes at any particular locus in the autonomous group. The general similarities between the autonomous and the *Ac*-controlled mutable loci are nevertheless striking.

EXTENT OF INSTABILITY OF GENIC EXPRESSION IN MAIZE

The discussion so far has mentioned a number of different mutable conditions at known loci in maize. In order to show that the phenomenon is much more prevalent than the particular cases described would indicate, some additional observations of changes in genic expression may be discussed briefly. Not only have *C, I, Bz, Wx, A_1,* and *A_2* become mutable during the course of these studies, but instability of other known dominants has been noted. These are *R, Pr, Yg, Pyd, Y,* and possibly *B* and *Pl*, although the evidence for the last two is observational and not genetic. Some previously unknown dominant loci have also become unstable. These are associated with various chlorophyll-determining factors, endosperm-starch-controlling factors, aleurone-color factors, growth-controlling factors, etc. Instability has arisen at the loci of some of the known recessives such as *wx, yg,* and the special case *a_1* described previously. Instability at the loci of the recessives *y* and *p* also appears to have arisen on several independent occasions. Genetic tests were not made, however, to determine the association of the instability with the known loci. Instability arising at recessive loci is recognized by mutations to or towards the expression of the dominant allele. It may be concluded, therefore, that many of the known recessive alleles are potentially capable of expressing action that is characteristic of the dominant alleles.

The expression of instability of various factors in maize is probably far more common than has been suspected in the past. Until recently, only a few such cases had been reported in the literature. One of the earliest recognized was that occurring at the *p* locus (pericarp and cob color, chromosome 1), studied by R. A. Emerson and his students (for literature citations, see Demerec, 1935) and recently being studied by Brink and his students (1951, and personal communication) and by Tavcar (personal communication). Reported cases of instability at the *a_1* and the *wx* loci have been mentioned previously in this discussion. In addition, Rhoades (1947) has studied instability at the *bt* locus (brittle endosperm, chromosome 5) and has reported two cases involving chlorophyll characters that appeared in the progeny of irradiated seed (Rhoades and Dempsey, 1950). Fogel (1950) has been investigating instability at the *R* locus (aleurone, plant, and pericarp color, chromosome 10); and Mangelsdorf (1948) has reported instability at the *Tu* locus (Tunicate, chromosome 4). It is believed that critical examination will uncover many such cases in maize, and that they will involve many different loci.

MUTABLE LOCI AND THE CONCEPT OF THE GENE

It will be noted that use of the term gene has been avoided in the foregoing discussion of instability. This does not imply a denial of the existence within chromosomes of units or elements having specific functions. The evidence for such units seems clear. The gene concept stems from studies of mutation. That heritable changes affecting a particular reaction, or the development of a particular character, in an organism arise repeatedly and are associated with a change of some kind occurring at one specific locus or within one specific region of a chromosome, has been established. This knowledge has been responsible for the development of a concept requiring unitary determiners. It cannot be denied, in the face of such evidence, that certain loci or regions in the chromosomes are associated in some manner with certain cellular reactions or with the development of particular phenotypic characters. This is not the major questionable aspect of current gene concepts. The principal questions relate to the mode of operation of the components at these loci, and the nature of the alterations that affect their constitution and their action. Within the organized nucleus, the modes of operation of units in the chromosomes, of whatever dimensions these may be, and the types of change that may result in specific alterations in their mode of action, are so little understood that no truly adequate concept of the gene can be developed until more has been discovered about the function of the various nuclear components. The author agrees with Goldschmidt that it is not possible to arrive at any clear understanding of the nature of a gene, or the nature of a change in a gene, from mutational evidence alone. At present, the most we know about any "gene mutation" is that a heritable change of some nature has occurred at a particular locus in a chromosome, and that any one locus is somehow concerned with a certain chemical reaction, or with a certain restricted phenotypic expres-

CHROMOSOME ORGANIZATION AND GENIC EXPRESSION 37

sion, or even with the control of a complex pattern of differentiation in the development of a tissue or organ. The various types of known mutation, each showing unitary inheritance, obviously reflect various levels of control of reactions and reaction paths. It is necessary to consider these various widely different levels of unitary control and how they may operate in the working nucleus, and also to consider the nature of the changes that can affect their operation. It is with the nature of such heritable changes, the conditions that induce them, and their consequences, that this report has been concerned. Various levels of unitary control, as witnessed by inheritance behavior, are evident from the study. That genes are present in the chromosomes and that they function to produce a specific type of reactive substance will be assumed in this discussion, even though such a restricted assumption may prove to be untenable. The knowledge gained from the study of mutable loci focuses attention on the components in the nucleus that function to control the action of the genes in the course of development. It is hoped that the evidence may serve to clarify some aspects of gene action and its control. Some of the interpretations of the author, based on this evidence, have been stated or implied at various points in the previous discussion, and may now be summarized.

The primary thesis states that instability arises from alterations that do not directly alter the genes themselves, but affect the functioning of the genic components at or near the locus of alteration. The particular class to which a mutable locus belongs is related to the particular kind of chromatin substance that is present at or near the genic component in the chromosome. It is this material and the changes that occur to it that control the types and the rates of action of the genic components. Thus the basic mechanism responsible for a change at a mutable locus is considered to be one that is associated with a structural alteration of the chromatin materials at the locus. The mechanism that brings about these changes is related to the mitotic cycle; and it may involve alterations of both sister chromatids at the given locus. Some of these alterations may immediately result in the expression of an altered phenotype, a "gene mutation." Others produce modifications controlling the type of events that will occur at the locus in future cell and plant generations. Still others produce changes of a more extensive type, such as duplications and deficiencies of segments of chromatin in the vicinity of the locus. With regard

to these conclusions, the evidence presented in the discussion of changes in state of c^{m-1} may be recalled.

BEHAVIOR AND ACTION OF *Ac* IN CELL AND PLANT GENERATIONS

The interpretations given above deal with the organization and the kinds of events that occur at genetically detectable loci in the chromosomes. The next level of consideration deals with mode of operation of the nucleus in controlling the course of events during development. Do these studies suggest a mode of operation, or at least one component in the operative system? It is believed that the behavior of *Ac* may be of importance in such a consideration. With reference to c^{m-1}, for example, it has been shown that mutations to *C* occur at particular times and places in development, under the control of the particular state and dose of *Ac*. It has also been shown that *Ac* is unstable, for changes arise affecting its dosage action, and changes also occur in its location in the chromosome complement. It has not been explained, however, that the time of occurrence of these changes at *Ac* during development is also a function of the particular state and dose of *Ac* that is present in a tissue. With any particular *Ac*, and any particular dose of this *Ac*, or with any combination of different isolates of *Ac*, the time when these events will occur to *Ac* is a function of the single or combined action of the *Ac* factors. An example of this may now be given.

One plant having an *Ac* factor at the same location in each homologue of a pair of chromosomes, and carrying *I, Sh, Bz, Wx*, and *Ds* (standard location) in both chromosomes 9, was crossed to a number of plants homozygous for *C, sh, bz,* and *wx* and carrying no *Ds* or *Ac*. On all the many ears resulting from these crosses, approximately 90 per cent of the kernels were sectorial with reference to the time of occurrence of *Ds* events. Pollen from this same plant was placed on silks of plants having other combinations of markers in chromosome 9, and similar types of sectorial kernels appeared. The sectoring was produced by segregations, occurring in the earliest nuclear divisions of the endosperm, that involved the controller of the time of occurrence of *Ds* events, that is, the Activator. The action of *Ac* in these different sectors resembled that occurring either (1) when no *Ac* is present, (2) when a sharply decreased dose of *Ac* is present, or (3) when a sharply increased dose of *Ac* is present. Illustrations that will make this relationship clear are shown in the photographs of Figures 17 to 22. In a definite fraction of the cases, a chromosome break at Ds was associated with segregations of Ac action. The mechanism responsible for this precise, somatically occurring segregation for Ac action was very likely the same as that responsible for the origins of germinal changes in action of Ac, as well as changes in its position in the chromosome complement. Such changes have been mentioned earlier in this report. It can be seen that if this same type of segregation occurred within some cells early enough in the development of a plant to be incorporated in a microspore or megaspore nucleus, the altered state or location of Ac, or both, would be recoverable in the plant that subsequently resulted from the functioning of the male or female gametophyte arising from such a spore. Such an early-timed event would allow for isolation and subsequent study of the transpositions and changes in state of Ac.

A study was initiated to determine the nature of the changes that occur at Ac by an analysis of the Ac constitutions in the gametes of plants having an Ac factor at the same location in each member of one pair of chromosomes, that is, in plants homozygous for Ac in allelic positions. In these plants the Ac factors were alike in state, and the homozygous condition was produced by self-pollination of plants having one Ac factor. Both chromosomes 9 in these plants carried the stable factors c, sh, and Wx. No Ds was present in these chromosomes. To test for Ac inheritance, these plants were crossed by plants having no Ac but carrying C, Sh, wx, and Ds in both chromosomes 9 (standard locations of Ds and identical states). If no changes had occurred to Ac, all the kernels on the resulting ears should be variegated. Similar variegation patterns, produced by sectors showing the c, sh, and Wx phenotype, should be present, because Ac would initiate

chromosome breaks at Ds in the C, Sh, wx, Ds-carrying chromosome 9 contributed by the male parent. With the exception of a few kernels, just such conditions were realized on these ears. A photograph of one such ear appears in Figure 23. It will be noted that the majority of kernels on this ear show very similar patterns of variegation. A few kernels that differ from the majority are completely colored, with no colorless sectors of any size. In them, no Ds breaks at all occurred. A few other atypical kernels show an altered timing of the breakage events at Ds. In them such events occurred either much earlier or very much later in development of the endosperm.

A study was made of the Ac constitution in plants derived from selections of all the different types of kernels appearing on such ears. In the plants coming from the kernels showing altered variegation patterns, it was necessary to determine the subsequent behavior of Ac; and in the plants coming from the nonvariegated kernels it was necessary to determine the presence or absence of Ac, and the presence or absence of Ds in the C, Sh, wx-carrying chromosome. The results of this study may be summarized. In the plants derived from the majority class of kernels, a single Ac factor was present. Its state was similar to that present in the parent plant (more than 25 cases studied). The Ac constitution in the plants derived from the nonvariegated kernels was most instructive. In 19 plants, no Ac was present. In 17 plants, two nonlinked Ac factors were present. In six plants, an Ac factor inherited as a single unit was present, but it gave a dose action equivalent to two doses of the Ac factor in the parent plant. In the plants derived from kernels that showed very late-occurring Ds events, either two nonlinked Ac factors were present (5 cases), or a single Ac factor was present giving a dose action greater than that of the Ac factor in the parent plant (3 cases).

FIGS. 17 to 22. Photographs of kernels illustrating the somatic segregations of Ac that may occur very early in the development of a kernel. These kernels arose from the cross of plants ($♀$) carrying C, bz and no Ds in each chromosome 9 and having no Ac factor, by plants carrying I, Bz, and Ds (standard location) in chromosome 9 and also carrying Ac. For phenotypes expected from breaks at Ds, see descriptions accompanying Figures 10 to 15. In Figures 17 and 18 there are 4 large sectors in each kernel: one is C bz (above in Figure 17, to right in Figure 18), one is I, non-variegated (to right in Figure 17, upper left in Figure 18), one is characterized by late occurring Ds breaks, producing speckles of the C bz genotype (left in Figure 17, lower left in Figure 18), and one shows that numerous Ds breaks occurred earlier in the development of the kernel (lower segment in Figure 17, middle segment in Figure 18). The kernel in Figure 19 has 3 sectors: one that is C bz, one with few specks of the C bz genotype and one with many specks of the C bz genotypes. Figure 20 shows a kernel with 3 sectors: one that is C bz, one that is wholly I Bz and one having many specks of the C bz genotype. Figure 21 shows a kernel with two sectors: one with many specks of the C bz genotype and one with few such specks. Figure 22 shows a kernel with five sectors: a large C bz sector (lower left), a large sector having many specks of C bz (upper), a large sector showing many larger C bz areas (upper right), a small sector with few C bz specks (middle), and a sector of I Bz with no C bz specks (lower right).

CHROMOSOME ORGANIZATION AND GENIC EXPRESSION   39

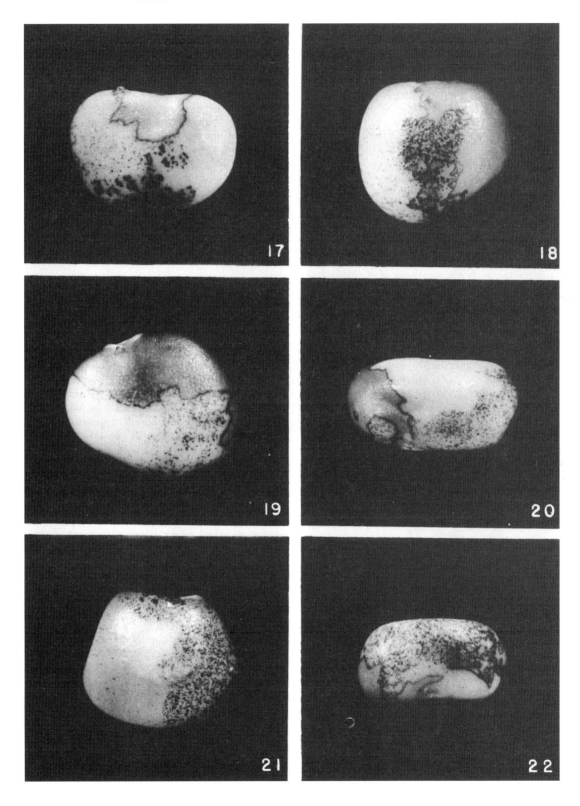

From this analysis it is clear that all the aberrant kernels on ears of the type shown in Figure 23 were produced because of some alteration of *Ac* that had occurred in cells of the parent plant. The reason that no *Ds* breaks were detected in some of these kernels is related either to the absence of *Ac* in the endosperm or to the presence of a marked increase in the dose of *Ac*. It will be recalled that the female parent contributed two gametophytic nuclei to the primary endosperm nucleus. If each nucleus carried two *Ac* factors, or a single *Ac* factor with a double-dose action, the endosperm would have either four *Ac* factors, or two *Ac* factors equivalent in action to four *Ac* factors. In such kernels, the high dose of *Ac* so delayed the time of occurrence of *Ds* breaks that none took place before the endosperm growth had been completed.

In order to verify the analyses of *Ac* constitution in some of these cases, tests were continued for another generation. For example, if a plant contains two nonlinked *Ac* factors, the gametic ratios approach 1 two-*Ac* : 2 one-*Ac* : 1 no-*Ac*—that is, a three-to-one ratio for the presence of *Ac*. On ears derived from crosses in which such plants ˌare used as male parents, the kernels with one or with two doses of *Ac* may be distinguished because of clearly seen differences in the time of response of *Ds* to *Ac* doses. Therefore, some of the kernels considered to have two *Ac* factors and others considered to have only one *Ac* factor were selected from the test ears. The plants grown from them were again tested for gametic ratio of *Ac*. In each case, verification was realized. The gametic ratios produced by the latter plants approached 1 one-*Ac* : 1 no-*Ac*, whereas those produced by the former approximated 3 with one or two *Ac* factors to 1 with no *Ac*.

The above-described series of tests, and still others that have been concerned with the time and type of changes occurring at *Ac*, have made it possible to understand the nature of its inheritance patterns. It has been found that, with any particular state or dose of *Ac*, the time of occurrence of changes of *Ac* is controlled by *Ac* itself. If, with a particular *Ac* state, the time of such changes is delayed until late in the development of the endosperm, then all the kernels should show this same late timing. This is known to occur with many of the isolates of *Ac*. As the photograph in Figure 23 has shown, however, a few aberrant kernels may be present on some of these ears. Some internal or external alteration in environmental conditions may have caused these few early-occurring changes at *Ac*. No attempts have been made, however, to study conditions that might alter the time of such changes.

If these tests for determining the inheritance behavior of *Ac* had not been made, considerable confusion might have arisen. This would certainly have been true had states of *Ac* giving relatively early changes been used in the initial inheritance studies. It must be stated that just such a situation has been observed. States of *Ac* giving aberrant gametic ratios have arisen. It is now realized that this is to be anticipated. It has been determined that the reason for the difference in patterns of inheritance between an *Ac* isolate that gives clear-cut mendelian gametic ratios and one of its modified derivatives, that gives aberrant gametic ratios, is related to the time in the development of the sporogenous or gametophytic cells at which such changes in *Ac* arise. With reference to the gametic constitutions that

FIG. 23. Photograph of an ear derived from a plant having two identical *Ac* factors located at allelic positions in an homologous pair of chromosomes. This plant carried *c* in each chromosome 9. The ♂ parent, having no *Ac*, introduced a chromosome 9 with *C* and *Ds* (standard location). The majority of the kernels are similarly variegated for sectors of the *c* genotype due to breaks at *Ds* that occurred in the *C Ds* chromosome during the development of the kernels. Note the few fully colored kernels in which *c* sectors are absent, and also the several kernels that show large sectors of the *c* genotype.

CHROMOSOME ORGANIZATION AND GENIC EXPRESSION   41

will be produced, the time when these changes occur is most critical. If they occur in somatic divisions before the meiotic mitoses, or in the male or female gametophytes, an apparently unorthodox inheritance pattern for *Ac* will result. If they occur late, that is, in the endosperm tissues—which act in this connection like a continuation of the development of the gametophyte—then no such confused pattern of inheritance will arise. The gametic constitutions will then closely approximate those predicted for mendelizing units. In the study of *Ac* inheritance, it was necessary to make selections for these latter states of *Ac*. A few exceptions with regard to the time of changes at *Ac* may occur in some cells, even with such selected states of *Ac*. It was the analysis of *Ac* constitutions in plants derived, in cell lineage, from those cells in which such exceptional timing of changes at *Ac* had occurred, that provided the information leading to appreciation of the somatic origins of altered states and locations of *Ac*.

Confused patterns of inheritance behavior of mutable loci have been described in the literature many times. The ratios obtained have often been so irregular that no satisfactory formulation of the nature of the inheritance patterns could be derived. This would be just as true of some of the autonomous mutable loci in maize if attention had not been given to altered states and their behavior. Two examples may illustrate this. Both a_1^{m-1} and a_2^{m-1}, when first discovered, produced many mutations and changes in state very early in the development of the plant. The plant, therefore, was sectorial for the altered conditions at these mutable loci. The sectors were present in the tassel. When pollen was collected without reference to the sectors present, and placed on the silks of plants carrying the stable a_1 or a_2 alleles, the kinds of kernels appearing on the resulting ears, and their frequencies, were not readily analyzable in terms of mendelian ratios. No such difficulty arises, however, when similar tests are made for gametic ratios in plants derived from those kernels on the original test ears that show only very late-occurring mutations. The inheritance pattern is now of the obvious mendelian type, for mutations and changes in state are mainly delayed until after meiosis and gamete formation. As with *Ac*, the selection of states of autonomous mutable loci that produce very late-occurring mutations makes it possible to examine the inheritance behavior of such loci, freed from the apparent confusion resulting from early-occurring modifications at the locus, which can distort the expected mendelian ratios.

IS THE BEHAVIOR OF *Ac* A REFLECTION OF A MECHANISM OF DIFFERENTIATION?

We now return to the original question. What is the significance of the somatically occurring changes at *Ac*, and the changes in state that occur at the autonomous mutable loci? Do they suggest the presence of nuclear factors that serve to control when and where certain decisive events will occur in the nucleus? With regard to *Ac*, it is known that the events leading to its loss, to increase or decrease of its dosage action, or to other changes involving its action or position in the chromosome complement, are related; and that they appear as the consequence of a mitotic event, controlled in time of occurrence by the state and dose of *Ac* itself. Sister nuclei are formed that differ with respect to *Ac* constitution, as the photographs of Figures 17 to 22 illustrate. Because of this somatically occurring event involving *Ac*, the *Ac*-controlled mutable loci will differ markedly as to the time when mutational events will occur at them, or as to whether or not any such events will occur at all in the cells arising from the sister cells. This precise timing of somatic segregations effects a form of differentiation, for it brings about changes in the control of occurrence and time of occurrence of genic action at other loci, and does so differentially in the progeny of two sister cells. This likewise applies to the autonomous mutable loci; but in these cases the controller of the time and place of appearance of genic activity is a component of the locus itself.

The process of differentiation is basically one involving patterns of action arising in sequential steps during development and affecting the types of activities of definitive cells. The ultimate expression of component parts of an organism represents the consequence of segregation mechanisms involving the various cellular components. The part played by any one component of the cell in this segregation system can not be divorced from that played by any other component. It is possible, however, to attempt to examine the various components in order to determine their respective relationships and the sequential events that involve them. Embryological studies have contributed much to our knowledge of the segregation of cytoplasmic components. The segregation of nuclear components is less well understood, although some outstanding examples are known. These examples show segregations or losses of obvious components of the nucleus—that is, of whole chromosomes or easily seen parts of chromosomes. The segregation or loss of smaller components, not readily visible on microscopic examination, may well be one of the mechanisms responsible for the nuclear aspects of control of the differentiation process. The phenomenon of variegation, as described here and observed in many other organisms, may be a reflection of such a segregation mechanism—exposed to view be-the timing of events leading to a specific type of genic action is "out-of-phase" in the developmental path. Variegation may represent merely an example of the usual process of differentiation that takes place at an abnormal time in development. Viewed in this way, it is possible to formulate an interpretation of the part played by the nuclear components in controlling the course of differentiation.

This interpretation considers that the nucleus is organized into definite units of action, and that the potentials for types of genic action in any one kind of cell differ from the potentials in another kind of cell. In other words, the functional capacities of the nuclei in different tissues or in different cells of a tissue are not alike. The differences are expressions of nonequivalence of nuclear components. This nonequivalence arises from events that occur during mitotic cycles. The differential mitotic segregations are of several types. Some involve controlling components, such as *Ac*, and produce sister nuclei that are no longer alike with respect to these components. As a result, the progeny of two such sister cells are not alike with respect to the types of genic action that will occur. Differential mitoses also produce the alterations that allow particular genes to be reactive. Other genes, although present, may remain inactive. This inactivity or suppression is considered to occur because the genes are "covered" by other nongenic chromatin materials. Genic activity may be possible only when a physical change in this covering material allows the reactive components of the gene to be "exposed" and thus capable of functioning.

A mechanism of differentiation that requires differences in nuclear composition in the various cells of an organism finds considerable support in the literature. The most conspicuous example is in *Sciara*, where a thorough cytological and genetical analysis has been made. (For reviews, see Metz, 1938, and Berry, 1941.) It is known, in this organism, at just what stage of development differences in nuclear composition will arise; and, with regard to the X chromosome, it is known what element in the chromosome controls the differential behavior. This element is at or near the centromere of the chromosome (Crouse, 1943). Furthermore, differential segre-

gations of the B-type or accessory chromosome have been found to occur in a number of plants. (For reviews of literature to 1949, see Müntzing, 1949.) Numerous other examples are known of differential segregation involving whole chromosome complements, certain types of chromosomes of a complement, or, occasionally, a certain component of a chromosome. (For literature citations see Melander, 1950; White, 1945, 1950; Berry, 1941.) Whether the differential segregations involve whole complements of chromosomes, individual chromosomes, individual parts of chromosomes that can be seen, or submicroscopic parts of chromosomes, may well be a matter of degree rather than type. Certainly, the evidence for differential segregation is not wholly negative.

With regard to mechanisms associated with differentiation and genic action, an additional factor may be mentioned. The part played by the doses of component elements in the chromosomes appears to be of considerable importance. First, a number of genetic factors associated with known loci produce measurable quantitative effects that are related to dose: the higher the dose the greater the effect. Such dosage actions, probably reflecting rates of reaction, are familiar to all geneticists, and some of them have been reviewed in this study of mutable loci. Dosage controls of the *Ac* type, affecting the time of action of certain other factors carried by the chromosomes, has been less well appreciated. A third type of dosage action has made itself evident in these studies. In some aspects, however, it resembles the action of different doses of *Ac*. In the study of the autonomous $a_2{}^{m-1}$ mutable locus, a number of mutants appeared, particularly on self-pollinated ears, showing a pale aleurone color. Study of the behavior of these pale mutants has revealed the following. Some of them produce pale-colored aleurone in one, two, and three doses and give no evidence of instability in the expression of the phenotype. (One and two doses are obtained by combinations of the pale-mutant allele with the stable a_2 allele.) That this stable expression may be deceptive is shown by the dosage effects of other similarly appearing pale-producing isolates derived from mutations at $a_2{}^{m-1}$. These may give pale aleurone color, and no indication of instability, in three and two doses; but with one dose something unexpected occurs. The kernels show a colorless aleurone in which mutations to deep aleurone color appear. Still other isolates give pale color in three doses, but in one or two doses produce the colorless background with deep-colored mutant areas. In these cases, it is clear that some of the mutations at $a_2{}^{m-1}$ giving pale color and appearing to be stable are stable only because of some dosage action produced by a mutation of the original $a_2{}^{m-1}$ to the pale-producing type.

The study of dose-provoked actions in the pale mutants mentioned above and those of Ac have given some indication of the importance of dosage action in affecting genic expression. The original isolate of $a_2{}^{m-1}$ did not give evidence of such striking dosage action. When present in one, two, or three doses, it gave rise to colorless kernels in which mutations, mainly to a deep color and occasionally to a pale color, appeared. The graded series of dosage action exhibited by the various pale mutants derived from $a_2{}^{m-1}$ is very much the same as the graded series exhibited by the various isolates of Ac. In these cases, it appears as if each isolate is composed of a specific number of reactive subunits and that the dosage expressions are related to the total number of such units that are present in the nucleus. Although these graded dosage effects may be visualized on a numerical basis, it is not claimed that such an interpretation is necessarily the correct one. The large differences in dosage expression exhibited by the various isolates of Ac, and also the various isolates of the pale mutants derived from $a_2{}^{m-1}$, nevertheless appear to follow such a scheme.

Why different doses of components of the chromosomes function as they do in controlling developmental processes takes us to another level of analysis that is not under consideration here. A relation to rates of particular reactions can be suspected. It is tempting to consider that changed environmental conditions may well alter otherwise-established rates of reaction, and thus initiate alterations in the nuclear components at predictable times, leading to strikingly modified phenotypic expression. Just such effects have been observed by students of developmental genetics. They have shown that alterations of environmental conditions at particular times in development can lead to predictable changes in the subsequent paths of differentiation.

CONSIDERATION OF THE CHROMOSOME ELEMENTS RESPONSIBLE FOR INITIATING INSTABILITY

It will be recalled that this study of the origin and behavior of mutable loci was undertaken because a large number of newly arisen mutable loci appeared in the progeny of plants in which an unusual sequence of chromosomal events had occurred—that is, the breakage-fusion-bridge cycle. Striking similarities in the patterns of behavior of these mutable loci were immediately noticed. It was the pattern of behavior, rather than the change in expression of the particular phenotypic character, that was obviously of importance. This pattern, revealed in all cases, stemmed from an event occurring at mitosis, which altered the time and frequency of mutations that would subsequently occur in the cells derived from those in which this event occurred. It was noticed that sister nuclei could differ in these respects—and sometimes reciprocally, as if the mitotic event had resulted in an increase in one nucleus of a component controlling the mutation time or frequency, and a decrease of this component in the sister nucleus. It was also noted that the change in phenotypic expression—that is, the mutation—likewise resulted from a mitotic event; and that the mutation itself and changes of the controller of the mutation process could result from the same mitotic event: one cell showing the mutation, the sister cell showing an altered condition with respect to control of future mutations in the cells derived from it.

Further, it may be recalled that the mechanism which resulted in the appearance of newly arisen mutable loci—that is, the breakage-fusion-bridge cycle involving chromosome 9—gave rise to numerous obvious alterations of the heterochromatic materials, in other chromosomes of the complement as well as in chromosome 9. It was also demonstrated that the effect of a known activator, Dt, located in the heterochromatin of the chromosome-9 short arm, and producing a very definite pattern of mutations of the otherwise stable a_1 locus in chromosome 3, could be recreated independently and on a number of different occasions in cells of a tissue in which the breakage-fusion-bridge cycle was in action. The combined observations and experiments point to elements in the heterochromatin as being the ones concerned with differential control of the times at which certain genes may become reactive. It is believed that somatic segregations of components of these elements may initiate the process of nuclear control of differentiation.

On the basis of these interpretations and those given in the previous section, it becomes apparent why a large number of newly arisen mutable loci appeared in the self-pollinated progeny of plants that had undergone the chromosome type of breakage-fusion-bridge cycle. This cycle induced alterations in the heterochromatin. These alterations changed the organization of the heterochro-

44   *BARBARA McCLINTOCK*

matic chromosome constituents and probably also, in many cases, the doses of their component elements. Changes were induced in these heterochromatic elements at times other than those at which they would normally occur during differentiation. This resulted in changes in the times in development when their action on specific chromatin material, associated with genic components of the chromosome, was expressed. The altered timing of their actions was consequently "out-of-phase" with respect to the timing that occurs during normal differentiation. This was made evident by the appearance of a "mutable locus." The "mutable locus" is thus a consequence of the alteration of an element of the heterochromatin produced by the breakage-fusion-bridge cycle. Once such an "out-of-phase" condition arises, others may subsequently appear because of the physical changes in the chromatin that occur at the mutable locus, leading, at times, to transpositions of this chromatin to new locations, as described earlier. In their new locations, these transposed chromatin elements continue their specific control of types of genic action but now affect the action of the genic components at the new locations.

RELATION OF "MUTABLE LOCI" TO "POSITION-EFFECT" EXPRESSIONS IN DROSOPHILA AND OENOTHERA

In a previous publication (McClintock, 1950) the author has suggested that the position-effect variegations in *Drosophila melanogaster* and the variegations observed in many other organisms, including those associated with the mutable loci here described, are essentially the same. An adequate discussion of the interrelations would require more space than can appropriately be given here. Attention will be drawn, therefore, only to a few relevant facts, which may serve to indicate why this conclusion has been reached. In the first place, a number of different types of position-effect expression are found in *Drosophila* (for review and literature citations, see Lewis, 1950). In maize, comparable types of instability expression have appeared. In *Drosophila*, some of the variegations appear to result from loss of segments of chromosomes. This applies to those cases where the expression of the dominant markers, carried by the chromosome showing the "position-effect" phenomenon, is absent in some sectors of the organism. The extent of the deficiency varies, but it includes in each case the region adjacent to the heterochromatic segment with which many of the variegation types of

position-effect expression are known to be associated. It may be recalled that such deficiencies are produced in maize when *Ds* is present.

That heterochromatic elements of the chromosomes of *Drosophila* undergo breakage events in somatic cells is suggested by the study of "somatic crossing over" in this organism (Stern, 1936). The appearance of the abnormally timed exchanges between chromosomes is conditioned by the presence of certain Minute factors, for example, $M(1)n$, much as the occurrence of structural aberrations at certain loci in maize (i.e., *Ds*, wherever it may be) is dependent on the presence of *Ac*.

Of particular significance for comparative purposes is the study of Griffen and Stone (1941) on the induction of changes in the position-effect expression in *Drosophila* of the white-eye variegation, w^{m5}. The w^{m5} case arose through an X-ray-induced translocation of the segment of the left end of the X chromosome at 3C2 (the w^+ locus) to the heterochromatic region of chromosome 4. Males carrying w^{m5} were X-rayed, and the progeny examined for changes in the variegation expression of the eye mottling. Many such changes were found. Studies of these cases were continued in order to determine the nature of the events associated with the changes. In all cases, the new modification in the phenotypic expression of the w^+ locus was found to be associated with a translocation, which placed the segment of the left end of the X chromosome, from 3C2 to the end of the arm, at a new location. In many cases, the new position was to a euchromatic region of another chromosome, and yet variegation persisted. Some of the new positions, however, gave rise to apparent "reversions" to a wild-type expression. Individuals having these "reversions" were X-rayed, and variegation types again appeared in the progeny. Here also the variegation was shown to be associated with a translocation involving the left end of the X chromosome at 3C2, from the location in the "reversion" stock to a new location—again, sometimes a euchromatic region. It may be suspected that the maintenance of variegation potentialities in all these cases was associated with the presence of a segment of heterochromatin of chromosome 4 that remained adjacent to the w^+ locus when the successive translocations occurred. This would not readily be detected in the salivary chromosomes. The presence of such "inserted" heterochromatin could be responsible for the continued expression of variegation at the w^+ locus in repeated translocations. If such was the case, then the resemblance to the maize cases, described in this report, is obvious. The appearance of "reversions," and the subsequent appearance of variegation after X-radiation of individuals carrying such "reversions," might seem to present a contradiction. On the basis of an analysis of the cases described by Griffen and Stone, the writer believes that no contradiction is involved. This analysis has suggested that the timing of variegation-producing events during development is, in part, a function of the relative distance of the translocated segment— i.e., the left end of the X chromosome—from the centromere of the chromosome that carries it: the farther removed the segment is from the centromere, the later in the development the variegation-producing events will occur. In the "reversions," this segment has been placed close to the end of one arm of a chromosome. The reappearance of variegation occurs when the segment is translocated to a position closer to a centromere. Another factor is also associated with the timing of the variegation-producing events. This is the Y chromosome. When the Y is absent, the areas of altered phenotype are larger than when it is present, indicating an earlier timing of the variegation-producing events. It may be noted in this connection that some of the cases of "reversions" are only apparent reversions. In XY constitutions they appear to give a stable wild-type expression but in XO constitutions, the eyes show a light speckling of the altered phenotype. With the latter constitution variegation occurs, but only very late in the development of the eye. The similarity of this effect of the Y chromosome to that of dosage action of Ac is apparent in these cases as well as in many others in *Drosophila* that have been examined.

In a recent report, Hinton and Goodsmith (1950) gave an analysis of induced changes at the bw^D (Dominant brown eye) locus in chromosome 2 of *Drosophila*. This case was considered to be a stable-type position effect. It arose originally through the insertion of an extra band next to the salivary-chromosome band where bw^+ is located. Males carrying bw^D were irradiated and crossed to wild-type females. The offspring (9,757 individuals) were examined for changes in the bw^D expression. Twenty-one individuals showing the wild-type expression appeared in the F_1, and progeny was obtained from one-third of them. A study of the inheritance behavior of each modification was undertaken, and a study was also made of the salivary-gland chromosomes. From these studies it was clear that the modifications

arose from changes that occurred in the vicinity of the bw^D locus and involved the inserted band. In four cases, restoration of the wild-type expression followed removal of this band. In two cases, it followed separation of this band from the bw^+ band by translocations. In one case, no obvious change in the salivary chromosomes was noted, but nevertheless a change in phenotypic expression had occurred. Of considerable importance, also, was the appearance, in some of these cases, of somatic instability of expression of the bw^+ phenotype. Variegation began to appear. It had never been observed in the brown-Dominant stock itself. It may be noted that the changes in the bw^D expression are associated with types of chromosomal alterations which are much the same as those proposed to account for changes in phenotypic expression at some mutable loci in maize.

A further resemblance between *Drosophila* and maize will be mentioned. In *Drosophila*, many of the translocations, inversions, and duplications are believed to be associated with the formation of dominant-lethal effects. In maize, a number of dominant lethals have arisen from transpositions of Ds. Some produce defective growth of the endosperm and embryo; others affect the development of the embryo but not the endosperm; and still others affect the capacity of the embryo to germinate, without affecting its morphological characters. Over half the newly arisen transpositions of Ds that are of this latter type have not produced viable plants, owing to lack of germination of the embryos in the kernels.

There are similarities between the maize cases and a case in *Drosophila pseudoobscura* described by Mampell (1943, 1945, 1946). In this *Drosophila* case, a heterochromatic element appeared to be associated with the initiation of instability at another locus, which in turn led to changes in chromosome organization and to numerous changes in genic action at various loci in the chromosome complement. These changes were expressed both somatically and germinally.

The position-effect behavior reported in *Oenothera* (Catcheside, 1939, 1947a, b) is much like that of Ds. The chromosomal events responsible for the observed types of change in phenotypic expression may be the same in the two organisms. In *Oenothera* as in maize, gross changes in chromosome constitution arise, such as duplications and deficiencies of segments of the chromosome involved. Similar cases in other organisms undoubtedly exist. It is probable, however, that

46   *BARBARA McCLINTOCK*

the lack of a critical mode of detection of a chromosome breakage mechanism has been responsible for the apparent delay in reporting such cases in connection with studies of somatic variegation and mutable loci. Also, because changes in state occur that involve reduction in the frequency of chromosome breaks, and because such breaks lead to lethal gametes, it is probable that states of a mutable locus producing some detectable breaks are rapidly eliminated from a population, leaving a state of the mutable locus that produces few or no such events to be propagated.

It has been argued that the variegation types of position effect in *Drosophila* usually do not give rise to germinal mutations, and that they belong, therefore, to a separate category of instability expression.   Since some variegation position effects do give rise to germinal changes, this argument could in any case be only partially applicable. However, whether or not germinal changes arise is not considered relevant in the interpretation developed here. The time and place of occurrence of such changes is related to controls, existing in the nucleus. The differentiation mechanism described above should effect controls that would exclude the germ lines from undergoing many changes, but should allow numerous alterations in the soma that would lead to altered patterns of genic expression. Whether or not a particular somatically expressed pattern of genic action—for example, the distribution of pigment—arises from mutations at a "mutable locus" or from the action of a particular "stable" allele of the locus cannot be decided by using the criterion of presence or absence of germinal mutations. The important consideration is when, where, and how the patterns of genic action are controlled and eventually expressed.

The combined evidence from many sources suggests that one should look first to the conspicuous heterochromatic elements in the chromosomes in search of the controlling systems associated with initiation of differential genic action in the various cells of an organism; and secondarily to other such elements, which are believed to be present along the chromosomes and to be either initially or subsequently involved in the events leading to differential genic action.   Evidence, derived from *Drosophila* experimentation, of the influences of various known modifiers on expression of phenotypic characters has led Goldschmidt (1949, 1951) to conclusions that are essentially similar to those given here.

The conclusions and speculations on nuclear, chromosomal, and genic organization and behavior included in this report are an outgrowth of studies of the instability phenomenon in maize. They are presented here for whatever value they may have in giving focus to thoughts regarding the basic genetic problems concerned with nuclear organization and genic functioning.   Until these problems find some adequate solution, our understanding and our experimental approach to many phenomena will remain obscured.

REFERENCES

BERRY, R. O., 1941, Chromosome behavior in the germ cells and development of the gonads in *Sciara ocellaris*. J. Morph. *68:* 547–576.

CATCHESIDE, D. G., 1939, A position effect in *Oenothera*. J. Genet. *38:* 345–352.
    1947a, The *P*-locus position effect in *Oenothera*. J. Genet. *48:* 31–42.
    1947b, A duplication and a deficiency in *Oenothera*. J. Genet. *48:* 99–110.

CROUSE, HELEN V., 1943, Translocations in *Sciara;* their bearing on chromosome behavior and sex determination. Res. Bull. Mo. Agric. Exp. Sta. *379:* 1–75.

DEMEREC, M., 1935, Mutable genes. Bot. Rev. *1:* 233–248.

EMERSON, R. A., 1921, Genetic evidence of aberrant chromosome behavior in maize. Amer. J. Bot. *8:* 411–424.

FOGEL, S., 1950, A mutable gene at the *R* locus in maize. Rec. Genet. Soc. Amer. *19:* 105.

GOLDSCHMIDT, R. B., 1949, Heterochromatic heredity. Hereditas, Suppl. *5:* 244–255.

GOLDSCHMIDT, R. B., HANNAH, A., and PITERNICK, L. K., 1951, The podoptera effect in *Drosophila melanogaster*. Univ. Calif. Publ. Zool. *55:* 67–294.

GRIFFEN, A. B., and STONE, W. S., 1941, The w^{m5} and its derivatives. Univ. Texas Publ. No. 4032: 190–200.

HINTON, T., and GOODSMITH, W., 1950, An analysis of phenotypic reversion at the brown locus in *Drosophila*. J. Exp. Zool. *114:* 103–114.

LAUGHNAN, JOHN R., 1949, The action of allelic forms of the gene *A* in maize. II. The relation of crossing over to mutations of A^b. Proc. Nat. Acad. Sci. Wash. *35:* 167–178.
    1950, Maize Genetics Coöperative News Letter *24:* 51–52.
    1951, Maize Genetics Coöperative News Letter *25:* 28–29.

LEWIS, E. B., 1950, The phenomenon of position effect. Advances in Genetics *3:* 73–115.

MAMPELL, K., 1943, High mutation frequency in *Drosophila pseudoöbscura*, Race B. Proc. Nat. Acad. Sci. Wash. *29:* 137–144.
    1945, Analysis of a mutator. Genetics *30:* 496–505.
    1946, Genic and non-genic transmission of mutator activity. Genetics *31:* 589–597.

MANGELSDORF, P. C., 1948, Maize Genetics Coöperative News Letter *22:* 21.

McCLINTOCK, B., 1941, The stability of broken ends of chromosomes in *Zea Mays*. Genetics *26:* 234–282.

CHROMOSOME ORGANIZATION AND GENIC EXPRESSION

1942, The fusion of broken ends of chromosomes following nuclear fusion. Proc. Nat. Acad. Sci. Wash. 28: 458-463.

1950, The origin and behavior of mutable loci in maize. Proc. Nat. Acad. Sci. Wash. 36: 344-355.

MELANDER, Y., 1950, Accessory chromosomes in animals, especially in Polycelis tenus. Hereditas 36: 19-38.

METZ, C. W., 1938, Chromosome behavior, inheritance and sex determination in Sciara. Amer. Nat. 72: 485-520.

MÜNTZING, A., 1949, Accessory chromosomes in Secale and Poa. Proc. Eighth Intern. Congr. Genetics. (Hereditas, Suppl. Vol.).

NUFFER, M. GERALD, 1951, Maize Genetics Coöperative News Letter 25: 38-39.

RHOADES, M. M., 1936, The effect of varying gene dosage on aleurone color in maize. J. Genet. 33: 347-354.

1938, Effect of the Dt gene on the mutability of the a_1 allele in maize. Genetics 23: 377-395.

1941, The genetic control of mutability in n Cold Spring Harb. Symposium Quant. Bic 138-144.

1945a, On the genetic control of mutability in n Proc. Nat. Acad. Sci. Wash. 31: 91-95.

1945b, Maize Genetics Coöperative News Lett 14.

1947, Maize Genetics Coöperative News Letter

1950, Maize Genetics Coöperative News Lett 49.

RHOADES, M. M., and DEMPSEY, E., 1950, Genetics Coöperative News Letter 24: 50.

SAGER, R., 1951, On the mutability of the waxy in maize. Genetics (in press).

STERN, C., 1936, Somatic crossing over and se tion in Drosophila melanogaster. Genetic 625-730.

WHITE, M. J. D., 1945, Animal Cytology and Evol Cambridge Univ. Press.

1950, Cytological studies on gall midges (Cecic dae). Univ. Texas Pub. No. 5007: 1-80.

Dissecting the Bacteriophage Life Cycle

Doermann A. 1952. **The Intracellular Growth of Bacteriophages. I. Liberation of Intracellular T4 by Premature Lysis with Another Phage or with Cyanide.** (Reprinted, with permission, from *J. Gen. Physiol.* **35:** 645–656 [©Rockefeller University Press].)

A.H. "Gus" Doermann and Salvador Luria take a break from the 1953 Symposium on "Viruses" and chat on the lawn of Blackford Hall. Demerec Laboratory is in the background.

BACTERIOPHAGE (KNOWN MORE FAMILIARLY AS phage) are viruses that prey on bacteria. Their use for genetic studies was promoted by Max Delbrück, a theoretical physicist who had worked with Max Born, Niels Bohr, and Wolfgang Pauli. Delbrück's interest in biology was stimulated by Bohr's "Light and Life" lecture on the implications of quantum physics for understanding life (Bohr 1933). Delbrück's first biological paper (with the *Drosophila* geneticist Timoféef-Ressovsky and the radiobiologist Zimmer) was on the nature of mutations induced by radiation (Timoféef-Ressovsky et al. 1935). This paper achieved great celebrity when Erwin Schrödinger used it as the basis for his discussion of the nature of the gene in his book *What Is Life?* (Schrödinger 1944). (This book has been regarded as an icon [Morange 1998], and several eminent molecular geneticists have acknowledged that it led them to join the quest for the understanding of the gene. Others have looked at its contents more skeptically. Linus Pauling wrote that its treatment of thermodynamics "...is vague and superficial to an extent that should not be tolerated even in a popular lecture [Pauling 1987], and Max Perutz wrote that "...what was true in his book was not original, and most of what was original was known not to be true even when the book was written" (Perutz 1987).)

In 1937, Delbrück came to the California Institute of Technology where he met Emory Ellis, who was already working on phage. Delbrück was elated to find that simple experiments on phage replication offered a route to understanding the gene. He and Ellis carried out the classic "one-step-growth" experiment that provided the basic system for analyzing phage growth (Ellis and Delbrück 1939). In 1940, Delbrück met Salvador Luria, a geneticist at Columbia's College of Physicians & Surgeons, and the two resolved to collaborate. It was here that Cold Spring Harbor entered the picture, for Demerec had invited Delbrück to take part in the 1941 Symposium. Luria joined him for the first of many summers that they spent there. Quite apart from his own research, Delbrück was determined to promote phage research widely. To do so, he forced standards on the field by insisting that everyone should work on *E. coli* strain B that was attacked by the "T" series phage which had been classified by Demerec and Fano (1945). In addition, he established the Phage Course at Cold Spring Harbor in 1945, which trained many of the leaders in molecular genetics (Brock 1990; Susman 1995).

Delbrück and Luria believed that an analysis of the gene required working with as simple a replicating system as possible, and Delbrück wrote that phage multiplication was "..so simple a phenomenon that answers cannot be hard to find" (Delbrück 1946). However, things were not proving so simple and there was a particularly embarrassing omission in what was known of the life of a phage. Nothing was known of what was going on in the "latent" period between infection and lysis, and yet understanding this was precisely the goal of the phage geneticists.

One of Delbrück's assistants on the first phage course was A.H. (Gus) Doermann, who later came to the Department of Genetics as a Carnegie Institution Fellow in 1947. Doermann's goal was to dispel "...our complete ignorance about the state and number of intracellular phage" (Doermann 1966), and his starting point was experiments done by Delbrück in 1940. Delbrück (1940) had shown that bacteria lysed rapidly when attacked by about 40 phage per cell, rather than the usual 3 or 4 per cell. Doermann found that T6 was particularly efficient at this "lysis from without," and he used it to lyse cells that had been infected with another phage. For example, Doermann infected a culture of cells with T4 and removed samples of the suspension at intervals. Cells in these aliquots were lysed with large numbers of T6. The production of infectious T4 phage was followed by plating the lysed samples on a strain of E. coli that was resistant to infection by T6. The only plaques formed were caused by T4, and, thus, he was able to detect T4 phage even in the presence of a very large excess of T6 used to induce lysis.

Doermann's first experiments were promising but not wholly successful because, it appeared, there was too much of a lag between addition of T6 and lysis of the cells. Because T4 phage replicated quickly relative to the time taken for lysis, the yield of T4 phage at lysis did not accurately represent the amount of phage at the time the T6 was added. To overcome this, Doermann added inhibitors along with the T6 so that T4 replication was blocked immediately even though the cells underwent lysis sometime after. This was done efficiently by both cyanide and 5-methyltryptophan. Doermann was encouraged when a very different method—disrupting the infected cells with sonic vibration—gave similar results (Anderson and Doermann 1952).

It was clear from Doermann's work that when a phage particle entered a cell, it underwent some transformation such that it was no longer infectious. The interval between infection and the first intracellular appearance of infectious phage particles was called the eclipse period, and what was happening during those ten minutes was still mysterious. Perhaps the infecting particle was in some state of limbo, no longer infectious but waiting for the bacterial cell metabolism to undergo whatever changes were necessary to begin phage replication. Alternatively, as Luria (1947) had suggested, a phage particle on entering the cell broke down into its constituent parts and began replicating. The end of the eclipse period, then, marked the first assembly of the newly synthesized phage components into intact phage.

But what did "breaking down into its constituent parts" mean? What were these parts and what did they do? The experiment that determined this was also carried out in the Department of Genetics and was destined to become one the most famous biological experiments of the twentieth century (pages 201–222).

A.H. Doermann was born in Blue Island, Illinois. He was an undergraduate at Wabash College and did his graduate studies at Stanford University where his advisor was George Beadle (who shared the Nobel Prize for Physiology or Medicine with Ed Tatum in 1958). Doermann joined Delbrück at Vanderbilt University, then spent several years in the Biology Division of Oak Ridge National Laboratory. His first faculty appointment was at the University of Rochester. There he was advisor to several graduate students who became key players in phage research, including Robert

Edgar, Richard Epstein, and Frank Stahl. Doermann moved to the Genetics Department at the University of Washington in 1964. He retired in 1984 and died in 1991.

Anderson T.F. and Doermann A.H. 1952. The intracellular growth of bacteriophages. II. The growth of T3 studied by sonic disintegration and by T6-cyanide lysis of infected cells. *J. Gen. Physiol.* **35:** 657–667.

Bohr N. 1933. Light and life. *Nature* **131:** 421–423, 457–459.

Brock T.D. 1990. *The emergence of bacterial genetics*, p. 127. Cold Spring Harbor Laboratory Press, Cold Spring Harbor, New York.

Delbrück M. 1940. The growth of bacteriophage and lysis of the host. *J. Gen. Physiol.* **23:** 643–660.

————. 1946. Experiments with bacterial viruses (bacteriophages). *Harvey Lect.* **41:** 161–187.

Demerec M. and Fano U. 1945. Bacteriophage-resistant mutants in *Escherichia coli*. *Genetics* **30:** 119–136.

Doermann A.H. 1952. The intracellular growth of bacteriophages. I. Liberation of intracellular T4 by premature lysis with another phage or with cyanide. *J. Gen. Physiol.* **35:** 645–656.

————. 1966. The eclipse in the bacteriophage life cycle. In *Phage and the origin of molecular biology* (ed. J. Cairns et al.), pp. 79–87. Cold Spring Harbor Laboratory, Cold Spring Harbor, New York (reprinted with additional material, 1992).

Ellis E.L. and Delbrück M. 1939. The growth of bacteriophage. *J. Gen. Physiol.* **22:** 365–384.

Luria S. 1947. Reactivation of irradiated bacteriophage by transfer of self-reproducing units. *Proc. Natl. Acad. Sci.* **33:** 253–264.

Morange M. 1998. *A history of molecular biology.* Harvard University Press, Cambridge, Massachusetts.

Pauling L. 1987. Schrödinger's contributions to chemistry and biology. In *Schrödinger: Centenary celebration of a polymath* (ed. C.W. Kilmister), pp. 225–233. Cambridge University Press, Cambridge, United Kingdom.

Perutz M.F. 1987. Erwin Schrödinger's *What is life?* and molecular biology. In *Schrödinger: Centenary celebration of a polymath* (ed. C.W. Kilmister), pp. 234–251. Cambridge University Press, Cambridge, United Kingdom.

Schrödinger E. 1944. *What is life? The physical aspect of the living cell.* Cambridge University Press, Cambridge, United Kingdom.

Susman M. 1995. The Cold Spring Harbor Phage Course (1945–1970): A 50th anniversary remembrance. *Genetics* **139:** 1101–1106.

Timoféeff-Ressovsky N.W., Zimmer K.G., and Delbrück M. 1935. Ueber die Natur der Genmutation ind der Genstruktur. *Nachrbl. Gesamte. Wiss. Göttingen* **13:** 190–245.

THE INTRACELLULAR GROWTH OF BACTERIOPHAGES

I. LIBERATION OF INTRACELLULAR BACTERIOPHAGE T4 BY PREMATURE LYSIS WITH ANOTHER PHAGE OR WITH CYANIDE

BY A. H. DOERMANN*, ‡

(*From the Department of Genetics, Carnegie Institution of Washington, Cold Spring Harbor*)

(Received for publication, August 20, 1951)

Direct studies of bacteriophage reproduction have been handicapped by the fact that the cell wall of the infected bacterium presents a closed door to the investigator in the period between infection and lysis. As a result it was impossible to demonstrate the presence of intracellular phage particles during this so called latent period, and, much less, to estimate their number or to describe them genetically. This barrier has now been penetrated. It is the purpose of the first two papers of this series to describe two methods for disrupting infected bacteria in such a way that the intracellular phage particles can be counted and their genetic constitution analyzed.

The first method used to liberate intracellular bacteriophage depends on the induction of premature lysis in infected bacteria by "lysis from without" which occurs when a large excess of phage particles is adsorbed on bacteria (1). It was found by nephelometric tests that T6 lysates are efficient in disrupting cells when moderately high multiplicities are used (2). The further observation was made that the addition of a large number of T6 particles to bacteria previously infected with T4, would, under some conditions, cause liberation of T4 particles before the expiration of the normal latent period of these cells. It therefore seemed hopeful that a method of reproducibly disrupting infected bacteria could be developed on the basis of this preliminary knowledge.

* The experiments described here were carried out while the author was a fellow of the Carnegie Institution of Washington. The author is indebted to Dr. M. Demerec and the staff of the Department of Genetics of the Carnegie Institution of Washington for providing facilities for this work. In particular the stimulating discussions with Dr. Barbara McClintock are gratefully acknowledged. The manuscript was prepared while the author held a fellowship in the Department of Biology of the California Institute of Technology. He is grateful to Dr. G. W. Beadle and the staff of that department for their interest in this work, and especially to Dr. Max Delbrück for criticism of the manuscript.

‡ Present address: Biology Division, Oak Ridge National Laboratory, Oak Ridge, Tennessee.

The first experiments in devising a method of this kind were made with phage T5 (3). It was found that T5 is liberated before the end of the latent period if the infected cells are exposed to a high excess of T6. However, the extremely low rate of adsorption of T5 coupled with difficulties in inactivation of unadsorbed phage by specific antisera indicated that this phage was a poor choice. Hence T4 was chosen because of its fast rate of adsorption and because of the availability of high titer antisera against it. The first experiments with T4, along with the T5 results, showed conclusively that, by itself, lysis from without is not sufficiently rapid for the purpose of this investigation. It is likely that phage growth continues after the addition of the lysing agent T6. Therefore the attempt was made to stop phage growth while T6 was allowed to accomplish lysis from without. Low temperature could not be used for this purpose

TABLE I

Composition of the Growth Medium

Material	Amount
	gm. per liter
KH_2PO_4	1.5
Na_2HPO_4 (anhydrous)	3.0
NH_4Cl	1.0
$MgSO_4 \cdot 7HOH$	0.2
Glycerol	10.0
Acid-hydrolyzed casein	5.0
dl-Tryptophan	0.01
Gelatin*	0.02
Tween-80	0.2

* To reduce surface inactivation of free phage particles (4).

since it also inhibits lysis from without. A search for a suitable metabolic inhibitor was therefore undertaken, and cyanide was eventually chosen as the most suitable one.

Materials and Methods

The experiments described here were carried out with the system T4r_{48}[1] growing at 37°C. in *Escherichia coli*, strain B/r/1. The latter is a T1 resistant, tryptophan-dependent mutant of B/r obtained from Dr. E. M. Witkin.

Two media were used in these experiments, namely the growth medium and the lysing medium. The composition of the growth medium used for both bacterial and phage cultures, is given in Table I. The lysing medium consists of growth medium with

---

[1] The subscript refers to a particular *r* mutant of T4 which arose by mutation and was numbered after the system of Hershey and Rotman (12) using the high subscript number to avoid confusion with the mutants already described by Hershey and Rotman.

the addition of one part in ten of a high titer T6 phage filtrate (concentration of T6 in lysing medium was *ca.* 4×10^9 particles per ml.) and cyanide brought to a final concentration of 0.01 M. Specially designed experiments showed that at this concentration the cyanide does not inactivate free phage particles, nor does the amount which reaches the plate affect titration by interference with plaque development.

T6 was used as the lysing phage because in several experiments it proved to be a more effective lysing agent than any of the other T phages tested. Since only single stocks of the phages were compared in the early experiments, the superiority of T6 over the other phages may have been due to a difference in the particular stock used, and not to an inherent difference among the phages. In fact, later experiments with different T6 stocks showed marked differences in lysing efficiency, and phage titer proved to be a poor criterion of lysing ability. The experiments described here were made with T6 stocks selected for their ability to induce lysis from without. The selections were made on the basis of nephelometric comparisons.

Platings were made in agar layer (0.7 per cent agar) poured over nutrient agar plates (1.3 per cent agar), and in order to assay T4 in the presence of high titer T6, the indicator strain, B/6, was used. B/6 is completely resistant to T6 (no host range mutants have so far been found which will lyse the strain used here) and gives full efficiency of plating (compared to B) with T4.

EXPERIMENTAL

Experiments with the Standard Lysing Medium.—The experimental procedure used consisted essentially of a one-step growth experiment (5) with certain modifications. B/r/1 cells in the exponential growth phase were concentrated by centrifugation to about 10^9 cells per ml. To these concentrated bacteria $T4r_{48}$ was added and this adsorption mixture was incubated for 1 to 2 minutes with aeration, allowing at least 80 per cent of the phage to be adsorbed to the bacteria. Then a 40-fold or larger dilution was made into growth medium containing anti-T4 rabbit serum. The serum inactivated most of the residual unadsorbed phage. After several minutes' incubation in the serum tube, a further dilution was made to reduce the serum concentration to one of relative inactivity. The resulting culture will be referred to as the *source culture* (SC). The entire experiment was carried out with the infected bacteria from SC. The titer of infected B/r/1 in this tube was approximately 10^5 cells per ml.

Simultaneously with the dilution into the tube containing serum, another dilution from the adsorption tube was made. From the latter an estimation of the unadsorbed phage was made by assaying the supernatant after sedimentation of the cells. This step permits calculation of the multiplicity of infection (5).

From SC a further dilution of 1:20 was made at some time before the end of the latent period. The resulting culture, containing approximately 5×10^3 infected bacteria per ml., was used for determining the normal end of the latent period and for estimating the average yield of phage per infected cell. It will be called the *control growth tube* (GT). In addition, a number of precisely timed 20-fold dilutions were made from SC into lysing medium. These were titrated after they had been incubated in the lysing medium for 30 minutes or longer. Serial platings from the lysing medium cultures over a longer period of time have shown that the phage titer remained constant after 30 minutes' incubation. The titer calculated from these platings, divided by the titer of infected bacteria given by the preburst control platings, gives the average yield per infected cell. As a working hypothesis, this yield was considered to be the average number of intracellular phage particles per bacterium at the time of dilution into the lysing medium. Dividing these numbers by the control burst size gives the fraction of the control yield found in the experimental lysing medium tubes.

The results of several typical experiments are shown in Fig. 1 in which the data are plotted on semilogarithmic coordinates. The fraction of the control yield found in a given experimental culture is plotted against the time at which the dilution is made into lysing medium. Curve 1 shows the results from a single experiment in which the bacteria were infected with an average of 7 phage particles each. Curve 2 is the composite result of four experiments in which the bacteria were infected with single phage particles. Curve 3 is the control one-step growth curve derived from the control growth tube platings in the four experiments of curve 2.

Several striking results can be seen in these experiments. First, it is clearly seen that during the early stages of the latent period the virus-host complex is inactivated by the cyanide-T6 mixture, and that not even the infecting particles are recovered. Even when 7 phage particles were adsorbed on each bacterium, less than two are recovered per cell at the earliest stage tested, and the shape of the curve suggests that if earlier stages had been tested, still fewer would have been recovered. In experiments with singly infected bacteria, the earliest tests indicated that less than one infected bacterium in 80 liberated any phage at all. A second point to be noted is that the multiplicity of infection appears to influence slightly the time at which phage particles can be recovered from the cell, and it continues to affect the fraction found in the bacteria at a given time. That this difference is a real one seems clear from the consistency among the points of curve 2. This result has been observed in each experiment, although the effect appeared to be less pronounced in some experiments made at the lower temperature of 30°C. Attention should also be drawn to the fact that the shape of the curves is clearly not exponential. In fact, it parallels with a delay of several minutes the approximately linear DNA increase observed in this system (6).

In connection with the preceding experiments a test was made to establish whether the cyanide concentration chosen was maximally effective in inhibiting phage synthesis. Using the described technique, but changing the cyanide concentration of the lysing medium from 0.01 M to 0.004 M and 0.001 M in parallel aliquots, no difference was detectable in the three lysing media. Thus 0.01 M cyanide is well beyond the minimum concentration necessary and was considered to be adequate for these experiments.

Action of Cyanide in the Absence of the Lysing Agent T6.—Cohen and Anderson (7) reported a loss of infectious centers when infected bacteria were incubated in the presence of the antimetabolite 5-methyltryptophan. Although the details

A. H. DOERMANN   649

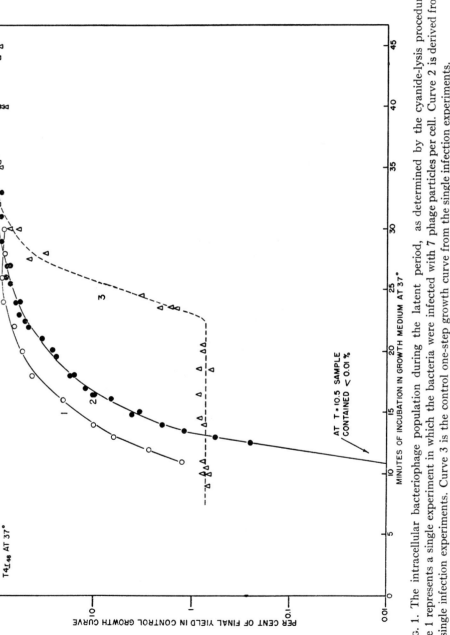

FIG. 1. The intracellular bacteriophage population during the latent period, as determined by the cyanide-lysis procedure. Curve 1 represents a single experiment in which the bacteria were infected with 7 phage particles per cell. Curve 2 is derived from four single infection experiments. Curve 3 is the control one-step growth curve from the single infection experiments.

650   INTRACELLULAR GROWTH OF BACTERIOPHAGES. I

of their experiments differed somewhat from those presented here, the loss of infectious centers in their experiments suggested testing whether cyanide could cause a similar loss of infected bacteria in the present procedure. An experiment was made which was identical with the standard cyanide lysis experiment

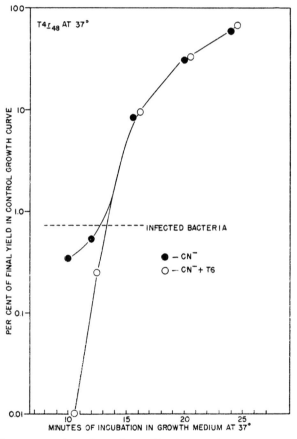

Fig. 2. The comparative effect of cyanide alone and cyanide plus T6 on singly infected bacteria at various stages in the latent period.

except that T6 was omitted from one set and included in a parallel set of lysing medium cultures (Fig. 2). As in the case of 5-methyltryptophan, it is seen that cyanide alone caused a loss of infectious centers when added in the early stages of phage growth, although the loss is less than that produced by cyanide and T6 together. Furthermore, in the second half of the latent period, comparison of the two media showed clearly and surprisingly that a definite *rise* in titer of infective centers occurred even when the lysing agent, T6, was omitted from the lysing medium. In fact, during the second half of the latent period, phage liberation is identical in the two media.

In order to see whether lysis is actually occurring and can account for the liberation of phage, a nephelometric experiment was made introducing CN⁻ at two points in the latent period. Three cultures of B/r/1 growing exponentially in growth medium were infected with $T4r_{48}$ (*ca.* fivefold multiplicity). One culture served as a control for normal lysis. To the second culture cyanide (0.01 M final concentration) was added 7.5 minutes after addition of the virus and to

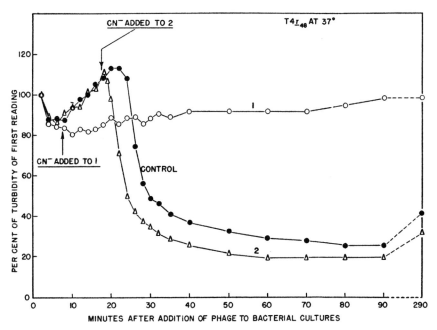

FIG. 3. The turbidity of T4-infected bacterial cultures as affected by addition of cyanide at two stages in the latent period.

the third tube 17.5 minutes after addition of the $T4r_{48}$. The turbidities of these three cultures were followed with a nephelometer designed like that described previously by Underwood and Doermann (8), but with four separate units which permit independent readings on the four tubes without removing any of them from the instrument. The results indicate that CN⁻ added to infected bacteria early during the latent period does not induce lysis (Fig. 3). From the plaque count experiment (Fig. 2) it is seen that a loss of infective centers does occur. This loss must therefore be due to some cause other than lysis of these cells. In the later stages of the latent period, the turbidimetric experiment indicates that lysis occurs promptly upon the addition of CN⁻ to the culture

652   INTRACELLULAR GROWTH OF BACTERIOPHAGES. I

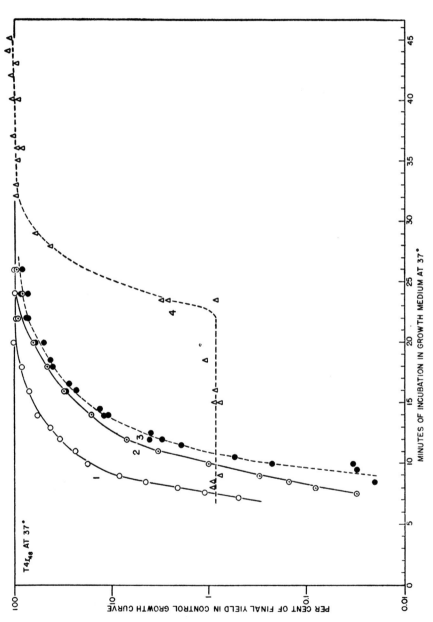

Fig. 4. The intracellular bacteriophage population during the latent period, as determined by premature lysis in the presence of 5-methyltryptophan. Curve 1 represents a multiple (fivefold) infection and curve 2 a single infection experiment with a different T4r mutant than that used in the other experiments described here. Curve 3 is a composite of three single infection experiments made with the same phage as was used in the cyanide experiments. Curve 4 is the standard one-step growth curve from the experiments which gave the data for curve 3.

(Fig. 3). The increase of infective centers in comparable cultures (Fig. 2) at these later stages is probably brought about by liberation of phage particles concurrent with this lysis.

Experiments Using 5-Methyltryptophan as the Metabolic Inhibitor.—In trying to find a suitable metabolic inhibitor for instantaneously stopping phage growth, a large number of experiments was done using the antimetabolite 5-methyltryptophan (5MT)[2] whose bacteriostatic action is blocked by tryptophan (9). The technique used was similar to the cyanide lysis procedure except that tryptophan was omitted from the lysing medium and 5MT was used in place of cyanide. The results (Fig. 4) are quite similar to the cyanide results in all respects except one. They are similar in failure to recover any phage particles during the early stages of the latent period, in the difference between single and multiple infection, and in the shapes of the curves. They are different, however, in that both the single and the multiple infection curves are moved to the left along the time scale by 3 to 4 minutes. This indicates that more phage is liberated per cell if lysis is induced in the presence of 5MT than if it is brought about in the presence of CN^-. This difference may be interpreted on the basis of two alternative hypotheses.

First, it might be suspected that CN^- penetrates the cell and reaches its site of inhibition more quickly than 5MT. This would allow more phage reproduction to go on between the time of exposure to the 5MT and the time at which the cell breaks open. In this event, a higher concentration of 5MT would enable penetration of an inhibitory amount in a shorter period of time, thus reducing the amount of phage found. To test this, the concentration of 5MT in the lysing medium was increased fivefold. No difference in the amount of phage liberated was found, suggesting that the rate of penetration of the poison is not limiting its effectiveness.

A second hypothesis is that the reaction blocked by 5MT may be one of the earlier ones involved in the synthesis of phage constituents. At the time of addition of 5MT many individual phages may already have acquired these constituents and thus be able to go on to maturity before lysis disperses the enzyme equipment of the infected cell. Cyanide, on the other hand, may block one of the terminal reactions in phage production, with the result that at a given time fewer individuals will have passed this reaction than will have passed the 5MT-inhibitable step. Consequently, fewer particles will be liberated when using cyanide than when 5MT is used.

DISCUSSION

Earlier experiments (2) and tests made of the lysing efficiency of the T6 stocks used here indicate that rapid lysis occurs when T6 stocks are added in

[2] Obtained through the courtesy of Dr. M. L. Tainter, Sterling-Winthrop Research Institute, Rensselaer, New York.

654   INTRACELLULAR GROWTH OF BACTERIOPHAGES. I

sufficient concentration to bacterial cultures. The very first experiments with bacteria infected with T5 (3) left no doubt that lysis from without by T6 will liberate T5 particles prematurely from infected bacteria. From the evidence contained in the present paper it cannot be definitely established whether the combined action of T6 and cyanide liberates all of the mature phage present in the cells. However, the fact that, during the terminal stages of intracellular development, the cyanide-lysis method yields as much phage as does spontaneous lysis, suggests that the cyanide method liberates all of the mature phage. Furthermore, during the second half of the latent period, exactly the same amount of phage is liberated by cyanide alone as by cyanide plus T6. This suggests that cyanide acts promptly in arresting phage growth. Otherwise one would expect to find a consistently higher number of phage particles in the cyanide medium than in the medium in which cyanide and T6 are combined. The experiments presented here therefore warrant the working hypothesis that mature intracellular phage is effectively liberated by the treatment described, and that the method gives a true picture of the intracellular phage population. The validity of this working hypothesis will be conclusively demonstrated for the phage T3 in the second paper of this series (10).

The bearing of the present experiments on our concept of phage reproduction might be discussed here. The finding that the original infecting particles are not recoverable from the cells during the first stages of the latent period appears at first sight surprising. Nevertheless some indirect evidence indicates that this is to be expected. The discovery that yields from mixedly infected bacteria may contain new combinations of the genetic material of the infecting types (11–13) suggests that some alteration of the infecting particles may occur. Furthermore, in mixed infections of bacteria with unrelated phages only one type is reproduced. The other type, although adsorbed on the cells, it not only prevented from multiplying but the infecting particle of that type is lost (5, 14). On the basis of multiple infection experiments with ultraviolet-inactivated phage particles, Luria (15) has proposed that reproduction of phage occurs by reproduction of subunits which are at some later stage assembled into complete virus particles. The failure to find infective phage particles within the infected cell in the early stages of reproduction agrees with what would have been predicted from these experiments.

The results of our experiments agree quite well with the scheme which Latarjet (16) suggested on the basis of x-ray inactivation studies of phage inside infected bacteria. Latarjet differentiated three segments of the latent period of phage growth. Using T2 he found that during the first segment of 6 to 7 minutes' duration, singly infected bacteria show the same inactivation characteristics as do unadsorbed phage particles. In the second period, from time 7 to time 13 minutes, the phage in infected cells became more resistant to x-rays, even during the first 2 minutes of this segment in which the inactivation curves still retain a single hit character. During the last 4 minutes of this period the curves take on a multiple hit character. In the final segment, from time 13 minutes to the end of the latent period, the curves retain the multiple hit character, but gradually regain the original x-ray sensitivity characteristic of free phage. These x-ray experiments suggest again that a rather drastic alteration occurs to the infecting particle, and that particles with the original characteristics are not found in the cell until the second half of the latent period. This is precisely what is observed in the results presented here. Our experiments were done with T4, but comparison seems legitimate since the two viruses are quite closely related (17).

Results of a similar nature to those discussed here were published by Foster (18). In studying the effect of proflavine on the growth of phage T2, Foster found that the time at which this poison was added influenced the amount of phage liberated by the bacteria. No phage is liberated from T2-infected cells (latent period 21 minutes) if proflavine is added during the first 12 minutes after infection even though lysis of the cells does occur at the normal time. When proflavine is added at later points in the latent period, lysis yields phage particles, the number depending on the time of proflavine addition. When the results of these single infection experiments are compared to the cyanide single infection experiments (Fig. 1) the results are seen to be quite similar. From other experiments Foster concluded that proflavine inhibits one of the final stages in the formation of fully infective phage. These facts, taken together, suggests that proflavine experiments were, in fact, measuring intracellular phage.

SUMMARY

A method is described for liberating and estimating intracellular bacteriophage at any stage during the latent period by arresting phage growth and inducing premature lysis of the infected cells. This is brought about by placing the infected bacteria into the growth medium supplemented with 0.01 M cyanide and with a high titer T6 lysate. It was found in some of the later experiments that the T6 lysate is essential only during the first half of the latent period. Cyanide alone will induce lysis during the latter part of the latent period.

Using this method on T4-infected bacteria it is found that during the first half of the latent period no phage particles, not even those originally infecting the bacteria, are recovered. This result is in agreement with the gradually emerging concept that a profound alteration of the infecting phage particle takes place before reproduction ensues. During the second half of the latent period mature phage is found to accumulate within the bacteria at a rate which is parallel to the approximately linear increase of intracellular DNA in this system. However, the phage production lags several minutes behind DNA production.

656 INTRACELLULAR GROWTH OF BACTERIOPHAGES. I

When 5-methyltryptophan replaced cyanide as the metabolic inhibitor, similar results were obtained. The curves were, however, displaced several minutes to the left on the time axis.

The results are compared with Latarjet's (16) data on x-radiation of infected bacteria and with Foster's data (18) concerning the effect of proflavine on infected bacteria. Essential agreement with both is apparent.

BIBLIOGRAPHY

1. Delbrück, M., *J. Gen. Physiol.*, 1940, **23**, 643.
2. Doermann, A. H., *J. Bact.*, 1948, **55**, 257.
3. Doermann, A. H., *Ann. Rep. Biol. Lab., Long Island Biol. Assn.*, 1946, 22.
4. Adams, M. H., *J. Gen. Physiol.*, 1948, **31**, 417.
5. Delbrück, M., and Luria, S. E., *Arch. Biochem.*, 1942, **1**, 111.
6. Cohen, S. S., *Bact. Rev.*, 1949, **13**, 1.
7. Cohen, S. S., and Anderson, T. F., *J. Exp. Med.*, 1946, **84**, 525.
8. Underwood, N., and Doermann, A. H., *Rev. Scient. Instr.*, 1947, **18**, 665.
9. Anderson, T. F., *Science*, 1945, **101**, 565.
10. Anderson, T. F., and Doermann, A. H., *J. Gen. Physiol.*, 1952, **35**, 657.
11. Delbrück, M., and Bailey, W. T., Jr., *Cold Spring Harbor Symp. Quant. Biol.*, 1946, **11**, 33.
12. Hershey, A. D., and Rotman, R., *Proc. Nat. Acad. Sc.*, 1948, **34**, 89.
13. Hershey, A. D., and Rotman, R., *Genetics*, 1949, **34**, 44.
14. Delbrück, M., *J. Bact.*, 1945, **50**, 151.
15. Luria, S. E., *Proc. Nat. Acad. Sc.*, 1947, **33**, 253.
16. Latarjet, R., *J. Gen. Physiol.*, 1948, **31**, 529.
17. Adams, M. H., *in* Methods in Medical Research, (J. H. Comroe, editor), Chicago, The Yearbook Publishers, Inc., 1950, **2**, 1.
18. Foster, R. A. C., *J. Bact.*, 1948, **56**, 795.

The Hershey–Chase Experiment

Hershey A. and Chase M. 1952. **Independent Functions of Viral Protein and Nucleic Acid in Growth of Bacteriophage.** (Reprinted, with permission, from *J. Gen. Physiol.* **36:** 39–56 [©Rockefeller University Press].)

The Hershey Laboratory in 1952. Martha Chase and Al Hershey are second and third from the left.

IT IS NOT ONLY IMPORTANT BIOLOGICAL OBJECTS that come in pairs; so also do the scientists whose names grace classic experiments or conceptual advances. The Hershey–Chase experiment reprinted here is one of the most notable of a group that includes Watson–Crick (DNA structure); Meselson–Stahl (DNA replication); Beadle–Tatum (one gene–one enzyme); Luria–Delbrück (fluctuation test); Embden–Meyerhof (glycolysis pathway); and Hardy–Weinberg (gene frequency equilibrium), to name just a few drawn from biology.

Alfred Hershey's professional career began in 1934 when he joined J. Bronfenbrenner, a pioneer of bacteriophage research, at Washington University in St. Louis to study the growth and nutritional requirements of enteric bacteria. In 1939, Hershey read the one-step-growth paper by Ellis and Delbrück (1939) which intimated that bacteriophage replication might be the key to understanding the gene. As Hershey put it, "That paper persuaded me that phage research had a future" (Hershey 1981). The future of bacteriophage genetics began properly in December, 1940, when Delbrück and Salvador Luria met in New York and agreed to spend the following summer together at Cold Spring Harbor. This was the first of many summers that they spent at Cold Spring Harbor, and the Laboratory became the center of a growing world of phage geneticists when Delbrück began the celebrated Phage Course in 1945. Delbrück and Luria visited Hershey in St. Louis in 1943. Twenty-six years later, these "two enemy aliens and one social misfit" (Hershey 1981) received the Nobel Prize for Physiology or Medicine.

Between 1941 and 1944, Hershey's phage research reflected the interests of Bronfenbrenner, but in 1946, Hershey reported his first genetic experiments using phage, describing the *r* mutants that produce rapid lysis of infected cells. By then, phage genetics had become central to Hershey's research program, and in 1949, he and Raquel Rotman published an important paper describing genetic recombination between phage infecting the same cell (Hershey and Rotman 1949). In 1950, Demerec recruited Hershey as a Carnegie Institution Fellow and, in the same year, Martha Chase came to Hershey's laboratory as a research assistant.

It was still not known what went on in the ten minutes between attachment of the phage particle to the bacterial cell and the detection of intracellular infectious particles. It was clear from the electron microscope pictures that the shell or envelope of a phage particle remained attached to the outside of the bacterial cell that was being infected (Luria et al. 1943). However, even though Herriott (1951) showed that these shells were DNA-free protein, there were cogent arguments not to be especially interested in the DNA, as opposed to the protein, of the phage (Hershey 1966). Herriot did make an analogy between phage infection and a hypodermic needle (Hershey 1966), but even as late as the summer of 1950 or 1951 the idea that the phage DNA alone entered the infected cell was thought to be a "...wildly comical possibility" (Anderson 1966).

The Hershey–Chase experiment, then, was designed to resolve which of the two components of the phage particle—DNA or protein—entered the cell. Bacteria that had been growing in culture medium with either radioactive ^{35}S or ^{32}P were infected with phage. The ^{35}S was incorporated in the protein of the new phage, whereas the ^{32}P was incorporated in the DNA. These radioactively labeled phage were used to infect a second bacterial culture, and at varying times after infection, a Waring blender was used to break the attachment of phage to cell. It was found that up to 80% of the ^{35}S, but less than 25% of the ^{32}P ,was stripped off. Thus, about 75% of the DNA had demonstrably entered the cells, while no more than 20% of the protein had done so. Since none of the protein label appeared in the progeny phage particles, but about 30% of the DNA label did so, Hershey and Chase concluded that the hereditary material of the phage was probably DNA.

It is striking that the conclusions of Hershey and Chase about the role of phage DNA were as tentative as those of Avery et al. (1944) on the role of DNA in pneumococcal transformation. Hershey and Chase wrote: "The chemical nature of the genetic part must wait, however, until some of the questions asked above have been answered." Although posterity has linked the blender experiments of 1951 with the transformation experiments of 1944, it is notable that Hershey and Chase did not cite Avery et al. (1944). It seems clear that the blender experiment was part of Hershey's quest to elucidate the process of phage replication, rather than an experiment designed to determine the chemical nature of the genetic material.

Alfred Hershey was born in Owosso, Michigan, on December 4, 1908. He graduated from Michigan State University in 1930 and received his Ph.D. from the same institution in 1934. Hershey moved to Washington University School of Medicine in 1934 and remained there until 1950, when he joined the staff of the Carnegie Institution of Washington's Department of Genetics at Cold Spring Harbor. Following the reorganization of the institutes on the Cold Spring Harbor campus, Hershey became Director of the Genetics Research Unit in 1962 and remained so until his retirement in 1974. He was elected a member of the National Academy of Sciences in 1958 and of the American Academy of Arts and Sciences in 1959. Among many honors, Hershey received the Albert Lasker Award in 1958 and the Kimber Genetics Award in 1965; he shared the 1969 Nobel Prize for Physiology or Medicine with Max Delbrück and Salvador Luria. He died on May 22, 1997.

Martha Chase was born on November 30, 1927, in Cleveland. She graduated from The College of Wooster, Ohio, in 1950 and was recommended to Demerec by Warren Spencer of the California Institute of Technology. She was appointed as research assistant to Hershey in August, 1950, and worked with him until she resigned in 1953.

Chase joined Doermann at Oak Ridge National Laboratory and then at the University of Rochester. In 1959, Chase moved to California where she began her Ph.D. with Giuseppe Bertani at the University of Southern California. After Bertani moved to Sweden, Chase continued her Ph.D. with Margaret Lieb. Her thesis "Reactivation of Phage P2 Damaged by Ultraviolet Light" was accepted in 1964. After receiving her Ph.D., Chase returned to her home in Cleveland, Ohio.

Anderson T.F. 1966. Electron microscopy of phages. In *Phage and the origins of molecular biology* (ed. J. Cairns et al.), pp. 63–78. Cold Spring Harbor Laboratory Press, Cold Spring Harbor, New York (reprinted with additional material, 1992).

Avery O.T., Macleod C.M., and McCarty M. 1944. Studies on the chemical nature of the substance inducing transformation of pneumococcal types. Induction of transformation by a desoxyribonucleic acid fraction isolated from pneumococcus Type III. *J. Exp. Med.* **79:** 137–158.

Ellis E.L. and Delbrück M. 1939. The growth of bacteriophage. *J. Gen. Physiol.* **22:** 137–158.

Herriott R.M. 1951. Nucleic acid-free T2 virus "ghosts" with specific biological action. *J. Bacteriol.* **61:** 752–754.

Hershey A. 1966. The injection of DNA into cells by phage. In *Phage and the origins of molecular biology* (ed. J. Cairns et al.), pp. 100–108. Cold Spring Harbor Laboratory Press, Cold Spring Harbor, New York (reprinted with additional material, 1992).

———. 1981. In *The Max Delbrück Laboratory dedication ceremony* (booklet). Cold Spring Harbor Laboratory Archives, Cold Spring Harbor, New York.

Hershey A. and Chase M. 1952. Independent functions of viral proteins and nucleic acid in growth of bacteriophage. *J. Gen. Physiol.* **36:** 39–56.

Hershey A. and Rotman R. 1949. Genetic recombination between host-range and plaque-type mutants of bacteriophage in single bacterial cells. *Genetics* **34:** 44–71.

Luria S.E., Delbrück M., and Anderson T.F. 1943. Electron microscope studies of bacterial viruses. *J. Bacteriol.* **46:** 57–77.

[Reprinted from THE JOURNAL OF GENERAL PHYSIOLOGY, September 20, 1952,
Vol. 36, No. 1, pp. 39–56]
Printed in U.S.A.

INDEPENDENT FUNCTIONS OF VIRAL PROTEIN AND NUCLEIC ACID IN GROWTH OF BACTERIOPHAGE*

BY A. D. HERSHEY AND MARTHA CHASE

(*From the Department of Genetics, Carnegie Institution of Washington, Cold Spring
Harbor, Long Island*)

(Received for publication, April 9, 1952)

The work of Doermann (1948), Doermann and Dissosway (1949), and Anderson and Doermann (1952) has shown that bacteriophages T2, T3, and T4 multiply in the bacterial cell in a non-infective form. The same is true of the phage carried by certain lysogenic bacteria (Lwoff and Gutmann, 1950). Little else is known about the vegetative phase of these viruses. The experiments reported in this paper show that one of the first steps in the growth of T2 is the release from its protein coat of the nucleic acid of the virus particle, after which the bulk of the sulfur-containing protein has no further function.

Materials and Methods.—Phage T2 means in this paper the variety called T2H (Hershey, 1946); T2*h* means one of the host range mutants of T2; UV-phage means phage irradiated with ultraviolet light from a germicidal lamp (General Electric Co.) to a fractional survival of 10^{-5}.

Sensitive bacteria means a strain (H) of *Escherichia coli* sensitive to T2 and its *h* mutant; resistant bacteria B/2 means a strain resistant to T2 but sensitive to its *h* mutant; resistant bacteria B/2*h* means a strain resistant to both. These bacteria do not adsorb the phages to which they are resistant.

"Salt-poor" broth contains per liter 10 gm. bacto-peptone, 1 gm. glucose, and 1 gm. NaCl. "Broth" contains, in addition, 3 gm. bacto-beef extract and 4 gm. NaCl.

Glycerol-lactate medium contains per liter 70 mM sodium lactate, 4 gm. glycerol, 5 gm. NaCl, 2 gm. KCl, 1 gm. NH$_4$Cl, 1 mM MgCl$_2$, 0.1 mM CaCl$_2$, 0.01 gm. gelatin, 10 mg. P (as orthophosphate), and 10 mg. S (as MgSO$_4$), at pH 7.0.

Adsorption medium contains per liter 4 gm. NaCl, 5 gm. K$_2$SO$_4$, 1.5 gm. KH$_2$PO$_4$, 3.0 gm. Na$_2$HPO$_4$, 1 mM MgSO$_4$, 0.1 mM CaCl$_2$, and 0.01 gm. gelatin, at pH 7.0.

Veronal buffer contains per liter 1 gm. sodium diethylbarbiturate, 3 mM MgSO$_4$, and 1 gm. gelatin, at pH 8.0.

The HCN referred to in this paper consists of molar sodium cyanide solution neutralized when needed with phosphoric acid.

* This investigation was supported in part by a research grant from the National Microbiological Institute of the National Institutes of Health, Public Health Service. Radioactive isotopes were supplied by the Oak Ridge National Laboratory on allocation from the Isotopes Division, United States Atomic Energy Commission.

40 VIRAL PROTEIN AND NUCLEIC ACID IN BACTERIOPHAGE GROWTH

Adsorption of isotope to bacteria was usually measured by mixing the sample in adsorption medium with bacteria from 18 hour broth cultures previously heated to 70°C. for 10 minutes and washed with adsorption medium. The mixtures were warmed for 5 minutes at 37°C., diluted with water, and centrifuged. Assays were made of both sediment and supernatant fractions.

Precipitation of isotope with antiserum was measured by mixing the sample in 0.5 per cent saline with about 10^{11} per ml. of non-radioactive phage and slightly more than the least quantity of antiphage serum (final dilution 1:160) that would cause visible precipitation. The mixture was centrifuged after 2 hours at 37°C.

Tests with DNase (desoxyribonuclease) were performed by warming samples diluted in veronal buffer for 15 minutes at 37°C. with 0.1 mg. per ml. of crystalline enzyme (Worthington Biochemical Laboratory).

Acid-soluble isotope was measured after the chilled sample had been precipitated with 5 per cent trichloroacetic acid in the presence of 1 mg./ml. of serum albumin, and centrifuged.

In all fractionations involving centrifugation, the sediments were not washed, and contained about 5 per cent of the supernatant. Both fractions were assayed.

Radioactivity was measured by means of an end-window Geiger counter, using dried samples sufficiently small to avoid losses by self-absorption. For absolute measurements, reference solutions of P^{32} obtained from the National Bureau of Standards, as well as a permanent simulated standard, were used. For absolute measurements of S^{35} we relied on the assays (±20 per cent) furnished by the supplier of the isotope (Oak Ridge National Laboratory).

Glycerol-lactate medium was chosen to permit growth of bacteria without undesirable pH changes at low concentrations of phosphorus and sulfur, and proved useful also for certain experiments described in this paper. 18-hour cultures of sensitive bacteria grown in this medium contain about 2×10^9 cells per ml., which grow exponentially without lag or change in light-scattering per cell when subcultured in the same medium from either large or small seedings. The generation time is 1.5 hours at 37°C. The cells are smaller than those grown in broth. T2 shows a latent period of 22 to 25 minutes in this medium. The phage yield obtained by lysis with cyanide and UV-phage (described in context) is one per bacterium at 15 minutes and 16 per bacterium at 25 minutes. The final burst size in diluted cultures is 30 to 40 per bacterium, reached at 50 minutes. At 2×10^8 cells per ml., the culture lyses slowly, and yields 140 phage per bacterium. The growth of both bacteria and phage in this medium is as reproducible as that in broth.

For the preparation of radioactive phage, P^{32} of specific activity 0.5 mc./mg. or S^{35} of specific activity 8.0 mc./mg. was incorporated into glycerol-lactate medium, in which bacteria were allowed to grow at least 4 hours before seeding with phage. After infection with phage, the culture was aerated overnight, and the radioactive phage was isolated by three cycles of alternate slow (2000 G) and fast (12,000 G) centrifugation in adsorption medium. The suspensions were stored at a concentration not exceeding 4 μc./ml.

Preparations of this kind contain 1.0 to 3.0 $\times 10^{-12}$ μg. S and 2.5 to 3.5 $\times 10^{-11}$ μg. P per viable phage particle. Occasional preparations containing excessive amounts of sulfur can be improved by absorption with heat-killed bacteria that do not adsorb the phage. The radiochemical purity of the preparations is somewhat uncertain, owing to the possible presence of inactive phage particles and empty phage membranes. The presence in our preparations of sulfur (about 20 per cent) that is precipitated by antiphage serum (Table I) and either adsorbed by bacteria resistant to phage, or not adsorbed by bacteria sensitive to phage (Table VII), indicates contamination by membrane material. Contaminants of bacterial origin are probably negligible for present purposes as indicated by the data given in Table I. For proof that our principal findings reflect genuine properties of viable phage particles, we rely on some experiments with inactivated phage cited at the conclusion of this paper.

The Chemical Morphology of Resting Phage Particles.—Anderson (1949) found that bacteriophage T2 could be inactivated by suspending the particles in high concentrations of sodium chloride, and rapidly diluting the suspension with water. The inactivated phage was visible in electron micrographs as tadpole-shaped "ghosts." Since no inactivation occurred if the dilution was slow

TABLE I

Composition of Ghosts and Solution of Plasmolyzed Phage

Per cent of isotope‖	Whole phage labeled with		Plasmolyzed phage labeled with	
	P^{32}	S^{35}	P^{32}	S^{35}
Acid-soluble..............................	—	—	1	—
Acid-soluble after treatment with DNase.......	1	1	80	1
Adsorbed to sensitive bacteria.................	85	90	2	90
Precipitated by antiphage...................	90	99	5	97

he attributed the inactivation to osmotic shock, and inferred that the particles possessed an osmotic membrane. Herriott (1951) found that osmotic shock released into solution the DNA (desoxypentose nucleic acid) of the phage particle, and that the ghosts could adsorb to bacteria and lyse them. He pointed out that this was a beginning toward the identification of viral functions with viral substances.

We have plasmolyzed isotopically labeled T2 by suspending the phage (10^{11} per ml.) in 3 M sodium chloride for 5 minutes at room temperature, and rapidly pouring into the suspension 40 volumes of distilled water. The plasmolyzed phage, containing not more than 2 per cent survivors, was then analyzed for phosphorus and sulfur in the several ways shown in Table I. The results confirm and extend previous findings as follows:—

1. Plasmolysis separates phage T2 into ghosts containing nearly all the sulfur and a solution containing nearly all the DNA of the intact particles.

2. The ghosts contain the principal antigens of the phage particle detectable by our antiserum. The DNA is released as the free acid, or possibly linked to sulfur-free, apparently non-antigenic substances.

42   VIRAL PROTEIN AND NUCLEIC ACID IN BACTERIOPHAGE GROWTH

3. The ghosts are specifically adsorbed to phage-susceptible bacteria; the DNA is not.

4. The ghosts represent protein coats that surround the DNA of the intact particles, react with antiserum, protect the DNA from DNase (desoxyribonuclease), and carry the organ of attachment to bacteria.

5. The effects noted are due to osmotic shock, because phage suspended in salt and diluted slowly is not inactivated, and its DNA is not exposed to DNase.

TABLE II

Sensitization of Phage DNA to DNase by Adsorption to Bacteria

Phage adsorbed to	Phage labeled with	Non-sedimentable isotope, *per cent*	
		After DNase	No DNase
Live bacteria..............................	S^{35}	2	1
" "	P^{32}	8	7
Bacteria heated before infection...............	S^{35}	15	11
" " " "	P^{32}	76	13
Bacteria heated after infection................	S^{35}	12	14
" " " "	P^{32}	66	23
Heated unadsorbed phage: acid-soluble P^{32} 70°.......	P^{32}	5	
80°.......	P^{32}	13	
90°.......	P^{32}	81	
100°.......	P^{32}	88	

Phage adsorbed to bacteria for 5 minutes at 37°C. in adsorption medium, followed by washing.

Bacteria heated for 10 minutes at 80°C. in adsorption medium (before infection) or in veronal buffer (after infection).

Unadsorbed phage heated in veronal buffer, treated with DNase, and precipitated with trichloroacetic acid.

All samples fractionated by centrifuging 10 minutes at 1300 *G*.

Sensitization of Phage DNA to DNase by Adsorption to Bacteria.—The structure of the resting phage particle described above suggests at once the possibility that multiplication of virus is preceded by the alteration or removal of the protective coats of the particles. This change might be expected to show itself as a sensitization of the phage DNA to DNase. The experiments described in Table II show that this happens. The results may be summarized as follows:—

1. Phage DNA becomes largely sensitive to DNase after adsorption to heat-killed bacteria.

2. The same is true of the DNA of phage adsorbed to live bacteria, and then

heated to 80°C. for 10 minutes, at which temperature unadsorbed phage is not sensitized to DNase.

3. The DNA of phage adsorbed to unheated bacteria is resistant to DNase, presumably because it is protected by cell structures impervious to the enzyme.

Graham and collaborators (personal communication) were the first to discover the sensitization of phage DNA to DNase by adsorption to heat-killed bacteria.

The DNA in infected cells is also made accessible to DNase by alternate freezing and thawing (followed by formaldehyde fixation to inactivate cellular enzymes), and to some extent by formaldehyde fixation alone, as illustrated by the following experiment.

Bacteria were grown in broth to 5×10^7 cells per ml., centrifuged, resuspended in adsorption medium, and infected with about two P^{32}-labeled phage per bacterium. After 5 minutes for adsorption, the suspension was diluted with water containing per liter 1.0 mM $MgSO_4$, 0.1 mM $CaCl_2$, and 10 mg. gelatin, and recentrifuged. The cells were resuspended in the fluid last mentioned at a concentration of 5×10^8 per ml. This suspension was frozen at $-15°C$. and thawed with a minimum of warming, three times in succession. Immediately after the third thawing, the cells were fixed by the addition of 0.5 per cent (v/v) of formalin (35 per cent HCHO). After 30 minutes at room temperature, the suspension was dialyzed free from formaldehyde and centrifuged at 2200 G for 15 minutes. Samples of P^{32}-labeled phage, frozen-thawed, fixed, and dialyzed, and of infected cells fixed only and dialyzed, were carried along as controls.

The analysis of these materials, given in Table III, shows that the effect of freezing and thawing is to make the intracellular DNA labile to DNase, without, however, causing much of it to leach out of the cells. Freezing and thawing and formaldehyde fixation have a negligible effect on unadsorbed phage, and formaldehyde fixation alone has only a mild effect on infected cells.

Both sensitization of the intracellular P^{32} to DNase, and its failure to leach out of the cells, are constant features of experiments of this type, independently of visible lysis. In the experiment just described, the frozen suspension cleared during the period of dialysis. Phase-contrast microscopy showed that the cells consisted largely of empty membranes, many apparently broken. In another experiment, samples of infected bacteria from a culture in salt-poor broth were repeatedly frozen and thawed at various times during the latent period of phage growth, fixed with formaldehyde, and then washed in the centrifuge. Clearing and microscopic lysis occurred only in suspensions frozen during the second half of the latent period, and occurred during the first or second thawing. In this case the lysed cells consisted wholly of intact cell membranes, appearing empty except for a few small, rather characteristic refractile bodies apparently attached to the cell walls. The behavior of intracellular P^{32} toward DNase, in either the lysed or unlysed cells, was not significantly different from that shown in Table III, and the content of P^{32} was only slightly less after lysis. The phage liberated during freezing and thawing was also titrated in this experiment. The lysis occurred without appreciable liberation of phage in suspensions frozen up to and including the 16th minute, and the 20 minute sample yielded only five per bacterium. Another sample of the culture formalinized at 30 minutes, and centrifuged without freezing, contained 66 per cent of the P^{32} in non-sedimentable form. The yield of extracellular phage at 30 minutes was 108 per bacterium, and the sedimented material consisted largely of formless debris but contained also many apparently intact cell membranes.

TABLE III

Sensitization of Intracellular Phage to DNase by Freezing, Thawing, and Fixation with Formaldehyde

	Unadsorbed phage frozen, thawed, fixed	Infected cells frozen, thawed, fixed	Infected cells fixed only
Low speed sediment fraction			
Total P^{32}	—	71	86
Acid-soluble	—	0	0.5
Acid-soluble after DNase	—	59	28
Low speed supernatant fraction			
Total P^{32}	—	29	14
Acid-soluble	1	0.8	0.4
Acid-soluble after DNase	11	21	5.5

The figures express per cent of total P^{32} in the original phage, or its adsorbed fraction.

We draw the following conclusions from the experiments in which cells infected with P^{32}-labeled phage are subjected to freezing and thawing.

1. Phage DNA becomes sensitive to DNAse after adsorption to bacteria in buffer under conditions in which no known growth process occurs (Benzer, 1952; Dulbecco, 1952).

2. The cell membrane can be made permeable to DNase under conditions that do not permit the escape of either the intracellular P^{32} or the bulk of the cell contents.

3. Even if the cells lyse as a result of freezing and thawing, permitting escape of other cell constituents, most of the P^{32} derived from phage remains inside the cell membranes, as do the mature phage progeny.

4. The intracellular P^{32} derived from phage is largely freed during spontaneous lysis accompanied by phage liberation.

We interpret these facts to mean that intracellular DNA derived from phage is not merely DNA in solution, but is part of an organized structure at all times during the latent period.

Liberation of DNA from Phage Particles by Adsorption to Bacterial Fragments.—The sensitization of phage DNA to specific depolymerase by adsorption to bacteria might mean that adsorption is followed by the ejection of the phage DNA from its protective coat. The following experiment shows that this is in fact what happens when phage attaches to fragmented bacterial cells.

TABLE IV

Release of DNA from Phage Adsorbed to Bacterial Debris

	Phage labeled with	
	S^{35}	P^{32}
Sediment fraction		
Surviving phage	16	22
Total isotope	87	55
Acid-soluble isotope	0	2
Acid-soluble after DNase	2	29
Supernatant fraction		
Surviving phage	5	5
Total isotope	13	45
Acid-soluble isotope	0.8	0.5
Acid-soluble after DNase	0.8	39

S^{35}- and P^{32}-labeled T2 were mixed with identical samples of bacterial debris in adsorption medium and warmed for 30 minutes at 37°C. The mixtures were then centrifuged for 15 minutes at 2200 G, and the sediment and supernatant fractions were analyzed separately. The results are expressed as per cent of input phage or isotope.

Bacterial debris was prepared by infecting cells in adsorption medium with four particles of T2 per bacterium, and transferring the cells to salt-poor broth at 37°C. The culture was aerated for 60 minutes, M/50 HCN was added, and incubation continued for 30 minutes longer. At this time the yield of extracellular phage was 400 particles per bacterium, which remained unadsorbed because of the low concentration of electrolytes. The debris from the lysed cells was washed by centrifugation at 1700 G, and resuspended in adsorption medium at a concentration equivalent to 3 × 10^9 lysed cells per ml. It consisted largely of collapsed and fragmented cell membranes. The adsorption of radioactive phage to this material is described in Table IV. The following facts should be noted.

1. The unadsorbed fraction contained only 5 per cent of the original phage particles in infective form, and only 13 per cent of the total sulfur. (Much of this sulfur must be the material that is not adsorbable to whole bacteria.)

2. About 80 per cent of the phage was inactivated. Most of the sulfur of this phage, as well as most of the surviving phage, was found in the sediment fraction.

3. The supernatant fraction contained 40 per cent of the total phage DNA (in a form labile to DNase) in addition to the DNA of the unadsorbed surviving phage. The labile DNA amounted to about half of the DNA of the inactivated phage particles, whose sulfur sedimented with the bacterial debris.

4. Most of the sedimentable DNA could be accounted for either as surviving phage, or as DNA labile to DNase, the latter amounting to about half the DNA of the inactivated particles.

Experiments of this kind are unsatisfactory in one respect: one cannot tell whether the liberated DNA represents all the DNA of some of the inactivated particles, or only part of it.

Similar results were obtained when bacteria (strain B) were lysed by large amounts of UV-killed phage T2 or T4 and then tested with P^{32}-labeled T2 and T4. The chief point of interest in this experiment is that bacterial debris saturated with UV-killed T2 adsorbs T4 better than T2, and debris saturated with T4 adsorbs T2 better than T4. As in the preceding experiment, some of the adsorbed phage was not inactivated and some of the DNA of the inactivated phage was not released from the debris.

These experiments show that some of the cell receptors for T2 are different from some of the cell receptors for T4, and that phage attaching to these specific receptors is inactivated by the same mechanism as phage attaching to unselected receptors. This mechanism is evidently an active one, and not merely the blocking of sites of attachment to bacteria.

Removal of Phage Coats from Infected Bacteria.—Anderson (1951) has obtained electron micrographs indicating that phage T2 attaches to bacteria by its tail. If this precarious attachment is preserved during the progress of the infection, and if the conclusions reached above are correct, it ought to be a simple matter to break the empty phage membranes off the infected bacteria, leaving the phage DNA inside the cells.

The following experiments show that this is readily accomplished by strong shearing forces applied to suspensions of infected cells, and further that infected cells from which 80 per cent of the sulfur of the parent virus has been removed remain capable of yielding phage progeny.

Broth-grown bacteria were infected with S^{35}- or P^{32}-labeled phage in adsorption medium, the unadsorbed material was removed by centrifugation, and the cells were resuspended in water containing per liter 1 mM $MgSO_4$, 0.1 mM $CaCl_2$, and 0.1 gm. gelatin. This suspension was spun in a Waring blendor (semimicro size) at 10,000 R.P.M. The suspension was cooled briefly in ice water at the end of each 60 second running period. Samples were removed at intervals, titrated (through antiphage serum) to measure the number of bacteria capable of yielding phage, and centrifuged to measure the proportion of isotope released from the cells.

The results of one experiment with each isotope are shown in Fig. 1. The data for S[35] and survival of infected bacteria come from the same experiment, in which the ratio of added phage to bacteria was 0.28, and the concentrations

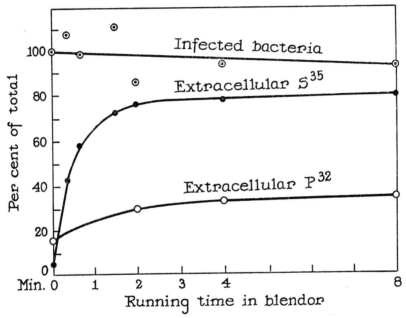

FIG. 1. Removal of S[35] and P[32] from bacteria infected with radioactive phage, and survival of the infected bacteria, during agitation in a Waring blendor.

of bacteria were 2.5×10^8 per ml. infected, and 9.7×10^8 per ml. total, by direct titration. The experiment with P[32]-labeled phage was very similar. In connection with these results, it should be recalled that Anderson (1949) found that adsorption of phage to bacteria could be prevented by rapid stirring of the suspension.

At higher ratios of infection, considerable amounts of phage sulfur elute from the cells spontaneously under the conditions of these experiments, though the elution of P[32] and the survival of infected cells are not affected by multiplicity of infection (Table V). This shows that there is a cooperative action among phage particles in producing alterations of the bacterial membrane which weaken the attachment of the phage. The cellular changes detected in

48   VIRAL PROTEIN AND NUCLEIC ACID IN BACTERIOPHAGE GROWTH

this way may be related to those responsible for the release of bacterial components from infected bacteria (Prater, 1951; Price, 1952).

A variant of the preceding experiments was designed to test bacteria at a later stage in the growth of phage. For this purpose infected cells were aerated in broth for 5 or 15 minutes, fixed by the addition of 0.5 per cent (v/v) commercial formalin, centrifuged, resuspended in 0.1 per cent formalin in water, and subsequently handled as described above. The results were very similar to those already presented, except that the release of P^{32} from the cells was slightly less, and titrations of infected cells could not be made.

The S^{35}-labeled material detached from infected cells in the manner described possesses the following properties. It is sedimented at 12,000 G, though less completely than intact phage particles. It is completely precipitated by

TABLE V

Effect of Multiplicity of Infection on Elution of Phage Membranes from Infected Bacteria

Running time in blendor	Multiplicity of infection	P^{32}-labeled phage		S^{35}-labeled phage	
		Isotope eluted	Infected bacteria surviving	Isotope eluted	Infected bacteria surviving
min.		*per cent*	*per cent*	*per cent*	*per cent*
0	0.6	10	120	16	101
2.5	0.6	21	82	81	78
0	6.0	13	89	46	90
2.5	6.0	24	86	82	85

The infected bacteria were suspended at 10^9 cells per ml. in water containing per liter 1 mM $MgSO_4$, 0.1 mM $CaCl_2$, and 0.1 gm. gelatin. Samples were withdrawn for assay of extracellular isotope and infected bacteria before and after agitating the suspension. In either case the cells spent about 15 minutes at room temperature in the eluting fluid.

antiphage serum in the presence of whole phage carrier. 40 to 50 per cent of it readsorbs to sensitive bacteria, almost independently of bacterial concentration between 2×10^8 and 10^9 cells per ml., in 5 minutes at 37°C. The adsorption is not very specific: 10 to 25 per cent adsorbs to phage-resistant bacteria under the same conditions. The adsorption requires salt, and for this reason the efficient removal of S^{35} from infected bacteria can be accomplished only in a fluid poor in electrolytes.

The results of these experiments may be summarized as follows:—

1. 75 to 80 per cent of the phage sulfur can be stripped from infected cells by violent agitation of the suspension. At high multiplicity of infection, nearly 50 per cent elutes spontaneously. The properties of the S^{35}-labeled material show that it consists of more or less intact phage membranes, most of which have lost the ability to attach specifically to bacteria.

2. The release of sulfur is accompanied by the release of only 21 to 35 per

cent of the phage phosphorus, half of which is given up without any mechanical agitation.

3. The treatment does not cause any appreciable inactivation of intracellular phage.

4. These facts show that the bulk of the phage sulfur remains at the cell surface during infection, and takes no part in the multiplication of intracellular phage. The bulk of the phage DNA, on the other hand, enters the cell soon after adsorption of phage to bacteria.

Transfer of Sulfur and Phosphorus from Parental Phage to Progeny.—We have concluded above that the bulk of the sulfur-containing protein of the resting phage particle takes no part in the multiplication of phage, and in fact does not enter the cell. It follows that little or no sulfur should be transferred from parental phage to progeny. The experiments described below show that this expectation is correct, and that the maximal transfer is of the order 1 per cent

Bacteria were grown in glycerol-lactate medium overnight and subcultured in the same medium for 2 hours at 37°C. with aeration, the size of seeding being adjusted nephelometrically to yield 2×10^8 cells per ml. in the subculture. These bacteria were sedimented, resuspended in adsorption medium at a concentration of 10^9 cells per ml., and infected with S^{35}-labeled phage T2. After 5 minutes at 37°C., the suspension was diluted with 2 volumes of water and resedimented to remove unadsorbed phage (5 to 10 per cent by titer) and S^{35} (about 15 per cent). The cells were next suspended in glycerol-lactate medium at a concentration of 2×10^8 per ml. and aerated at 37°C. Growth of phage was terminated at the desired time by adding in rapid succession 0.02 mM HCN and 2×10^{11} UV-killed phage per ml. of culture. The cyanide stops the maturation of intracellular phage (Doermann, 1948), and the UV-killed phage minimizes losses of phage progeny by adsorption to bacterial debris, and promotes the lysis of bacteria (Maaløe and Watson, 1951). As mentioned in another connection, and also noted in these experiments, the lysing phage must be closely related to the phage undergoing multiplication (*e.g.*, T2H, its *h* mutant, or T2L, but not T4 or T6, in this instance) in order to prevent inactivation of progeny by adsorption to bacterial debris.

To obtain what we shall call the maximal yield of phage, the lysing phage was added 25 minutes after placing the infected cells in the culture medium, and the cyanide was added at the end of the 2nd hour. Under these conditions, lysis of infected cells occurs rather slowly.

Aeration was interrupted when the cyanide was added, and the cultures were left overnight at 37°C. The lysates were then fractionated by centrifugation into an initial low speed sediment (2500 G for 20 minutes), a high speed supernatant (12,000 G for 30 minutes), a second low speed sediment obtained by recentrifuging in adsorption medium the resuspended high speed sediment, and the clarified high speed sediment.

The distribution of S^{35} and phage among fractions obtained from three cultures of this kind is shown in Table VI. The results are typical (except for the excessively good recoveries of phage and S^{35}) of lysates in broth as well as lysates in glycerol-lactate medium.

The striking result of this experiment is that the distribution of S^{35} among the fractions is the same for early lysates that do not contain phage progeny, and later ones that do. This suggests that little or no S^{35} is contained in the mature phage progeny. Further fractionation by adsorption to bacteria confirms this suggestion.

Adsorption mixtures prepared for this purpose contained about 5×10^9 heat-killed bacteria (70°C. for 10 minutes) from 18 hour broth cultures, and

TABLE VI

Per Cent Distributions of Phage and S^{35} among Centrifugally Separated Fractions of Lysates after Infection with S^{35}-Labeled T2

Fraction	Lysis at $t = 0$ S^{35}	Lysis at $t = 10$ S^{35}	Maximal yield	
			S^{35}	Phage
1st low speed sediment......................	79	81	82	19
2nd " " "	2.4	2.1	2.8	14
High speed "	8.6	6.9	7.1	61
" " supernatant......................	10	10	7.5	7.0
Recovery.................................	100	100	96	100

Infection with S^{35}-labeled T2, 0.8 particles per bacterium. Lysing phage UV-killed *h* mutant of T2. Phage yields per infected bacterium: <0.1 after lysis at $t = 0$; 0.12 at $t = 10$; maximal yield 29. Recovery of S^{35} means per cent of adsorbed input recovered in the four fractions; recovery of phage means per cent of total phage yield (by plaque count before fractionation) recovered by titration of fractions.

about 10^{11} phage (UV-killed lysing phage plus test phage), per ml. of adsorption medium. After warming to 37°C. for 5 minutes, the mixtures were diluted with 2 volumes of water, and centrifuged. Assays were made from supernatants and from unwashed resuspended sediments.

The results of tests of adsorption of S^{35} and phage to bacteria (H) adsorbing both T2 progeny and *h*-mutant lysing phage, to bacteria (B/2) adsorbing lysing phage only, and to bacteria (B/2*h*) adsorbing neither, are shown in Table VII, together with parallel tests of authentic S^{35}-labeled phage.

The adsorption tests show that the S^{35} present in the seed phage is adsorbed with the specificity of the phage, but that S^{35} present in lysates of bacteria infected with this phage shows a more complicated behavior. It is strongly adsorbed to bacteria adsorbing both progeny and lysing phage. It is weakly adsorbed to bacteria adsorbing neither. It is moderately well adsorbed to bacteria adsorbing lysing phage but not phage progeny. The latter test shows that the S^{35} is not contained in the phage progeny, and explains the fact that the S^{35} in early lysates not containing progeny behaves in the same way.

The specificity of the adsorption of S^{35}-labeled material contaminating the phage progeny is evidently due to the lysing phage, which is also adsorbed much more strongly to strain H than to B/2, as shown both by the visible reduction in Tyndall scattering (due to the lysing phage) in the supernatants of the test mixtures, and by independent measurements. This conclusion is further confirmed by the following facts.

TABLE VII

Adsorption Tests with Uniformly S^{35}-Labeled Phage and with Products of Their Growth in Non-Radioactive Medium

Adsorbing bacteria	Per cent adsorbed				
	Uniformly labeled S^{35} phage		Products of lysis at $t=10$	Phage progeny (Maximal yield)	
	+ UV-h	No UV-h			
	S^{35}	S^{35}	S^{35}	S^{35}	Phage
Sensitive (H).....................	84	86	79	78	96
Resistant (B/2)...................	15	11	46	49	10
Resistant (B/2h)................	13	12	29	28	8

The uniformly labeled phage and the products of their growth are respectively the seed phage and the high speed sediment fractions from the experiment shown in Table VI.

The uniformly labeled phage is tested at a low ratio of phage to bacteria: +UV-h means with added UV-killed h mutant in equal concentration to that present in the other test materials.

The adsorption of phage is measured by plaque counts of supernatants, and also sediments in the case of the resistant bacteria, in the usual way.

1. If bacteria are infected with S^{35} phage, and then lysed near the midpoint of the latent period with cyanide alone (in salt-poor broth, to prevent readsorption of S^{35} to bacterial debris), the high speed sediment fraction contains S^{35} that is adsorbed weakly and non-specifically to bacteria.

2. If the lysing phage and the S^{35}-labeled infecting phage are the same (T2), or if the culture in salt-poor broth is allowed to lyse spontaneously (so that the yield of progeny is large), the S^{35} in the high speed sediment fraction is adsorbed with the specificity of the phage progeny (except for a weak non-specific adsorption). This is illustrated in Table VII by the adsorption to H and B/2h.

It should be noted that a phage progeny grown from S^{35}-labeled phage and containing a larger or smaller amount of contaminating radioactivity could not be distinguished by any known method from authentic S^{35}-labeled phage, except that a small amount of the contaminant could be removed by adsorption to bacteria resistant to the phage. In addition to the properties already mentioned, the contaminating S^{35} is completely precipitated with the phage by antiserum, and cannot be appreciably separated from the phage by further fractional sedimentation, at either high or low concentrations of electrolyte. On the other hand, the chemical contamination from this source would be very small in favorable circumstances, because the progeny of a single phage particle are numerous and the contaminant is evidently derived from the parents.

The properties of the S^{35}-labeled contaminant show that it consists of the remains of the coats of the parental phage particles, presumably identical with the material that can be removed from unlysed cells in the Waring blendor. The fact that it undergoes little chemical change is not surprising since it probably never enters the infected cell.

The properties described explain a mistaken preliminary report (Hershey et al., 1951) of the transfer of S^{35} from parental to progeny phage.

It should be added that experiments identical to those shown in Tables VI and VII, but starting from phage labeled with P^{32}, show that phosphorus is transferred from parental to progeny phage to the extent of 30 per cent at yields of about 30 phage per infected bacterium, and that the P^{32} in prematurely lysed cultures is almost entirely non-sedimentable, becoming, in fact, acid-soluble on aging.

Similar measures of the transfer of P^{32} have been published by Putnam and Kozloff (1950) and others. Watson and Maaløe (1952) summarize this work, and report equal transfer (nearly 50 per cent) of phosphorus and adenine.

A Progeny of S^{35}-Labeled Phage Nearly Free from the Parental Label.—The following experiment shows clearly that the obligatory transfer of parental sulfur to offspring phage is less than 1 per cent, and probably considerably less. In this experiment, the phage yield from infected bacteria from which the S^{35}-labeled phage coats had been stripped in the Waring blendor was assayed directly for S^{35}.

Sensitive bacteria grown in broth were infected with five particles of S^{35}-labeled phage per bacterium, the high ratio of infection being necessary for purposes of assay. The infected bacteria were freed from unadsorbed phage and suspended in water containing per liter 1 mM $MgSO_4$, 0.1 mM $CaCl_2$, and 0.1 gm. gelatin. A sample of this suspension was agitated for 2.5 minutes in the Waring blendor, and centrifuged to remove the extracellular S^{35}. A second sample not run in the blendor was centrifuged at the same time. The cells from both samples were resuspended in warm salt-poor broth at a concentration of 10^8 bacteria per ml., and aerated for 80 minutes. The cultures were then lysed by the addition of 0.02 mM HCN, 2×10^{11} UV-killed T2, and 6 mg. NaCl per ml. of culture. The addition of salt at this point causes S^{35} that would otherwise be eluted (Hershey et al., 1951) to remain attached to the bacterial debris. The lysates were fractionated and assayed as described previously, with the results shown in Table VIII.

The data show that stripping reduces more or less proportionately the S^{35}-content of all fractions. In particular, the S^{35}-content of the fraction containing most of the phage progeny is reduced from nearly 10 per cent to less than 1 per cent of the initially adsorbed isotope. This experiment shows that the bulk of the S^{35} appearing in all lysate fractions is derived from the remains of the coats of the parental phage particles.

Properties of Phage Inactivated by Formaldehyde.—Phage T2 warmed for 1 hour at 37°C. in adsorption medium containing 0.1 per cent (v/v) commercial formalin (35 per cent HCHO), and then dialyzed free from formalde-

TABLE VIII

Lysates of Bacteria Infected with S^{35}-Labeled T2 and Stripped in the Waring Blendor

Per cent of adsorbed S^{35} or of phage yield:	Cells stripped		Cells not stripped	
	S^{35}	Phage	S^{35}	Phage
Eluted in blendor fluid..............................	86	—	39	—
1st low-speed sediment...............................	3.8	9.3	31	13
2nd " " "	(0.2)	11	2.7	11
High-speed "	(0.7)	58	9.4	89
" " supernatant.........................	(2.0)	1.1	(1.7)	1.6
Recovery...	93	79	84	115

All the input bacteria were recovered in assays of infected cells made during the latent period of both cultures. The phage yields were 270 (stripped cells) and 200 per bacterium, assayed before fractionation. Figures in parentheses were obtained from counting rates close to background.

hyde, shows a reduction in plaque titer by a factor 1000 or more. Inactivated phage of this kind possesses the following properties.

1. It is adsorbed to sensitive bacteria (as measured by either S^{35} or P^{32} labels), to the extent of about 70 per cent.

2. The adsorbed phage kills bacteria with an efficiency of about 35 per cent compared with the original phage stock.

3. The DNA of the inactive particles is resistant to DNase, but is made sensitive by osmotic shock.

4. The DNA of the inactive particles is not sensitized to DNase by adsorption to heat-killed bacteria, nor is it released into solution by adsorption to bacterial debris.

5. 70 per cent of the adsorbed phage DNA can be detached from infected cells spun in the Waring blendor. The detached DNA is almost entirely resistant to DNase.

These properties show that T2 inactivated by formaldehyde is largely incapable of injecting its DNA into the cells to which it attaches. Its behavior in the experiments outlined gives strong support to our interpretation of the corresponding experiments with active phage.

DISCUSSION

We have shown that when a particle of bacteriophage T2 attaches to a bacterial cell, most of the phage DNA enters the cell, and a residue containing at least 80 per cent of the sulfur-containing protein of the phage remains at the cell surface. This residue consists of the material forming the protective membrane of the resting phage particle, and it plays no further role in infection after the attachment of phage to bacterium.

These facts leave in question the possible function of the 20 per cent of sulfur-containing protein that may or may not enter the cell. We find that little or none of it is incorporated into the progeny of the infecting particle, and that at least part of it consists of additional material resembling the residue that can be shown to remain extracellular. Phosphorus and adenine (Watson and Maaløe, 1952) derived from the DNA of the infecting particle, on the other hand, are transferred to the phage progeny to a considerable and equal extent. We infer that sulfur-containing protein has no function in phage multiplication, and that DNA has some function.

It must be recalled that the following questions remain unanswered. (1) Does any sulfur-free phage material other than DNA enter the cell? (2) If so, is it transferred to the phage progeny? (3) Is the transfer of phosphorus (or hypothetical other substance) to progeny direct—that is, does it remain at all times in a form specifically identifiable as phage substance—or indirect?

Our experiments show clearly that a physical separation of the phage T2 into genetic and non-genetic parts is possible. A corresponding functional separation is seen in the partial independence of phenotype and genotype in the same phage (Novick and Szilard, 1951; Hershey et al., 1951). The chemical identification of the genetic part must wait, however, until some of the questions asked above have been answered.

Two facts of significance for the immunologic method of attack on problems of viral growth should be emphasized here. First, the principal antigen of the infecting particles of phage T2 persists unchanged in infected cells. Second, it remains attached to the bacterial debris resulting from lysis of the cells. These possibilities seem to have been overlooked in a study by Rountree (1951) of viral antigens during the growth of phage T5.

SUMMARY

1. Osmotic shock disrupts particles of phage T2 into material containing nearly all the phage sulfur in a form precipitable by antiphage serum, and capable of specific adsorption to bacteria. It releases into solution nearly all

the phage DNA in a form not precipitable by antiserum and not adsorbable to bacteria. The sulfur-containing protein of the phage particle evidently makes up a membrane that protects the phage DNA from DNase, comprises the sole or principal antigenic material, and is responsible for attachment of the virus to bacteria.

2. Adsorption of T2 to heat-killed bacteria, and heating or alternate freezing and thawing of infected cells, sensitize the DNA of the adsorbed phage to DNase. These treatments have little or no sensitizing effect on unadsorbed phage. Neither heating nor freezing and thawing releases the phage DNA from infected cells, although other cell constituents can be extracted by these methods. These facts suggest that the phage DNA forms part of an organized intracellular structure throughout the period of phage growth.

3. Adsorption of phage T2 to bacterial debris causes part of the phage DNA to appear in solution, leaving the phage sulfur attached to the debris. Another part of the phage DNA, corresponding roughly to the remaining half of the DNA of the inactivated phage, remains attached to the debris but can be separated from it by DNase. Phage T4 behaves similarly, although the two phages can be shown to attach to different combining sites. The inactivation of phage by bacterial debris is evidently accompanied by the rupture of the viral membrane.

4. Suspensions of infected cells agitated in a Waring blendor release 75 per cent of the phage sulfur and only 15 per cent of the phage phosphorus to the solution as a result of the applied shearing force. The cells remain capable of yielding phage progeny.

5. The facts stated show that most of the phage sulfur remains at the cell surface and most of the phage DNA enters the cell on infection. Whether sulfur-free material other than DNA enters the cell has not been determined. The properties of the sulfur-containing residue identify it as essentially unchanged membranes of the phage particles. All types of evidence show that the passage of phage DNA into the cell occurs in non-nutrient medium under conditions in which other known steps in viral growth do not occur.

6. The phage progeny yielded by bacteria infected with phage labeled with radioactive sulfur contain less than 1 per cent of the parental radioactivity. The progeny of phage particles labeled with radioactive phosphorus contain 30 per cent or more of the parental phosphorus.

7. Phage inactivated by dilute formaldehyde is capable of adsorbing to bacteria, but does not release its DNA to the cell. This shows that the interaction between phage and bacterium resulting in release of the phage DNA from its protective membrane depends on labile components of the phage particle. By contrast, the components of the bacterium essential to this interaction are remarkably stable. The nature of the interaction is otherwise unknown.

8. The sulfur-containing protein of resting phage particles is confined to a

56   VIRAL PROTEIN AND NUCLEIC ACID IN BACTERIOPHAGE GROWTH

protective coat that is responsible for the adsorption to bacteria, and functions as an instrument for the injection of the phage DNA into the cell. This protein probably has no function in the growth of intracellular phage. The DNA has some function. Further chemical inferences should not be drawn from the experiments presented.

REFERENCES

Anderson, T. F., 1949, The reactions of bacterial viruses with their host cells, *Bot. Rev.*, **15**, 464.

Anderson, T. F., 1951, *Tr. New York Acad. Sc.*, **13**, 130.

Anderson, T. F., and Doermann, A. H., 1952, *J. Gen. Physiol.*, **35**, 657.

Benzer, S., 1952, *J. Bact.*, **63**, 59.

Doermann, A. H., 1948, *Carnegie Institution of Washington Yearbook, No. 47*, 176.

Doermann, A. H., and Dissosway, C., 1949, *Carnegie Institution of Washington Yearbook, No. 48*, 170.

Dulbecco, R., 1952, *J. Bact.*, **63**, 209.

Herriott, R. M., 1951, *J. Bact.*, **61**, 752.

Hershey, A. D., 1946, *Genetics*, **31**, 620.

Hershey, A. D., Roesel, C., Chase, M., and Forman, S., 1951, *Carnegie Institution of Washington Yearbook, No. 50*, 195.

Lwoff, A., and Gutmann, A., 1950, *Ann. Inst. Pasteur*, **78**, 711.

Maaløe, O., and Watson, J. D., 1951, *Proc. Nat. Acad. Sc.*, **37**, 507.

Novick, A., and Szilard, L., 1951, *Science*, **113**, 34.

Prater, C. D., 1951, Thesis, University of Pennsylvania.

Price, W. H., 1952, *J. Gen. Physiol.*, **35**, 409.

Putnam, F. W., and Kozloff, L., 1950, *J. Biol. Chem.*, **182**, 243.

Rountree, P. M., 1951, *Brit. J. Exp. Path.*, **32**, 341.

Watson, J. D., and Maaløe, O., 1952, *Acta path. et microbiol. scand.*, in press.

Assembly Line Genes

Demerec M. and Hartman Z. 1956. **Tryptophan Mutants in *Salmonella typhimurium*.**
(Reprinted, with permission, from *Carnegie Inst. Wash. Publ.* **612:** 5–33.)

Milislav Demerec and Andre Lwoff discuss "Viruses,"
the subject of the 1953 Cold Spring Harbor
Symposium.

D EMEREC AND HARTMAN DEVELOPED AND exploited a new gene mapping strategy based on phage transduction of bacterial genes. When they applied it to genes involved in tryptophan synthesis, they found that the genes were arranged on the chromosome in the same order as the biochemical steps that they control.

If there is one tool that distinguishes the practitioner of genetics from other biologists, it is linkage. For more than 85 years, geneticists have used linkage analysis to produce genetic maps showing the positions of genes on chromosomes and to locate a gene by determining its position relative to known positions. Linkage was first detected by Bateson and Punnett when crossing pea plants. They found that two genes—one for flower color and the other for pollen shape—appeared to be "coupled"; that is, the two genes did not obey Mendel's Second Law of Independent assortment (Bateson et al. 1905). Bateson and Punnett had no explanation for their results, and none was forthcoming until Morgan discovered the same phenomenon in *Drosophila*.

Morgan suggested that genes were *genetically* coupled because they were *physically* coupled by being on the same chromosome; the occasional breakdown of coupling was due to recombination—somehow genes were exchanged between chromosomes—a process supported by the work of cytogeneticists like Belling. Morgan proposed that the frequency with which recombination took place was related to the physical distance separating the genes on the chromosome (Morgan 1911) and proposed that this could be used for mapping the positions of genes relative to each other. Sturtevant, then an undergraduate research assistant in Morgan's laboratory, took the data home and overnight produced the first genetic map, of three genes on the *Drosophila* X-chromosome (Sturtevant 1913; Allen 1978).

Determining linkage between genetic loci requires detecting recombinants, and this becomes increasingly difficult as the distance between genes gets smaller and recombination occurs less frequently. For very small distances, very large numbers of matings have to be screened to detect the very small number of recombinants. This became possible only when bacterial genetics developed and it was realistic to screen many millions of crosses (Brock 1990). The ultimate linkage analysis was done by Seymour Benzer, who mapped mutations in the *rII* locus of phage T4 to a resolution of single nucleotides (Benzer 1961). Bacterial geneticists also exploited some special features of bacteria and phage. For example, during mating in *E. coli*, a chromosome passes from one cell to the other, beginning with one end, so that genes enter the recipient cell in their order on the chromosome. Thus, the time at which a gene is transferred is correlated with its position. Wollman and Jacob (1955) interrupted the mating at various times and so produced a gene map.

In the paper reprinted here, Demerec and Hartman (1956) use the phenomenon of transduction, discovered by Zinder and Lederberg (1952) for mapping. The latter had shown that bacteriophage P22 can carry small numbers of bacterial genes on a chromosomal fragment from one *Salmonella typhimurium* cell to another, and that these can recombine with the host cell chromosome. Demerec realized that this could be used for gene mapping and, about 1954, he began an intensive project mapping genes involved in tryptophan synthesis, a project that Hayes (1964) described as "...a brilliant and revealing series of investigations.... ."

The basic strategy in these experiments was to infect the "donor" bacteria with the transducing phage, isolate the phage, infect the "recipient" bacteria, plate out the infected bacteria on selective media, and count the recombinants. Demerec and Hartman (1956) worked with the same system as Zinder and Lederberg—phage PLT-22 and *Salmonella typhimurium*. A set of ten *Salmonella* mutants that required tryptophan for growth were isolated, and pair-wise transductions using all ten mutants and wild-type cells were carried out. The power that bacteria and phage brought to genetics is shown by the numbers involved in these experiments. Transduction is a rare event occurring once in 100,000 infected cells, but Demerec and Hartman routinely added 1,000,000,000 phage to 200,000,000 bacteria. They found, depending on the transduction, as many as several hundred recombinants not requiring tryptophan when these infected bacteria were plated out on a minimal medium. This indicated that two genes were different although close together.

Brenner (1955) had sorted the ten mutants into four groups, A through D, depending on which step was affected in the pathway to tryptophan; *try-2* and *try-4* formed Group B, blocking the first step leading from anthranilic acid to an unknown intermediate, whereas *try-1, try-6, try-7, try-9, try-10,* and *try-11* comprised Group D, blocking the last step from indole to tryptophan. When Demerec analyzed the data, he found, for example, that transductions between *try-2* and *try-4* did not yield wild-type cells (less than 20 recombinants), whereas transductions between *try-2* and *try-4* and any of the mutants from the last group did produce cells that could synthesize tryptophan (between 44 and 542 recombinants). In fact, the genetic data were consistent with the biochemical findings.

Demerec then turned to mapping the mutations, using *try* mutants that also carried the *cysB-12* mutation that is linked with the *try* locus. Now, they used cells with two *try* mutations and the *cysB-12* mutation—so-called three-point crosses. The data from these experiments enabled Demerec and Hartman to determine that the order of the groups of *try* mutations is *tryD-tryC-tryB-tryA-cysB*. From further experiments, Demerec and Hartman were able to determine the order of alleles within groups; *try-4-try-2* in Group B and *try-1-(try-10,try-11)-try-7-try-9-try-6* in Group D.

At the same time that Demerec and Zlata Hartman were mapping the *try* mutations, Philip Hartman was performing a similar analysis for mutations in the histidine locus of *Salmonella* (Hartman 1956). His findings were very similar. He, too, found that there was a correlation between mutations grouped by genetic analysis and by biochemical activity. What is more, he and Demerec saw that the *order of the loci on the chromosome corresponded with the order of the reactions in the biochemical pathway* (Demerec and Hartman 1959). Demerec called this an "assembly line" arrangement and, together with Philip Hartman (1956), suggested that this close association came about as a consequence of evolutionary selection for the most efficient arrangement of genes. This was an early hint of the operon concept, developed in full and so successfully by Jacob and Monod who cited Demerec's Cold Spring Harbor Symposium presentation (Demerec 1956) in their classic paper (Jacob and Monod 1961). It transpired later that the order of genes on the chromosome did not invariably correlate with the steps of the biochemical pathway, but the clustering of genes that are coordinately controlled is common.

Biographical information on Milislav Demerec can be found on page 78.

Zlata (Demerec) Hartman was a research assistant in the Demerec laboratory and, from 1957 to 1996, continued as a part-time research assistant in Philip Hartman's laboratory in the Department of Biology, the Johns Hopkins University. She took part in research that characterized and mapped 1500 histidine mutants in Salmonella typhimurium, *as well as selecting strains with low spontaneous reversion rates for use in the Ames test for assessing mutagenicity of environmental chemicals. Zlata Hartman collaborated also with Ernest Bueding in a study of hycanthone, an antischistosomal agent, which showed that it was possible to separate its desirable pharmacological activity from its mutagenic activity.*

Allen G.E. 1978. *Thomas Hunt Morgan: The man and his science*. Princeton University Press, Princeton, New Jersey.

Bateson W., Saunders E.R., and Punnett R.C. 1905. Experimental studies in the physiology of heredity. *Reports to the Evolution Committee of the Royal Society* **2:** 1–131.

Benzer S. 1961. On the topography of the genetic fine structure. *Proc. Natl. Acad. Sci.* **47:** 403–415.

Brenner S. 1955. Tryptophan biosynthesis in *Salmonella typhimurium. Proc. Natl. Acad. Sci.* **41:** 862–863.

Brock T.D. 1990. Transduction. In *The emergence of bacterial genetics*, pp. 189–263. Cold Spring Harbor Laboratory Press, Cold Spring Harbor, New York.

Demerec M. 1956. A comparative study of certain gene loci in *Salmonella. Cold Spring Harbor Symp. Quant. Biol.* **21:** 113–121.

Demerec M. and Hartman P.E. 1959. Complex loci in microorganisms. *Annu. Rev. Microbiol.* **13:** 377–406.

Demerec M. and Hartman Z. 1956. Tryptophan mutants of *Salmonella typhimurium. Carnegie Inst. Wash. Publ.* **612:** 5–33.

Hartman P.E. 1956. Linked loci in the control of consecutive steps in the primary pathway of histidine synthesis in *Salmonella typhimurium. Carnegie Inst. Wash. Publ.* **612:** 35–61.

Hayes W. 1964. *The genetics of bacteria and their viruses*, p. 130. Wiley, New York.

Jacob F. and Monod J. 1961. Genetic regulatory mechanisms in the synthesis of proteins. *J. Mol. Biol.* **3:** 318–356.

Morgan T.H. 1911. Random segregation versus coupling in Mendelian inheritance. *Science* **34:** 384.

Sturtevant A.H. 1913. The linear arrangement of six sex-linked factors in *Drosophila*, as shown by their mode of association. *J. Exp. Zool.* **14:** 43–59.

Wollman E.L. and Jacob F. 1955. Sur le mécanisme du transfert de matériel génétique au cours de la recombination chez *Escherichia coli* K12. *C.R. Acad. Sci.* **240:** 2449–2451.

Zinder N. and Lederberg P. 1952. Genetic exchange in *Salmonella. J. Bacteriol.* **64:** 679–699.

2

TRYPTOPHAN MUTANTS IN SALMONELLA TYPHIMURIUM

M. Demerec and Zlata Hartman

The main purpose of this paper is to present data on linkage relations among four loci controlling tryptophan synthesis and one locus associated with cystine synthesis in the genome of Salmonella typhimurium. Since the technique used in this work is new, the data will be reported in greater detail than would otherwise be warranted. On the basis of evidence published earlier (Demerec and Demerec, 1956), we reached the conclusion that transduction is accomplished in the following manner. A phage particle carries a fragment of a chromosome from the bacterium in which it was raised to the bacterium which it infects. This fragment may synapse with the homologous region of the chromosome of the recipient bacterium. When it synapses it participates in the chromosomal division of the recipient bacterium; crossing over occurs between the fragment and the bacterial chromosome, resulting in a recombination of genetic markers present in the region involved. Thus, by means of transduction experiments one may bring about duplication of small regions of a bacterial chromosome; and, when known genetic markers are available, the genetic structure of such regions may be studied by the application of crossover analyses similar to those worked out for Drosophila, maize, and other organisms.

In addition to the authors, Wynn Westover and Ernest L. Lahr took part in the experimental work to be discussed.

Materials

Nine of the tryptophan-requiring mutants used in our experiments were isolated in strain LT-2 of S. typhimurium, and one (try-7) in strain LT-7. Both these strains were obtained from Dr. Norton Zinder. Originally, eleven mutants were isolated, and were designated try-1, try-2, try-3, ..., try-11; but the try-5 stock was lost before the analysis was begun.

The ten tryptophan requirers originated independently of one another. They were isolated by the penicillin method. In our early isolation experiments ultraviolet irradiation was used to increase the frequency of mutants, but we soon realized that the spontaneous-mutation rate was high enough so that irradiation could be omitted from the procedure. Thus try-1, -2, -3, -4, -6, -7, and -8 were obtained from cultures treated with UV, and try-9, -10, and -11 were

5

Carnegie Institution of Washington Publication 612, Genetic Studies with Bacteria (1956), pages 5-33

6   GENETIC STUDIES WITH BACTERIA

isolated from cultures that had received no treatment. In strains carrying two
mutant markers, each was derived by independent selection. In the cases of
strains tryD-10 cysB-12, tryA-8 cysB-12, and tryD-11 cysB-18, the cys character was isolated first; in tryB-2 cysB-45 and tryB-2 cysD-50, try was obtained
first; and in tryA-8 cysB-40 and tryA-8 cysB-41, cys was derived from tryA-8
cysB-12+.

Methods

For all the transduction experiments reported in this paper we used temperate phage PLT-22. In each experiment a mixture consisting of approximately
2×10^8 bacteria per ml and 1×10^9 phage particles per ml was kept in a water
bath at 37° C for 5 to 15 minutes, and then 0.5 ml was plated on each of four
plates. The values given in the tables (at the end of this paper) represent total
numbers of colonies appearing on four plates. In cases where one or two plates
out of a set of four could not be scored, because of contamination or some other
accident, the count of an incomplete set was adjusted to correspond to the value
expected for four plates.

In order to increase the efficiency of transduction, Davis' minimal medium
was enriched by the addition of 0.01 per cent of broth powder. When supplement-
ing either the minimal or the enriched minimal medium with one or another of
the amino acids required, we used 20 µg of the amino acid per ml of the medium.

Grouping of the Mutant Markers

Biochemical blocks. Brenner (1955) established that the ten try mutants
could be divided into four distinct groups, according to the positions of the blocks
they represent in the pathway of tryptophan biosynthesis (fig. 1). The try-8 mutant (group A) is not able to synthesize anthranilic acid; try-2 and -4 (group B)

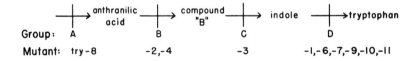

Fig. 1. Diagram showing sequence of biochemical steps in tryptophan synthesis, positions of the blocks associated with the different
groups of mutants, and the mutants belonging to each group.

are unable to transform anthranilic acid into a still-unidentified intermediate
compound called "B," and they accumulate anthranilic acid; try-3 (group C) cannot transform compound "B" into indole, and it accumulates compound "B";
try-1, -6, -7, -9, -10, and -11 (group D) are unable to transform indole into
tryptophan, and they accumulate a mixture of compound "B" and indole. Group D

TRYPTOPHAN MUTANTS 7

mutants feed mutants of groups A, B, and C; and group B mutants feed group A. The mutants of group D require tryptophan for growth, those of groups B and C are able to grow on either tryptophan or indole, and the group A mutant can grow on tryptophan, indole, or anthranilic acid.

Transduction. Table 1 gives data obtained in transduction experiments involving bacteria carrying the different try markers and phages raised on these or on wild-type bacteria. The ten tryptophan mutants clearly fall into four groups on the basis of transduction results, even though the data have not been corrected for differences in the ease with which different bacterial strains are transduced by the same phage or for differences in the efficiency of transduction of the same strain of bacteria by different phages. Such corrections would have changed the values recorded in the various columns, but would not have affected the groupings. For example, as shown by the numbers of transductions of try-8 and of try-3 bacteria effected by + phage (i.e., phage grown on wild-type bacteria), the efficiency of transduction is almost ten times as great in try-8 as in try-3; and therefore, to make the data comparable, the values obtained in the try-3 experiments would have to be multiplied by a factor of 9.7. But inasmuch as these data are not to be used to calculate relations between members of the same group, it seems better to give the actual rather than adjusted values.

The data of table 1 indicate that try-1, -6, -7, -9, -10, and -11 belong in one group, try-2 and -4 in another group, and try-3 and try-8 each by itself. Thus it is evident that the grouping arrived at by transduction tests is in agreement with that based on the biochemical findings. As pointed out earlier (Demerec and Demerec, 1956), we assume that all mutants belonging to any one such functional group are alleles, most of them nonidentical but some possibly identical. Accordingly, we conclude that a gene locus takes in a section of chromosome, that changes occurring at different sites within that section give rise to different nonidentical alleles, and that the transduction technique is a very sensitive method of detecting recombination between the sites of a locus.

In the experiments reported in table 1, no cases of transduction were observed in tests between try-10 and try-11, from which it might be concluded that these two try mutants represent changes at the same site. Any decision about their relationship, however, must be deferred until they are compared with respect to various biological properties such as spontaneous and induced mutability, and growth on media containing different concentrations of tryptophan.

Linkage Relations

It is evident that transduction techniques can enable us to detect linkage only between loci that are situated close enough together on the chromosome to be included in one fragment carried by an infecting phage particle, for only in such cases is it possible to recover classes of recombinants derived by simultaneous transduction of the two genetic markers. For example, if two markers, a and b, occur together in bacteria (a b) which are infected with phage from the wild-type strain ($\underline{a}^+$ $\underline{b}^+$), the recovery of a wild-type class indicates simultaneous transduction of both markers, and is evidence of linkage. Similarly, if $\underline{a}$ is present in the infected bacteria and $\underline{b}$ is carried by the transducing phage, then a $\underline{b}$ class represents the simultaneous transduction of both markers (fig. 2). Simultaneous transfer of two markers could also result if two independent transductions occurred in one bacterium; but in our material transductions are so infrequent (about one per 5×10^6 infecting particles for $\underline{try}$ and for $\underline{cys}$) as to make it unlikely that such simultaneous transductions would occur at all in our experiments (expected frequency, about one per 2.5×10^{11} infecting particles). So we can assume that double transduction indicates close linkage between the loci involved.

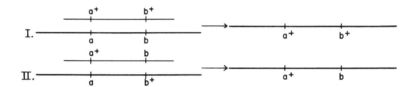

Fig. 2. Classes resulting from simultaneous transduction of two closely linked genetic markers ($\underline{a}$ and $\underline{b}$): (I) when both the markers are located in the bacterial chromosome; (II) when one marker, $\underline{a}$, is located in the bacterial chromosome and the other, $\underline{b}$, in the transducing segment.

For technical reasons it is easier to detect linkage if the markers being tested occur in the same genome than if one is brought into an experiment by the recipient and another by the donor. Therefore, in order to facilitate the detection of linkages, we have developed a number of stocks with two markers by isolating new mutants in bacteria that already carried a mutation.

Experiments with double mutants. Experiments in which bacteria with two markers were transduced by phage from wild-type donors revealed linkage between the $\underline{cysB}$ locus and the $\underline{tryA}$, B, and D loci; and experiments with $\underline{cysB}$ recipients and $\underline{tryC}$ donors showed that $\underline{tryC}$ also is linked with $\underline{cysB}$. Table 2 summarizes the data from experiments in which bacteria carrying a $\underline{try}$ and a $\underline{cys}$ marker were transduced by wild-type phage. Samples containing a mixture of about 1×10^8 bacteria and 5×10^8 phage particles per ml were plated on three sets of plates (four to each set) containing media on which only certain classes of bacteria would be able to grow. Set (1) contained minimal medium, on which only wild-type ($\underline{try}^+$ $\underline{cys}^+$) bacteria could form colonies; set (2), minimal medium with tryptophan added, which would support both wild-type and tryptophan-requiring ($\underline{try}$ $\underline{cys}^+$) colonies; and set (3), minimal medium with cysteine added, on which wild-type and cystine-requiring ($\underline{try}^+$ $\underline{cys}$) colonies could grow. By printing the colonies that appeared on (2) and (3) plates onto minimal medium, it was possible to identify the wild-type class on these plates. Table 2 gives the average number of colonies per experiment of the different

TRYPTOPHAN MUTANTS   9

classes of bacteria observed on a set of four plates; and figure 3 is a schematic
representation of the crossing over responsible for the origin of these classes.
It is evident that between 18 and 44 per cent of the colonies obtained on medium
containing either tryptophan or cysteine were wild type, that is, they resulted
from simultaneous transductions of both markers used in the experiments.

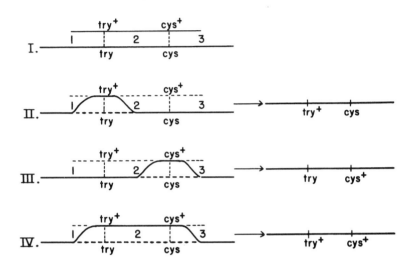

Fig. 3. Schematic representation of crossing over responsible for
the origin of classes obtained when try cys bacteria were transduced by
wild-type phage. Lower lines represent the bacterial chromosome; upper lines, the chromosomal segment introduced by the phage. I: parental constitution, indicating chromosomal regions (1, 2, 3). II, III, IV:
left, diagrams of crossing over; right, classes derived.

Three-point tests. Since the cysB locus is closely linked with the try loci,
and since a large proportion of the classes arising from recombinations between
try markers and a cysB marker can be readily recognized, we have been able to
use three-point tests to determine the order of the four try loci and the cysB
locus on a linkage map. To verify the reliability of these tests, markers representing three loci were often used in several combinations; and to test the variability of the results, the experiments were repeated several times. Data from
all the experiments will be presented here.

1. Experiments involving tryB, tryD, and cysB: Table 3 shows the data of
two experiments in which tryD-10 cysB-12 bacteria were transduced by phage
grown on tryB-2 bacteria, and an analysis of these data is given in table 4. It is
evident from the analysis that the linkage order tryD—cysB—tryB is impossible,
because the value for the tryB-2 cysB-12 class is too high for a four-crossover
class. Further, the data indicate that, of the two remaining possibilities, tryB—

tryD—cysB is the less likely, for the difference between the ratios 2-3/2-4 and 1-3/1-4 is considerably greater than would be expected on the basis of evidence presented in table 2. It can be concluded from these results, therefore, that the probable order of the three loci is tryD—tryB—cysB. More definite evidence in support of this choice was obtained from reciprocal experiments in which tryB-2 bacteria were infected with phage from tryD-10 cysB-12 bacteria (tables 5 and 6). The data show that the cysB class was much too large to be accounted for by quadruple crossing over, and eliminate the possibility that the order might be tryB—tryD—cysB. They fit well with the conclusion that the order is tryD—tryB—cysB.

Similar results were obtained when different nonidentical alleles of the tryB, tryD, and cysB loci were used in experiments. The data of experiments with tryB-2, tryD-11, and cysB-18 have been published (Demerec and Demerec, 1956), and data for tryB-4, tryD-11, and cysB-18 are given in tables 7 and 8.

Our method of determining the order of loci could be subjected to a critical test in experiments in which the cysB and tryB markers were together in one strain and the tryD marker was in the other strain. In such experiments the relative proportions of derived cysB, tryB, and tryD classes could be expected to be quite different from those found in the experiments with cysB and tryD together in one of the participating strains. The results are given in tables 9 and 10, and they are in complete agreement with expectation based on the assumption that the order of loci is tryD—tryB—cysB.

2. Experiments involving tryA, tryD, and cysB: Table 11 gives the data of an experiment in which tryD-10 cysB-12 bacteria were infected with tryA-8 phage. In this experiment the tryA-8 class (which cannot synthesize anthranilic acid) was detected on the minimal medium as well as on anthranilic acid medium, because the tryD bacteria used here as recipient accumulated a sufficient quantity of indole to support growth of the tryA-8 class. Analysis of the classes obtained (table 12) eliminates the order tryD—cysB—tryA, because the tryA-8 cysB-12 class was too large to be due to quadruple crossing over, and indicates that of the two remaining possibilities the order tryD—tryA—cysB is in better agreement with the data. The results of reciprocal experiments, using tryA-8 bacteria and tryD-10 cysB-12 phage, are given in table 13. An analysis of these data similar to the one given in table 6 shows that the order is tryD—tryA—cysB.

Data from experiments in which tryA and cysB occurred together and tryD alone are given in tables 14 and 15. They are comparable with the data of table 11, which were analyzed in table 12. Similar analyses applied to the data of tables 14 and 15 confirm the conclusion that the order of these loci is tryD—tryA—cysB.

3. Experiments involving tryA, tryB, and cysB: Data of three tryA-8 cysB-12 (×) tryB-2 experiments are given in table 16, and those of a reciprocal experiment in table 18. In this last experiment an anthranilic acid class (tryA-8) was detected on the minimal medium because the tryB-2 bacteria used as recipient accumulated a sufficient quantity of anthranilic acid to support its growth.

Results of the analysis shown in table 19 eliminate the possibility that the order might be tryB—cysB—tryA, because of the size of the tryA-8 class. In addition, the observed frequencies of +, tryA-8, and cysB-12 classes indicate tryB—tryA—cysB as the probable order. This conclusion is supported by the analysis of the data from the first experiments, shown in table 17. In particular, the size of the + class makes the tryA—tryB—cysB order improbable. Data from an experiment using tryB-4 in place of tryB-2 are in full agreement with this interpretation (table 20), as are the data from experiments in which tryB and cysB were together in one strain (table 21).

4. Experiments involving tryC, tryD, and cysB: Data from three tryD-10 cysB-12 (×) tryC-3 experiments are given in table 22, and those of two reciprocal experiments in table 23. The order of these three markers could be determined by applying the patterns of analysis used in tables 4 and 6. The conclusion derived was that the order is tryD—tryC—cysB. Results of experiments in which tryC-3 bacteria were infected with tryD-11 cysB-18 phage (table 24) are in agreement with that conclusion.

5. Experiments involving tryA, tryC, and cysB: Tables 25 and 26 give the data of experiments with markers representing these three loci. The analysis follows the patterns demonstrated in tables 17 and 19, and shows the order to be tryC—tryA—cysB.

6. Experiments involving tryB, tryC, and cysB: Table 27 gives the data of two tryB-2 cysB-45 (×) tryC-3 experiments, and table 28 presents the analysis of these data. The large size of the + class in comparison with the cys and try classes eliminates the order tryB—tryC—cysB, and the ratio between the + and cys classes suggests the order tryC—tryB—cysB. This is supported by the data from the reciprocal experiments (table 29). Analysis of these data (table 30) shows that the cys class was too small to be due to crossing over in regions 1-3, which would be expected if the order were tryC—cysB—tryB.

Conclusions regarding the order of the try loci. The data presented here show that all four try loci are on the same side of the cysB locus; and the relative arrangements indicated by the different three-point tests are as follows:

1. tryD—tryB—cysB
2. tryD—tryA—cysB
3. tryB—tryA—cysB
4. tryD—tryC—cysB
5. tryC—tryA—cysB
6. tryC—tryB—cysB

It is evident from (1) and (3) that the order of three of the try loci is tryD—tryB—tryA, and from (4) and (6) that tryC is between tryD and tryB. Therefore we conclude that the order of these five loci is tryD—tryC—tryB—tryA—cysB.

Distances between loci. From the data presented in the preceding sections it is possible to estimate the relative distances between the loci in question. The genetic markers of the five loci with which we have been dealing divide the chromosome section into six regions, as follows:

$$\underline{\quad 1 \quad} tryD \underline{\quad 2 \quad} tryC \underline{\quad 3 \quad} tryB \underline{\quad 4 \quad} tryA \underline{\quad 5 \quad} cysB \underline{\quad 6 \quad}$$

We used in every three-point experiment, in addition to a cysB marker, two other markers each representing one of the try loci. Thus, in every experiment in which appropriate classes could be recognized it was possible to observe crossovers in four regions of this chromosome section. We have to keep in mind, however, that in transduction experiments we are dealing with crossing over between a short duplication and a whole chromosome, and that single-crossover classes cannot be detected because only those classes that represent an even number of crossovers are able to survive (Demerec and Demerec, 1956). In our calculations we shall consider only certain of the double-crossover classes observed in the experiments.

In the tryD-10 cysB-12 (×) tryC-3 experiment (table 22) the markers were distributed as follows:

$$\underline{\quad 1 \quad} tryD\text{-}10 \underline{\quad 2 \quad} tryC\text{-}3 \underline{\quad (3+4+5) \quad} cysB\text{-}12 \underline{\quad 6 \quad}$$

In this case the cys class, with an average value of 59, represented crossing over in the regions 1-2, and the try class, with a value of 776, represented crossing over in regions 2-6 and in regions (3+4+5)-6. For our calculations we are interested in the size of the (3+4+5)-6 crossover class. This cannot be determined directly, but evidence from other experiments indicates that the 2-6 class could not be larger than the 1-2 class; and so by subtracting the value for class 1-2 from the observed value for the combined classes 2-6 and (3+4+5)-6 (776 − 59) we obtain a fair estimate of the minimum value for class (3+4+5)-6 (717). Now, in both class 1-2 and class (3+4+5)-6, one of the two crossovers occurred in an end section (either 1 or 6), and it is reasonable to assume that the lengths of these end sections in terms of crossover units were about equal, since they were limited only by the length of the transducing fragment. Accordingly, it can be concluded that the relative sizes of the 1-2 and (3+4+5)-6 classes were a function of the relative lengths of regions 2 and (3+4+5), and thus that region (3+4+5) is about twelve times as long as region 2.

Using the data of experiment tryD-10 cysB-12 (×) tryB-2 (table 4), we can make a similar calculation from the values for classes 1-(2+3) and (4+5)-6, and it indicates that region (4+5) is about twelve times as long as region (2+3). Using data from table 12, it can be calculated that region 5 is about three times as long as region (2+3+4); and from the data of table 25 it can be estimated that region 5 is about six times as long as region (3+4).

Although all these estimates are mere approximations, they indicate that regions 2, 3, and 4 are short in proportion to region 5, and, indeed, that region 5 is considerably longer than the total length of these three regions together. It seems very likely that the four try loci are adjacent to one another, and that a region still genetically unknown to us is present between the tryA and cysB loci on this section of the bacterial chromosome.

TRYPTOPHAN MUTANTS   13

Order of Nonidentical Alleles within a Locus

Of the ten try mutant markers being analyzed here, two are nonidentical alleles of the tryB locus (tryB-2 and -4) and six are nonidentical alleles of the tryD locus (tryD-1, -6, -7, -9, -10, and -11). Since we have one of the tryB alleles and two of the tryD alleles available in stocks in combination with a cysB allele (tryB-2 cysB-45, tryD-10 cysB-12, and tryD-11 cysB-18), we could carry out three-point tests to determine the order of some of the alleles within these two try loci.

In the case of the tryB alleles, there are two possible orders with relation to the cysB locus, namely, tryB-2—tryB-4—cysB and tryB-4—tryB-2—cysB; and in experiments involving the three markers we should expect to obtain recombinants resulting from crossovers in four regions: region 1 to the left of the try locus, region 2 between the two try alleles, region 3 between the try and cys loci, and region 4 to the right of the cys locus. The results of the three-point tests already described, involving two try loci and the cysB locus, in which region 2 represented the distance between the two try loci, showed that region to be considerably shorter than region 1, 3, or 4; and thus we should expect region 2 in these experiments involving two tryB alleles to be still shorter. This expectation was fulfilled.

Table 31 gives the data of three tryB-2 cysB-45 (X) tryB-4 experiments, and table 32 the results of reciprocal experiments. The + and cys classes provide the information needed. Figures 4 and 5 are diagrams of the crossovers that would produce these two classes, assuming the order of the alleles to be B-2—B-4 or B-4—B-2. In the tryB-2 cysB-45 (X) tryB-4 experiments (table 31), the average values were 12 for the + class and 31 for the cys class. These results agree better with the assumption that the order is B-4—B-2 (fig. 4, II) than with the assumption that it is B-2—B-4 (fig. 4, I), because, judging by data obtained in previous experiments (table 2), we should expect the 2-3 crossover class (cys) to be approximately two to three times as large as the 2-4 class (+). The data of the reciprocal experiments (table 32) also support this conclusion, since they suggest that the cys class resulted from quadruple crossovers (fig. 5, II) rather than from double crossovers. From the results of these experiments we conclude that the order of the markers is tryB-4—tryB-2—cysB.

By applying to the data of tables 33 and 34 the method of analysis used in establishing the order of the tryB-4 and -2 alleles, it can be shown that tryD-1 is to the left of tryD-10; and from the data of tables 35 to 40 we can conclude that tryD-9, -7, and -6 are to the right of either tryD-10 or tryD-11.

From the data given in tables 35, 37, and 39 the distances between tryD-10 (or -11) and tryD-9, -7, and -6 can be estimated and the approximate order of these alleles determined. To illustrate these calculations we shall use the data of table 35, relating to tryD-10, tryD-9, and cysB-12. The average value for the try class, which represents recombinants resulting from 3-4, 2-4, and 1-4 crossovers, is 2877; the average value for the cys class, representing 1-2 crossovers,

14   GENETIC STUDIES WITH BACTERIA

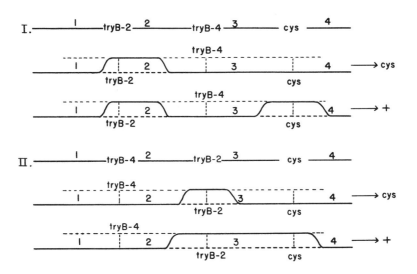

Fig. 4. Diagram indicating crossing over that would produce cys
or + classes in experiments involving two nonidentical alleles of a try
locus and a cys marker, using tryB-2 cysB-45 bacteria and tryB-4
phage, and assuming the order of markers to be (I) B-2—B-4—cys
or (II) B-4—B-2—cys.

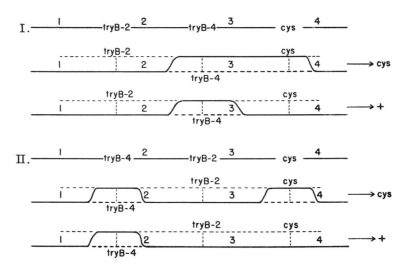

Fig. 5. Diagram indicating crossing over that would produce cys
or + classes in experiments involving two nonidentical alleles of a try
locus and a cys marker, using tryB-4 bacteria and tryB-2 cysB-45
phage, and assuming the order of markers to be (I) B-2—B-4—cys or
(II) B-4—B-2—cys.

TRYPTOPHAN MUTANTS   15

is 75; and that for the + class, representing 1-2-3-4 crossovers, is 6. The distance between D-10 and D-9 will be estimated as the ratio of the value for crossing over in regions 1-2 to that for crossing over in regions 3-4. Since regions 1 and 4 are end regions, it seems justifiable to assume that crossing over in these two regions occurs with equal frequency. Thus our calculation gives us a value for crossing over in region 2—that is, between tryD-10 and -9—in terms of percentage of the value for crossing over in region 3—that is, between tryD-9 and cysB-12. An estimate of the value for the 3-4 class can be obtained by deducting from the value for the try class (2877) the values for the 2-4 and 1-4 classes. Class 2-4 should not be larger than class 1-2, which is 75, and according to the data given in table 2 the 3-4 class can be estimated as 82 per cent of the sum of the 3-4 and 1-4 classes. Thus, in our calculations, the value for the 3-4 class amounts to about 2300, so that the crossover value for region 2 is about 3.2 per cent of the value for region 3, and we take 3.2 as the value for the distance between tryD-10 and -9. Similar calculations give the distance between tryD-11 and -7 as 0.55, and that between tryD-11 and -6 as 4.2. From these operations it may be concluded that the order of the alleles of the tryD locus is: -1, (-10, -11), -7, -9, -6.

Four-point tests. Using the same double-marker stocks as in the three-point tests, it was possible to conduct experiments in which four-point recombination could be observed. The results of two such experiments are given here, to show that the frequencies of the recombination classes, when four markers were involved, were in good agreement with expectation based on the assumption that crossing over was responsible for their origin. Through analyses of these results, the relative order of the cysB markers used in the experiment could be determined.

Tables 41 and 43 give the data of experiments employing tryD-11 cysB-18 and tryA-8 cysB-12. In analyzing these data (tables 42 and 44), only two possible orders were considered, since the data agreed with the earlier conclusions regarding the relative positions of the loci tryD, tryA, and cysB. From the values for the + and tryA-8 classes obtained in the two sets of experiments it can be concluded that the order of the two cysB markers is cysB-18—cysB-12.

Tables 45 to 48 give a similar set of data for experiments involving tryD-11 cysB-18 and tryB-2 cysB-45, from which it can be concluded that the order of these two cysB markers is cysB-18—cysB-45.

Discussion

This study showed that by transduction techniques it is possible to determine the order of loci on a bacterial chromosome and to estimate the distances between them. It revealed that the four try loci we have investigated in Salmonella are very closely linked, probably adjacent to one another, and that cysB, which is the next known locus in this chromosome section, is at a considerable distance from the nearest try locus (tryA).

Furthermore, the study demonstrated in the case of the try loci, as in the other Salmonella material analyzed in our laboratory, that the genetic changes at one gene locus which lead to independent occurrences of phenotypically similar mutants usually originate at different sites within the locus, and that the transduction method has a very high "resolving power" for the detection of crossing over between such sites. In our collection of ten try mutant markers, five or six are nonidentical alleles of the tryD locus, two others are nonidentical alleles of the tryB locus, and the remaining two are referable to mutations in the tryA and tryC loci, respectively.

Unquestionably, the most striking discovery of this study is the "assembly line" arrangement of the try loci on the linkage map: the fact that their order on the chromosome coincides with their sequence in respect to biochemical blocks in the chain of reactions leading to the synthesis of tryptophan. Hartman (1956) found a similar arrangement in the case of four hi loci. In both instances we have a situation in which a section of chromosome containing four loci is concerned with the control of a complex series of different, but related, metabolic reactions. This spatial relationship of the four loci suggests that they may have a common origin; in other words, that they originated through differentiation among the sites of a single precursor locus and an increase in number of the differentiated sites by duplication. On the other hand, the reactions controlled by the various members of the tryptophan series are so different that it is difficult to visualize a differentiation process which would change the action of a site of one locus in such a way as to give it the characteristics now observed in the locus adjacent to it.

It is not necessary, however, to assume a common origin in order to explain how a certain group of loci might become adjacent and arranged in a particular sequence, provided that this arrangement increased the efficiency of their performance and thus conferred selective advantage on the organisms in which it was present. In fact, given time, any two loci which were adjacent to each other because of common origin would eventually be separated, if a separating mechanism existed—and there is ample evidence of the existence of mechanisms that can separate not only adjacent loci, but also different sites within the same locus. Thus it is to be expected that in the course of evolutionary readjustment different loci or even different sites of a locus would become randomly distributed in a genome if some force did not either keep or bring certain of them together. It does not seem likely that this force would be a mechanical one, of a sort to prevent breaks between certain loci, since we have good evidence that breaks occur both between the four tryptophan loci and within them. A low frequency of occurrence of such breaks would only extend the time required for the attainment of randomness; and, since time is not a limiting factor in evolutionary processes, an inhibiting mechanism could not be responsible for the nonrandom distribution of certain loci. As already suggested, a more promising hypothesis to explain such nonrandomness is that it possesses selective advantage and that evolutionary forces operate in bringing about the gene arrangements best fitted to the

TRYPTOPHAN MUTANTS   17

species. Such forces must also operate to keep the sites of individual loci together; otherwise it is difficult to see how a locus could include upward of a hundred sites, as is indicated in the case of the histidine loci (Hartman, 1956). It seems likely, in fact, that the associative forces operating within a locus are of considerably greater magnitude than those effective between loci, because a gene functions as a unit and this functioning would probably be upset by any considerable change in either the number or the order of sites.

These considerations lead us to conclude that the close association in a definite sequence of the four tryptophan loci and the four histidine loci is of evolutionary advantage in Salmonella. Workers with Neurospora have analyzed four tryptophan loci in that organism which appear to be analogous to the four Salmonella loci with regard to the functions they control. These four Neurospora loci, however, are neither adjacent to one another nor linked in the order of sequence of their biosynthetic blocks, but are in different linkage groups, probably in different chromosomes (Barratt et al., 1954). This shows that a gene arrangement which is advantageous in Salmonella, and presumably in bacteria as such, possesses no apparent advantage for Neurospora; and it indicates that selective advantage as related to gene arrangement may be different in different organisms.

Summary

The detailed data presented and discussed show that four tryptophan loci are linked together, and that their order on a linkage map coincides with their sequence in respect to biochemical blocks in the chain of reactions they control.

Literature Cited

Barratt, R. W., D. Newmeyer, D. D. Perkins, and L. Garnjobst 1954 Map construction in Neurospora crassa. Advances in Genet. 6: 1-93.

Brenner, S. 1955 Tryptophan biosynthesis in Salmonella typhimurium. Proc. Nat. Acad. Sci. U. S. 41: 862-863.

Demerec, M., and Z. E. Demerec 1956 Analysis of linkage relationships in Salmonella by transduction techniques. Brookhaven Symposia Biol. No. 8: 75-84.

Hartman, P. E. 1956 Linked loci in the control of consecutive steps in the primary pathway of histidine synthesis in Salmonella typhimurium. In Genetic Studies with Bacteria, Carnegie Inst. Wash. Pub. 612, pp. 35-61.

TABLE 1

Transduction experiments with the try markers: numbers of colonies derived from transduced cells appearing on enriched minimal agar plated with about 2×10^7 infected bacteria per plate. Each figure represents the number of colonies observed on four plates. Multiplicity of infection, 5; adsorption of phage, 5 to 15 minutes at 37° C. Colonies were counted after 2 days' incubation at 37° C.

Recipient	Donor										
	try-1	-6	-7	-9	-10	-11	-3	-2	-4	-8	+
try-1.......	0	66	203	104	219	208	291	706	458	418	1264
-6.......	141	0	11	60	21	182	188	179	234	100	1617
-7.......	21	2	0	10	19	22	444	537	435	107	717
-9.......	26	8	41	0	101	66	310	361	247	437	1456
-10.....	4	2	7	12	0	0	270	628	602	206	1822
-11.....	22	1	23	22	0	0	280	240	315	497	1406
-3.......	166	50	30	75	88	107	0	139	111	123	336
-2.......	542	375	126	320	295	440	344	0	18	66	3074
-4.......	173	120	44	213	145	235	163	20	0	85	2257
-8.......	144	123	138	560	345	111	133	125	44	0	3264

TABLE 2

Average numbers of colonies per experiment of the different transduction classes recovered when bacteria carrying a try and a cys marker were infected with phage from the wild type (+). Media used: (1) minimal; (2) minimal with tryptophan added; (3) minimal with cysteine added.

Recipient bacteria	No. of expts.	1	2		3	
		+	+	try	+	cys
tryA-8 cysB-12	2	150	123	224	261	312
tryB-2 cysB-45	5	253	234	935	213	272
tryD-10 cysB-12	6	336	337	1512	351	877
tryD-11 cysB-18	3	143	141	523	193	391

TRYPTOPHAN MUTANTS   19

TABLE 3

Results of two transduction experiments in which <u>tryD-10</u> <u>cysB-12</u> bacteria were treated with <u>tryB-2</u> phage [<u>tryD-10</u> <u>cysB-12</u> (×) <u>tryB-2</u>]. Data show numbers of colonies of different classes appearing on a total of four plates. Experiment I, September 9, 1954; experiment II, December 7, 1954.

(ind = indole; min = Davis' minimal enriched with 0.01 per cent dehydrated broth.)

Medium	Phenotype	I	II	Total	Av.
min	+	16	14	30	15
ind	+	16	6	22	11
	ind	235	274	509	255
cys	+	21	6	27	14
	cys	103	89	192	96
try	+	14	49	63	32
	ind	231	325	556	278
	try	947	1809	2756	1378
cys & ind	+	9	11	20	10
	ind	236	260	496	248
	cys	113	120	233	117
	cys-ind	823	491	1314	657

TABLE 4

Analysis of the data given in table 3, showing the average values for the various transduction classes obtained in the <u>tryD-10</u> <u>cysB-12</u> (×) <u>tryB-2</u> experiments, and indicating the regions where crossing over would have had to occur to produce the observed phenotypes.

Genotype	Value	Crossing-over regions if order is:		
		tryD—tryB—cysB	tryD—cysB—tryB	tryB—tryD—cysB
+......................	16	1-2-3-4	1-3	2-4
tryB-2	260	1-4	1-4	1-4
cysB-12	107	1-2	1-2	2-3
tryB-2 cysB-12 ..	657	1-3	1-2-3-4	1-3
tryD-10 tryD-10 tryB-2 } ..	1378	{ 3-4 2-4	2-3 2-4	3-4 1-2-3-4

20 GENETIC STUDIES WITH BACTERIA

TABLE 5

tryB-2 (×) tryD-10 cysB-12 experiments

I and II, January 14, 1955; III and IV, February 16, 1955

Medium	Phenotype	I	II	III	IV	Total	Av.
min	+	277	317	284	290	1168	292
cys	+	228	313	306	280	1127	282
	cys	55	104	153	161	473	118

TABLE 6

Analysis of the data given in table 5

Genotype	Value	Crossing-over regions if order is:	
		tryD—tryB—cysB	tryB—tryD—cysB
+	287	2-3	1-2
cysB	118	2-4	1-2-3-4

TABLE 7

tryD-11 cysB-18 (×) tryB-4 experiment, December 7, 1954

Medium	Phenotype	No.
min	+	6
ind	+	7
	ind	192
cys	+	2
	cys	32
try	+	3
	ind	192
	try	709
cys & ind	+	4
	ind	148
	cys	21
	cys-ind	492

TRYPTOPHAN MUTANTS   21

TABLE 8

tryB-4 (×) tryD-11 cysB-18 experiment, November 10, 1955

Medium	Phenotype	No.
min	+	231
cys	+	259
	cys	69

TABLE 9

tryD-10 (×) tryB-2 cysB-45 experiment, October 18, 1955

Medium	Phenotype	No.
min	+	37
ind	+	52
	ind	300
cys	+	47
	ind	445
	cys	27
	cys-ind	69
cys & ind	+	40
	ind	447
	cys	12
	cys-ind	90

TABLE 10

Analysis of the data given in table 9, indicating crossing-over regions
on the assumption that the order of loci is tryD—tryB—cysB

Genotype	Value	Crossing-over regions
+	44	1-2
tryB-2	397	1-3
cysB-45	18	1-2-3-4
tryB-2 cysB-45	80	1-4

22   GENETIC STUDIES WITH BACTERIA

TABLE 11

tryD-10 cysB-12 (×) tryA-8 experiment, August 22, 1955

(anth = anthranilic acid.)

Medium	Phenotype	No.
min	+	16
	anth	169
anth	+	16
	anth	129
cys	+	27
	anth	198
	cys	215
	cys-anth	339
try	+	21
	anth	199
	try	825
cys & anth	+	15
	anth	136
	cys	198
	cys-anth	196

TABLE 12

Analysis of the data given in table 11

Genotype	Value	Crossing-over regions if order is:		
		tryD—tryA—cysB	tryD—cysB—tryA	tryA—tryD—cysB
+	19	1-2-3-4	1-3	2-4
tryA-8	166	1-4	1-4	1-4
cysB-12	207	1-2	1-2	2-3
tryA-8 cysB-12	268	1-3	1-2-3-4	1-3
tryD-10 tryD-10 tryA-8 }	825	{ 3-4 2-4	2-3 2-4	3-4 1-2-3-4

TABLE 13

tryA-8 (×) tryD-10 cysB-12 experiments, August 22, 1955

Medium	Phenotype	I	II	Total	Av.
min	+	487	425	912	456
cys...............	+	439	513	952	476
	cys	187	186	373	187

TRYPTOPHAN MUTANTS   23

TABLE 14
tryD-1 (×) tryA-8 cysB-12 experiment, October 4, 1954

Medium	Phenotype	No.
min	+	125
	anth	326
anth......................	+	124
	anth	278
cys	+	90
	anth	314
	cys	2
cys & anth	+	92
	anth	242
	cys	3
	cys-anth	61

TABLE 15
tryA-8 cysB-12 (×) tryD-1 experiment, August 15, 1955

Medium	Phenotype	No.
min	+	30
anth........................	+	49
	anth	537
cys	+	20
	cys	100
try	+	52
	anth	518
	try	65

TABLE 16
tryA-8 cysB-12 (×) tryB-2 experiments. I, September 10, 1954; II, October 12, 1954;
III, August 15, 1955.

Medium	Phenotype	I	II	III	Total	Av.
min	+	64	68	45	177	59
anth.............	+	49	59	59	167	56
	anth	1319	1378	1104	3801	1267
cys	+	67	75	42	184	61
	cys	116	144	62	322	107
try	+	37	61	43	141	47
	anth	1188	1414	1079	3681	1227
	try	73	77	87	237	79

TABLE 17

Analysis of the data given in table 16

Genotype	Value	Crossing-over regions if order is:		
		tryB—tryA—cysB	tryB—cysB—tryA	tryA—tryB—cysB
+..........................	56	2-4	2-4	1-2-3-4
tryA-8....................	1247	3-4	2-3	3-4
cysB-12	107	2-3	3-4	1-2
tryB-2 tryA-8 tryB-2 }	79	{ 1-4 1-2-3-4	1-4 1-3	1-4 2-4

TABLE 18

tryB-2 (×) tryA-8 cysB-12 experiment, October 5, 1954

Medium	Phenotype	No.
min	+	43
	anth	393
anth	+	37
	anth	320
cys	+	47
	anth	360
	cys	0
	cys-anth	30
cys & anth	+	39
	anth	457
	cys	0
	cys-anth	28

TABLE 19

Analysis of the data given in table 18

Genotype	Value	Crossing-over regions if order is:		
		tryB—tryA—cysB	tryB—cysB—tryA	tryA—tryB—cysB
+.......................	42	1-2	1-2	2-3
tryA-8	383	1-3	1-2-3-4	1-3
cysB-12	0	1-2-3-4	1-3	2-4
tryA-8 cysB-12....	29	1-4	1-4	1-4

TRYPTOPHAN MUTANTS 25

TABLE 20
tryB-4 (×) tryA-8 cysB-12 experiment, October 5, 1954

Medium	Phenotype	No.
min	+	54
	anth	427
anth	+	76
	anth	378
cys	+	58
	anth	397
	cys	0
	cys-anth	54
cys & anth	+	63
	anth	443
	cys	0
	cys-anth	70

TABLE 21
tryA-8 (×) tryB-2 cysB-45 experiments. I, January 12, 1955; II and III, September 1, 1955.

Medium	Phenotype	I	II	III	Total	Av.
min	+	134	295	262	691	230
cys	+	146	266	225	637	212
	cys	73	103	78	254	85

TABLE 22
tryD-10 cysB-12 (×) tryC-3 experiments. I, January 14, 1955; II, February 16, 1955; III, February 28, 1955.

Medium	Phenotype	I	II	III	Total	Av.
min	+	28	39	12	79	26
	ind	295	223	157	675	225
ind	+	18	29	10	57	19
	ind	213	300	128	641	214
cys	+	28	47	9	84	28
	ind	197	196	151	544	181
	cys	60	99	29	188	63
	cys-ind	475	331	256	1062	354
cys & ind	+	24	33	12	69	23
	ind	202	202	126	530	177
	cys	53	84	25	162	54
	cys-ind	358	565	255	1178	393
try	+	48	32	18	98	33
	ind	258	230	142	630	210
	try	1044	798	485	2327	776

26   GENETIC STUDIES WITH BACTERIA

TABLE 23
tryC-3 (×) tryD-10 cysB-12 experiments. I, January 14, 1955; II, February 16, 1955.

Medium	Phenotype	I	II	Total	Ave.
min	+	230	135	365	183
cys	+	143	89	232	116
	cys	26	25	51	26

TABLE 24
tryC-3 (×) tryD-11 cysB-18 experiments, July 15, 1955

Medium	Phenotype	I	II	Total	Av.
min	+	332	240	572	286
cys	+	361	346	707	354
	cys	83	55	138	69

TABLE 25
tryA-8 cysB-12 (×) tryC-3 experiment, August 15, 1955

Medium	Phenotype	No.
min	+	60
anth	+	54
	anth	726
cys	+	67
	cys	136
try	+	53
	anth	846
	try	47

TABLE 26
tryC-3 (×) tryA-8 cysB-12 experiments. I, January 14, 1955; II, February 4, 1955.

Medium	Phenotype	I	II	Total	Av.
min	+	130	517	647	324
anth	+	159	448	607	304
	anth	471	1556	2027	1014
cys	+	126	495	621	311
	cys	6	28	34	17
cys & anth	+	103	545	648	324
	anth	417	1308	1725	863
	cys	16	56	72	36
	cys-anth	11	78	89	45

TRYPTOPHAN MUTANTS   27

TABLE 27

tryB-2 cysB-45 (×) tryC-3 experiments. I, January 14, 1955; II, January 25, 1955.

Medium	Phenotype	I	III	Total	Av.
min	+	11	37	48	24
cys	+	19	27	46	23
	cys	82	50	132	66
try	+	16	25	41	21
	try	612	554	1166	583

TABLE 28

Analysis of the data given in table 27

Genotype	Value	Crossing-over regions if order is:		
		tryC—tryB—cysB	tryC—cysB—tryB	tryB—tryC—cysB
+	23	2-4	2-4	1-2-3-4
cysB	66	2-3	3-4	1-2
tryB 〕		3-4	2-3	3-4
tryC 〉	583	1-4	1-4	1-2
tryB tryC 〕		1-2-3-4	1-3	2-4

TABLE 29

tryC-3 (×) tryB-2 cysB-45 experiments. I, January 14, 1955; II, January 25, 1955.

Medium	Phenotype	I	II	Total	Av.
min	+	50	30	80	40
cys	+	54	23	77	39
	cys	8	8	16	8

TABLE 30

Analysis of the data given in table 29

Genotype	Value	Crossing-over regions if order is:	
		tryC—tryB—cysB	tryC—cysB—tryB
+	40	1-2	1-2
cys	8	1-2-3-4	1-3

28   GENETIC STUDIES WITH BACTERIA

TABLE 31

tryB-2 cysB-45 (×) tryB-4 experiments, January 28, 1955

Medium	Phenotype	I	II	III	Total	Av.
min	+	9	13	11	33	11
try	+	8	12	12	32	11
	try	1836	1874	1724	5434	1811
cys	+	14	14	13	41	14
	cys	23	38	31	92	31

TABLE 32

tryB-4 (×) tryB-2 cysB-45 experiments. I to III, January 28, 1955; IV and V, April 27, 1955.

Medium	Phenotype	I	II	III	IV	V	Total	Av.
min	+	29	28	30	66	62	215	43
cys	+	33	36	32	76	67	244	49
	cys	3	5	2	10	9	29	6

TABLE 33

tryD-10 cysB-12 (×) tryD-1 experiments. I, August 16, 1954; II, March 10, 1955.

Medium	Phenotype	I	II	Total	Av.
min	+	25	26	51	26
try	+	15	32	47	24
	try	1850	2090	3940	1970
cys	+	29	41	70	35
	cys	122	141	263	132

TABLE 34

tryD-1 (×) tryD-10 cysB-12 experiments, March 25, 1955

Medium	Phenotype	I	II	III	Total	Av.
min	+	359	362	389	1110	370
cys	+	455	425	425	1305	435
	cys	28	37	39	104	35

TRYPTOPHAN MUTANTS   **29**

TABLE 35

tryD-10 cysB-12 (×) tryD-9 experiments. I, August 16, 1954; II and III, March 17, 1955.

Medium	Phenotype	I	II	III	Total	Av.
min	+	7	5	8	20	7
try	+	5	2	5	12	4
	try	3520	2430	2680	8630	2877
cys	+	4	11	2	17	6
	cys	81	62	81	224	75

TABLE 36

tryD-9 (×) tryD-10 cysB-12 experiments, February 10, 1955

Medium	Phenotype	I	II	III	Total	Av.
min	+	138	140	155	433	144
cys	+	121	171	121	413	138
	cys	30	40	27	97	32

TABLE 37

tryD-11 cysB-18 (×) tryD-7 experiments, March 16, 1955

Medium	Phenotype	I	II	Total	Av.
min	+	4	2	6	3
try	+	1	3	4	2
	try	1260	1030	2290	1145
cys	+	3	4	7	4
	cys	19	11	30	15

TABLE 38

tryD-7 (×) tryD-11 cysB-18 experiments, March 8, 1955

Medium	Phenotype	I	II	III	IV	Total	Av.
min	+	30	41	28	27	126	32
cys	+	38	29	30	21	118	30
	cys	18	10	19	13	60	15

30   GENETIC STUDIES WITH BACTERIA

TABLE 39

tryD-11 cysB-18 (✕) tryD-6 experiments, March 14, 1955

Medium	Phenotype	I	II	Total	Av.
min	+	8	15	23	12
try	+	3	4	7	4
	try	1680	1740	3420	1710
cys	+	9	12	21	11
	cys	54	56	110	55

TABLE 40

tryD-6 (✕) tryD-11 cysB-18 experiments, March 25, 1955

Medium	Phenotype	I	II	III	Total	Av.
min	+	107	125	127	359	120
cys	+	105	93	113	311	104
	cys	339	326	365	1030	343

TABLE 41

tryD-11 cysB-18 (✕) tryA-8 cysB-12 experiments. I, January 20, 1955; II, February 16, 1955.

Medium	Phenotype	I	II	Total	Av.
min	+	3	0	3	2
anth	+	2	2	4	2
	anth	41	18	59	30
cys	+	3	5	8	4
	cys	790	284	1074	537
try	+	2	0	2	1
	anth	39	18	57	29
	try	114	56	170	85
cys & anth	+	4	0	4	2
	anth	48	8	56	28
	cys	746	238	984	492
	cys-anth	3914	1322	5236	2618

TRYPTOPHAN MUTANTS   31

TABLE 42

Analysis of the data given in table 41, assuming order of markers to be
(a) tryD-11—tryA-8—cysB-12—cysB-18, (b) tryD-11—tryA-8—cysB-18—cysB-12

Genotype	Value	Crossing-over regions if order is:	
		a	b
+	2	1-2-4-5	1-2-3-4
tryA-8	29	1-3-4-5	1-4
cysB-12 cysB-18 cysB-12 cysB-18	515	1-2-3-5 1-2 1-2-3-4	1-2-3-5 1-2 1-2-4-5
tryA-8 cysB-12 tryA-8 cysB-18 tryA-8 cysB-12 cysB-18	2618	1-5 1-3 1-4	1-5 1-3 1-3-4-5

TABLE 43

tryA-8 cysB-12 (×) tryD-11 cysB-18 experiments. I, October 26, 1954; II, January 17, 1955.

Medium	Phenotype	I	II	Total	Av.
min	+	1	0	1	0.5
anth	+	1	0	1	0.5
	anth	9	3	12	6
cys	+	2	0	2	1
	cys	320	104	424	212
try	+	1	0	1	0.5
	anth	10	4	14	7
	try	3	2	5	2.5

TABLE 44

Analysis of the data given in table 43, assuming order of markers to be
(a) tryD-11—tryA-8—cysB-12—cysB-18, (b) tryD-11—tryA-8—cysB-18—cysB-12

Genotype	Value	Crossing-over regions if order is:	
		a	b
+......................................	0.6	2-4	2-3-4-5
tryA-8	6.5	3-4	4-5
cysB-12 cysB-18 }............ cysB-12 cysB-18	212	2-3 2-5 2-3-4-5	2-3 2-5 2-4
tryD-11 tryD-11 tryA-8 }...............	2.5	1-4 1-2-3-4	1-3-4-5 1-2-4-5

TABLE 45

tryD-11 cysB-18 (×) tryB-2 cysB-45 experiments. I, December 8, 1954; II, January 20, 1955.

Medium	Phenotype	I	II	Total	Av.
min	+	0	0	0	0
ind	+	0	1	1	0.5
	ind	1	37	38	19
cys	+	0	1	1	0.5
	ind	15	7	22	11
	cys	99	82	181	91.
	cys-ind	735	484	1219	610
try................	+	0	0	0	0
	ind	5	15	20	10
	try	22	24	46	23
cys & ind	+	3	0	3	1.5
	ind	8	21	29	15
	cys	75	75	150	75
	cys-ind	551	947	1498	749

TABLE 46

Analysis of the data given in table 45, assuming order of markers to be
(a) tryD-11—tryB-2—cysB-18—cysB-45, (b) tryD-11—tryB-2—cysB-45—cysB-18

Genotype	Value	Crossing-over regions if order is:	
		a	b
+	0.5	1-2-3-4	1-2-4-5
tryB-2	14	1-4	1-3-4-5
cysB-18 cysB-45 cysB-18 cysB-45	83	1-2 1-2-3-5 1-2-4-5	1-2 1-2-3-5 1-2-3-4
tryD-11 tryD-11 tryB-2	23	3-4 2-4	4-5 2-3-4-5
tryB-2 cysB-18 tryB-2 cysB-45 tryB-2 cysB-18 cysB-45	680	1-3 1-5 1-3-4-5	1-3 1-5 1-4

TABLE 47

tryB-2 cysB-45 (×) tryD-11 cysB-18 experiments, March 8, 1955

Medium	Phenotype	No.
min	+	1
ind	+	1
	ind	36
cys	+	1
	cys	220
try	+	2
	ind	40
	try	3

TABLE 48

Analysis of the data given in table 47, assuming order of markers to be
(a) tryD-11—tryB-2—cysB-18—cysB-45, (b) tryD-11—tryB-2—cysB-45—cysB-18

Genotype	Value	Crossing-over regions if order is:	
		a	b
+	1.25	2-3-4-5	2-4
tryB-2	38	4-5	3-4
cysB-18 cysB-45 cysB-18 cysB-45	220	2-5 2-3 2-4	2-5 2-3 2-3-4-5
tryD-11 tryD-11 tryB-2	3	1-3-4-5 1-2-4-5	1-4 1-2-3-4

First Glimpses of DNA Repair

Witkin E.M. 1956. **Time, Temperature, and Protein Synthesis: A Study of Ultraviolet-induced Mutation in Bacteria.** (Reprinted from *Cold Spring Harbor Symp. Quant. Biol.* **21:** 123–140.)

Evelyn Witkin with William Hayes (left) and Norton Zinder (right) in 1953 at the "Viruses" Symposium.

For almost 50 years, the DNA double helix has been the foundation of our thinking about genes and their behavior, so much so that a world without the intellectual framework of the chemical nature of the gene has become inconceivable. The difficulties of thinking about genes in such a world are evident in Richard Goldschmidt's writings (Goldschmidt 1938), most notably (and notoriously) in his contribution to the 1951 Cold Spring Harbor Symposium. Here, he criticized the "...extrapolation from the mutant action to the existence of the original gene" (Goldschmidt 1951) and based his argument against the particulate concept of the gene on the existence of "position effects"—instances where a normal gene behaves like a mutant when it is close to a chromosomal rearrangement.

Mutations, in the absence of knowledge of the physical nature of the gene, were a black box, observable only through their effect on phenotype. As H.J. Muller put it, "...the path of the chemical action in the production of a mutation is still very much a mystery...," although "Obviously the cause of mutation lies buried in the vicissitudes of the 'molecular chaos'" (Muller 1950). One of the contributors to the mystery of mutations was Demerec. In 1946, he had exposed *E. coli* to ultraviolet radiation and measured the appearance of bacteriophage-resistant mutants at various times after exposure (Demerec 1946). Remarkably, mutants continued to appear long after the cells had been irradiated. That is, the more divisions cells underwent after irradiation, the higher the mutation yield. This "delayed effect" did not fit well with the idea that radiation induced mutations only at the time of exposure; how then could mutations continue to appear when the source of radiation had gone?

Demerec used the *E. coli B/r* strain in these experiments, a radiation-resistant mutant that had been isolated in the summer of 1944 by Evelyn Witkin (Witkin 1947). Witkin, then a graduate student with Dobzhansky, is likely one of the very few students to have been inspired to become a geneticist by the work of Trofim Lysenko (Soyfer 1994), although she rapidly concluded he was "...either a charlatan or an ignoramus, or quite possibly both" (Witkin 1994). She came back to Cold Spring Harbor in 1945 to complete her Ph.D. thesis on the nature of the *B/r* mutation and then returned to mutagenic effects of UV irradiation (Witkin 1947). Witkin and others showed that various post-irradiation treatments (temperature, for example) affected the delayed appearance of mutations, and that these treatments were effective if given prior to the cells' first post-irradiation division. In the paper reprinted here, Witkin made a definitive examination of the six hypotheses proposed to account for the delayed appearance of mutations.

Witkin adopted a technique for detecting mutants that had been developed by Demerec and Cahn (1953). They had treated bacterial auxotrophs—strains that required specific growth supplements, such as tryptophan, for growth—with UV light; controlled the number of cell divisions the cells underwent; and looked for cells that no longer required the supplement. Demerec and Cahn showed that the number of mutants was a function of the number of divisions undergone by the cells. Witkin was particularly interested in the timing of the appearance of mutants, and she examined this in post-irradiation cultures grown in media of varying degrees of richness. Her findings led, as she wrote, to an "impasse," for they pointed inescapably to the fact that all six hypotheses were wrong.

Witkin realized, however, that although the delayed appearance of mutants was correlated with the number of post-irradiation cell divisions, the delay might be a consequence of the richness of the culture medium, rather than of the cell divisions. She set out to examine in more detail how delayed appearance was related to the growth conditions of the cells prior to the first post-irradiation division. Her results showed that exposure of cells to a rich medium for only one to one and a half hours following irradiation led to high levels of mutants, no matter how the cultures were treated subsequently. Switching treatment after this period was without effect; it was the medium present in the first hour that determined the final yield of mutants. Witkin found that adding a pool of amino acids to the minimal medium was as effective as using complete medium. These findings suggested that protein synthesis was important, and this was confirmed by using chloramphenicol, an inhibitor of protein synthesis. When this was added to the cells during the first hour of post-irradiation culture, the yield of mutants fell dramatically. Finally, high and low temperatures—both expected to inhibit protein synthesis—had the same effect of lowering the yield of mutants, a phenomenon that came to be called mutation frequency decline (MFD).

Witkin's interpretation of MFD was, with hindsight, straightforward. She distinguished between processes involved in a "sensitive" period, the end of which was probably marked by the completion of DNA replication, and a second process, "X," changes in which affected the yield of mutants. Although she discussed three possibilities, Witkin appeared to favor the idea that "...X is the process of repair of genetic damage." She continued "... if repair is accomplished before the end of the sensitive period (DNA duplication?) the damaged cells survive, with a high probability of surviving as a mutant. If repair is not accomplished within the critical period, the damage is lethal."

The next steps in the elucidation of "X" as a DNA repair process came with the work of Setlow and Carrier (1964) on excision repair. It had been shown earlier that UV exposure led to the linking of two thymidines adjacent to each other in the DNA strand (Wacker et al. 1962). If left, these thymine dimers interfered with DNA transcription; excision repair cut out the dimers and replaced the thymidines. The kinetics of excision repair were similar to those for MFD and, even more significantly, mutant cells deficient for excision repair did not show MFD, and mutants lacking the MFD response removed thymine dimers very slowly. A little later, Witkin (1966) demonstrated that her process "X" was excision repair of pyrimidine dimers in genes coding two particular transfer RNA molecules.

Full understanding of the nature of MFD took many years (Witkin 1994; Friedberg 1997). A key finding was that the process is strand specific and that it affects pyrimidine dimers on the DNA strand that is used to make messenger RNA—the template DNA strand (Bockrath and Palmer 1977). Furthermore, excision repair occurs much more rapidly on a DNA strand that is actively being transcribed to make RNA than elsewhere (Bohr et al. 1985; Mellon et al. 1987). This makes good sense—this damage will immediately lead to defective proteins, and clearing it out has to be done as quickly as possible. This preferential repair of the template DNA strand comes about through the action of a protein-transcription-repair coupling factor (TRCF), which appears to recognize places where the transcription machinery has been stopped by DNA damage (Selby and Sancar 1991). And, in a nice closing of the circle, Selby et al. (1991) showed that Witkin's 1966 *mfd⁻* mutant lacking the MFD response also lacked TRCF; thus, TRCF is coded for by the *mfd* gene.

Evelyn Witkin was born in New York City, March 9, 1921. She was awarded her B.A. from New York University, majoring in zoology, in 1941 and began graduate studies with Theodosius Dobzhansky at Columbia University. Her interests changed from Drosophila *genetics to bacterial genetics and she spent the summer of 1944 at Cold Spring Harbor, where she isolated a radiation-resistant mutant of* E. coli. *Witkin was awarded her Ph.D. in 1947 and remained at the Carnegie Institution Department of Genetics at Cold Spring Harbor until 1955. She then moved to the State University of New York's Downstate Medical Center in Brooklyn. In 1971 she was appointed Professor of Biological Sciences at Douglass College, Rutgers University, where she was named Barbara McClintock Professor of Genetics in 1979. Witkin moved to the Waksman Institute at Rutgers University in 1983, becoming Barbara McClintock Professor Emerita in 1991. Among her many awards are membership in the National Academy of Sciences (1977); Fellow of the American Association for the Advancement of Science (1980); American Women of Science Award for Outstanding Research; and Fellow, American Academy of Microbiology.*

Bockrath R.C. and Palmer J.E. 1977. Differential repair of premutational UV-lesions at tRNA genes in *E. coli*. *Mol. Gen. Genet.* **156:** 133–140.

Bohr V.A., Smith C.A., Okumoto D.S., and Hanawalt P.C. 1985. DNA repair in an active gene: Removal of pyrimidine dimers from the DHFR gene of CHO cells is much more efficient than in the genome overall. *Cell* **40:** 359–369.

Demerec M. 1946. Induced mutations and possible mechanisms of the transmission of heredity in *Escherchia coli*. *Proc. Natl. Acad. Sci.* **32:** 36–46.

Demerec M. and Cahn E. 1953. Studies of mutability in nutritionally deficient strains of *Escherchia coli*. I. Genetic analysis of five auxotrophic strains. *J. Bacteriol.* **65:** 27–36.

Friedberg E.C. 1997. The emergence of excision repair. In *Correcting the blueprint of life: An historical account of the discovery of DNA repair mechanisms*, pp. 63–101. Cold Spring Harbor Laboratory Press, Cold Spring Harbor, New York.

Goldschmidt R.B. 1938. *Physiological genetics*. McGraw-Hill, New York.

—————. 1951. Chromosomes and genes. *Cold Spring Harbor Symp. Quant. Biol.* **16:** 1–11, p. 1.

Mellon I., Spivak G., and Hanawalt P.C. 1987. Selective removal of transcription-blocking DNA damage from the transcribed strand of the mammalian DHFR gene. *Cell* **51:** 241–249.

Muller H.J. 1950. The development of the gene theory. In *Genetics in the 20th century* (ed. L.C. Dunn), pp. 77–79. Macmillan, New York.

Selby C.P. and Sancar A. 1991. Gene- and strand-specific repair in vitro: Partial purification of a transcription-repair coupling factor. *Proc. Natl. Acad. Sci.* **88:** 8232–8236.

Selby C.P., Witkin E.M., and Sancar A. 1991. *Escherichia coli mfd* mutant deficient in "mutation frequency decline" lacks strand-specific repair: In vitro complementation with purified coupling factor. *Proc. Natl. Acad. Sci.* **88:** 11574–11578.

Setlow R.B. and Carrier W.L. 1964. The disappearance of thymine dimers from DNA: An error-correcting phenomenon. *Proc. Natl. Acad. Sci.* **51:** 226–231.

Soyfer V.N. 1994. *Lysenko and the tragedy of Soviet science*. Rutgers University Press, New Brunswick, New Jersey.

Wacker A., Dellweg H., and Jacherts D. 1962. Thymine dimerization and survival of bacteria. *J. Mol. Biol.* **4:** 410–412.

Witkin E.M. 1947. Genetics of resistance to radiation in *Escherichia coli*. *Genetics* **32:** 221–248.

—————. 1956. Time, temperature, and protein synthesis: A study of ultraviolet-induced mutation in bacteria. *Cold Spring Harbor Symp. Quant. Biol.* **21:** 123–140.

—————. 1966. Radiation-induced mutations and their repair. *Science* **152:** 1345–1353.

—————. 1994. Mutation frequency decline revisited. *BioEssays* **16:** 437–444.

Time, Temperature, and Protein Synthesis: A Study of Ultraviolet-Induced Mutation in Bacteria[1]

Evelyn M. Witkin

Department of Medicine, State University of New York College of Medicine at New York City, Brooklyn, New York[2]

About ten years ago (Demerec, 1946), it was reported that the yield of bacteriophage-resistant mutants induced by ultraviolet light in *Escherichia coli* increases as the irradiated population is allowed to multiply, until ten or twelve divisions are accomplished. Since that time, it has been observed repeatedly that maximal yields of many kinds of induced bacterial mutants, including auxotrophs (Davis, 1948), prototrophs (Davis, 1950; Demerec and Cahn, 1953) and streptomycin-independent variants (Labrum, 1953), are obtained only under conditions permitting more or less extensive periods of cell division after exposure to ultraviolet light. This "delayed appearance" of induced bacterial mutants has become one of the hardy perennials among the unsolved problems of microbial genetics.

Six possible mechanisms have been postulated to account for this phenomenon. Delayed mutation, phenotypic lag and segregation of multiple nuclei or multiple chromosomal strands were proposed by Demerec (1946). Newcombe and Scott (1949) added to the list the possible effect of irregularity of the growth pattern of irradiated cells, and Ryan (1954) suggested that delayed onset of division of newly induced mutant cells could account for the facts, in some cases. It has also been proposed that, in experiments with bacteriophage-resistant mutants, a methodological artefact may be responsible for exaggeration of the scope of delay (Newcombe, 1953). The complexity of the problem is exemplified by the multiplicity of the hypotheses offered to explain it, by the fact that they are by no means mutually exclusive, and by the possibility that the patterns of delay observed for different mutations may have their basis in different mechanisms. The hypotheses of segregation and irregularity of growth patterns have been satisfactorily eliminated, at least as primary factors, (Demerec and Cahn, 1953), but there has been conflicting evidence as to the part played by phenotypic lag (Newcombe and Scott, 1949; Ryan, 1955), and no decisive evidence in support of the possibility that induced mutations are often delayed in their occurrence for many generations (see, however, Witkin, 1951; Newcombe, 1953; Kaplan, 1952).

In the decade that has passed since the hypothesis of delayed mutation was proposed to account for the "delayed appearance" of induced mutants, it has become increasingly certain that the time interval between the absorption of radiant energy and the production of stable genetic changes can no longer be regarded as infinitesimal. The evidence that ultraviolet-induced mutations are irreversibly established only after an appreciable delay has come primarily from studies of postirradiation effects on the frequency of induced mutants, such as those of visible light (Kelner, 1949; Novick and Szilard, 1949; Newcombe and Whitehead, 1950), temperature (Witkin, 1953), and metabolites and metabolic inhibitors (Wainwright and Nevill, 1955a and b). Recent work (Anderson and Billen, 1955; Hollaender and Stapleton, 1956) suggests that mutations induced by ionizing radiations may also be susceptible to the influence of posttreatments. Information as to the duration of the period of susceptibility to postirradiation influences, limited as it is at the present time, does not, however, support the view that genetic instability persists for a number of cell generations after irradation. The period of susceptibility to postirradiation temperature (Witkin, 1953; Berrie, 1953) for a variety of induced mutations in *E. coli* corresponds to a fraction of the time required for the first cell division after exposure to ultraviolet light. Newcombe (1955) has shown a relation between the period of susceptibility to photoreactivation in *Streptomyces* and the timing of the first nuclear division after irradiation. Thus, the evidence derived from studies of posttreatments suggests that the metabolic pathway leading from the absorption of energy to the establishment of induced mutations is associated with the first postirradiation cell division.

The experiments to be described in this report were begun in an effort to determine decisively the basis of the "delayed appearance" of ultraviolet-induced prototrophs, with the particular goal of obtaining information about the timing of the induced mutation process.

Materials and Methods

Bacterial strains. Most of the experiments were done with a tryptophane-requiring substrain (*try3*) of strain LT2 of *Salmonella typhimurium*, obtained from Dr. M. Demerec. Other auxotrophic substrains, each requiring a single growth factor, and

[1] This study was conducted in part under a grant from the National Science Foundation.
[2] Part of the work described in this report was done at the Carnegie Institution of Washington, Cold Spring Harbor, N. Y., while the author was a Staff Member of the Department of Genetics.

124   EVELYN M. WITKIN

derived either from *S. typhimurium* LT2 or from *E. coli* B/r, were also used. Spontaneous rates of mutations to prototrophy were 10^{-8} or less, and all data presented in tables are corrected for spontaneous mutants.

Transducing Bacteriophage. In the experiments involving transduction, the temperate phage PLT22, obtained from Dr. N. Zinder, was used.

Culture Media. Difco nutrient broth and Difco nutrient agar were used as complete media. Synthetic medium "A" (Davis, 1950), consisting of inorganic salts and glucose, was used as minimal medium for strains derived from *S. typhimurium*, and was solidified with 1.5 per cent agar. The minimal medium used for strains derived from *E. coli* was medium "E," developed at Yale University by Vogel and Bonner (personal communication). Partial enrichment of minimal medium consisted of the addition of small amounts of nutrient broth, or small amounts of the specific growth factor required by the auxotrophic strain in use. The detailed description of the semi-enriched media will be given in connection with specific experiments.

Irradiation Procedure. Cultures to be irradiated were grown for 18 to 24 hours at 37°C, with aeration, in nutrient broth, unless otherwise specified. Ten-ml lots of culture were centrifuged, washed in saline or phosphate buffer, and resuspended in the same volume of saline or buffer for irradiation. The titer of the washed suspensions was usually between 1 and 3×10^9 cells per ml. The suspensions were irradiated in 7-ml portions in uncovered Petri dishes, with constant mechanical agitation during the exposure. The source of ultraviolet light was a G. E. Germicidal lamp. The dose used for *Salmonella* was 300 ergs/mm², and for *E. coli* B/r was 600 ergs/mm², giving survivals of between 10 and 20 per cent. Yellow light was used exclusively during and following irradiation to avoid uncontrolled photoreactivation.

Chloramphenicol was obtained from Charles Pfizer and Co.

EXPERIMENTAL RESULTS

PART I. THE "DELAYED APPEARANCE" OF
INDUCED AUXOTROPHS

Induction delay compared with transduction delay. When strain *try3* of *S. typhimurium* is irradiated with ultraviolet light, induced prototrophs are obtained. The same genetic change can be brought about by infecting *try3* cells with transducing bacteriophage initially grown on an established stock of the ultraviolet-induced prototroph (*try3*⁺ᵘᵛ). In these experiments, a comparison was made between the pattern of "delayed appearance" of the induced prototrophs (induction delay) and that of the prototrophs arising by transduction (transduction delay).

Since transduction is a parasexual process in which genetic properties are transferred from one bacterial strain to another (Zinder and Lederberg,

1952), the transduction from *try3* to *try3*⁺ᵘᵛ is essentially a recombination, and involves no mutagenic action. The same genetic change induced by ultraviolet light, however, is brought about by whatever events take place when one genetic specificity is altered by mutation to another. Thus, a comparison of the pattern of transduction delay with that of induction delay could provide critical evidence concerning the nature of the delay mechanism. Specifically, if the delayed appearance of induced mutants is primarily due to delayed mutation, it would be expected that prototrophs arising by transduction should exhibit no comparable delay. On the other hand, if phenotypic lag is the principal mechanism of induction delay, a similar delay should be found for transduction, since in both cases a physiological adjustment of the cell from tryptophane-dependence to independence must be effected.

In these experiments, a washed suspension of *try3* was irradiated, and the irradiated suspension divided into two portions. To the first portion (A), transducing phage (grown on a previously isolated *try3*⁺ᵘᵛ stock) was added; to the second portion (B), the same amount of nontransducing phage (the same phage strain, grown on *try3* itself, and thus incapable of transducing *try3* to prototrophy) was added. The two suspensions were thus identical in their previous growth history, in their exposure to ultraviolet light and in their state of infection with phage (except for the difference in the previous host of the infecting phage). Portion A is capable of giving rise to induced prototrophs and to prototrophs arising by transduction; portion B can give rise only to induced prototrophs. The pattern of delayed appearance of the prototrophs arising from each of the suspensions was measured by the method of increasing partial enrichment (Demerec and Cahn, 1953). Immediately after the addition of phage, the suspensions were placed in a 37°C water bath for ten minutes to allow adsorption to take place, and 0.1-ml aliquots were then spread on the surface of a series of minimal agar plates, some unsupplemented and others partially enriched with nutrient broth, the concentrations of nutrient broth in the plates ranging from 0.05 per cent to 2.5 per cent. Six platings from each suspension were made on each of the enrichment levels, and all of these plates were incubated for either 24 or 48 hours. The irradiated *try3* cells were able to undergo varying amounts of residual division on these plates, the amount of division being strictly dependent upon the enrichment level. In each experiment, the amount of residual division was determined by washing the cells from the surfaces of two plates of each enrichment level with 10 ml of saline, and assaying the wash fluid. These plate washes were performed after 24 hours of incubation, by which time the residual growth was completed, but before the colonies of prototrophs were large enough to be visible, or to contribute significantly to the total population on the plates. As a further precaution, the assays were plated on both minimal and complete media, so that differential counts could be used to eliminate the contribution of the prototrophs. After 48 hours of incubation, the colonies on the other four plates in each set were counted, and these data were used to construct "expression curves," relating the number of residual divisions undergone by the irradiated population (as determined by the level of enrichment) to the number of induced prototrophs (for suspension B), or to the number of prototrophs arising by transduction (for suspension A). Although suspension A gave rise to prototrophs by induction as well as by transduction, the frequency of the latter was much higher than that of the former under the conditions of these experiments, so that the induced mutants contributed very little to the plate counts in this series, and, in any case, their number was determined from the data in series B and subtracted from the A counts. Table 1 gives the results of a typical experiment, and the curves derived from a number of such experiments are shown in Figure 1.

It will be noted that the prototrophs arising by transduction are fully "expressed" on plates supporting only about two residual divisions, while the induced prototrophs continue to increase in numbers until an enrichment level supporting about six divisions is used. When corrections are made for the effects of irregularity of onset of division, as measured by the "checkerboard" respreading technique (Witkin and Thomas, 1955), it is found that the full yield of transductions is actually obtained after only one division, since these measurements show that only half the cells have actually divided when the population as a whole has doubled in number. A similar correction has very little effect on the pattern of delay shown by the induced mutants. The patterns of transduction delay and induction delay were compared for prototrophs derived from two other auxotrophs (adenineless and lysineless), and it was found in both cases that the maximum yield of transductions was obtained in one division, while the induced prototrophs continued to increase in numbers until enrichment levels supporting six residual divisions and at least ten residual divisions, for the two strains respectively, were used.

These results support the following conclusions concerning the mechanism of induction delay: 1) the "delayed appearance" of induced prototrophs cannot be primarily determined by phenotypic lag, which does not exceed one division; 2) irregularity of the growth pattern of irradiated cells cannot be primarily responsible for induction delay (nor can any other nongenetic effect of ultraviolet light), since the same irradiated suspension was used under comparable conditions of growth for the determination of transduction and induction delay patterns; 3) nuclear segregation can

TABLE 1. THE PATTERN OF "DELAYED APPEARANCE" OF PROTOTROPHS ARISING BY TRANSDUCTION COMPARED WITH THAT OF ULTRAVIOLET-INDUCED PROTOTROPHS

Culture used: 24-hour aerated broth culture of *try3*, washed and resuspended in buffer.

Ultraviolet treatment: 7 ml. of washed suspension irradiated with 300 ergs/mm² (survival 17%); irradiated suspension divided into two portions, A and B.

Bacteriophage: Transducing phage (previous host *try3⁺ᵘᵛ*) added to A; nontransducing phage (previous host *try3*) added to B; multiplicity of infection, ca. 5.

Number of viable cells plated: 4×10^7 per plate.

Enrichment of Minimal Agar	Number of Residual Div.	Number of Prototrophs per Plate	
		A, trans. + ind.	B, induced
% broth			
0.05	0.4	39, 48	0, 2
0.25	1.8	196, 226	4, 4
0.75	4.1	258, 267	17, 25
1.25	5.2	290, 276	72, 64
2.5	6.3	312, 288	79, 81

also be eliminated as a significant factor in induction delay, if the assumption is made that the probability of occurrence of either transduction or induced mutation in more than one nucleus of a multinucleate cell is negligible. The results are consistent with only two of the hypotheses that have been proposed, both attributing unique properties to the newly induced or potential mutant cell: delayed mutation, and delayed onset of division of cells in which a mutation has been actually or potentially produced. The next series of experiments was designed to test these possibilities.

The Clonal Increase of Newly Induced Prototrophs Compared with the Growth Rate of Previously Isolated Induced Prototrophs. Newcombe and Scott (1949) recognized that the possibility of delayed mutation as the major basis of induction delay carried with it the implication that mature

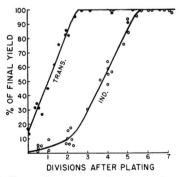

FIGURE 1. Expression curves of prototrophs obtained from *try3* by transduction and by induction with ultraviolet light.

126   EVELYN M. WITKIN

colonies containing induced mutants should also contain the descendents of the unmutated sister cells of the original mutant. In other words, induced mutants should rarely be found as pure clones of the mutant type when the colonies in which they occur are detected without selection. Because of the difficulty of detecting most kinds of induced mutants in the absence of selective techniques, there has been only limited work done to analyze the composition of clones in which they arise. Ten such clones containing induced bacteriophage-resistant mutants in *E. coli* have been analyzed (Newcombe, 1953), six of which proved to be pure clones, and four were mixtures containing both sensitive and resistant cells. Since no special precautions were taken to control nuclear multiplicity, it is probable that some or all of the mixed clones arose simply through nuclear segregation. This seems all the more likely in view of the work with induced lactose-negative mutants in *E. coli* (Witkin, 1951; Newcombe, 1953), in which the composition of clones can be read directly from the colony color on indicator medium. In this case, when special methods were used to insure that the colonies containing induced lactose-negative mutants were derived from single nuclei, they were found to be pure clones of the mutant type much more frequently than mixed clones containing both lactose-positive and lactose-negative cells. These data, limited as they are, do not support the hypothesis of persistent genetic instability. They suggest, rather, that most of the induced mutations are established before the completion of the first postirradiation division. Since the frequency of induced prototrophs obtained from *try3* is relatively low, their detection without the use of a selective medium

did not seem practical. An indirect approach, using the respreading technique (Newcombe, 1949) was taken.

Three series of platings were made on the surface of semi-enriched (2.5 per cent broth) minimal medium: (1) the first series of plates was seeded with about 1×10^7 irradiated *try3* cells (which, if incubated without disturbance, would give rise to about 30 induced prototroph colonies); (2) the second series was seeded with about 30 *try3^{+uv}* cells (from a previously isolated induced prototroph) which had been irradiated with the same dose, and (3) the third series was seeded with a mixture of about 30 *try3^{+uv}* cells irradiated with the same dose plus about 10^7 ultraviolet-killed *try3* cells which had been more heavily irradiated than the others, and could not give rise to induced prototrophs. All three sets of plates were incubated, and at periodic intervals, four plates from each set were withdrawn from the incubator and respread with 0.1 ml of saline to redistribute the individual members of growing microcolonies. Since the medium is selective for prototrophs, a reading of these plates after 48 hours of reincubation will show, for the first series, the rate of increase of the average number of mutant cells per clone containing any newly induced mutants. The second and third series, for purposes of comparison, will show the growth rate of the established prototroph on the same medium, with or without a heavy background of ultraviolet-killed cells. Table 2 shows the results. It is obvious, first of all, that the growth of the established mutant is not affected by the presence of a heavy background of ultraviolet-killed cells, and either control series can be used as a standard for the growth rate of the established mutant. It will be seen also that the rate of increase of the average number of newly induced mutant cells per clone in which mutation has occurred lags behind the growth rate of the established mutant significantly, but to an extent which is not inconsistent either with the hypothesis that most of the induced mutations occur after a delay of several generations, or with the hypothesis that the potential mutant cell is more than slightly delayed in its onset of multiplication. Nuclear segregation could account for a discrepancy of the magnitude observed (Ryan, Fried and Schwartz, 1954). Since these experiments are based upon the average rate of increase of induced mutants, they do not rule out the possibility that a small fraction of mutations may be delayed, nor that a small fraction of potential mutants may start to multiply late, but there can be no doubt that neither of these mechanisms is characteristic of the majority of the induced mutations, nor can either of them account for the "delayed appearance" of the induced mutants as characterized by the expression curve.

The Dissociation of Residual Division from Induced Mutation Frequency. The results of the foregoing experiments led quite inescapably to the

TABLE 2. THE GROWTH RATE OF THE ESTABLISHED INDUCED PROTOTROPH (*try3^{+uv}*) COMPARED WITH THE CLONAL INCREASE OF NEWLY INDUCED PROTOTROPHS

A. Plates seeded with irradiated *try3* (1×10^7 viable cells).

B. Plates seeded with irradiated *try3^{+uv}* (30 viable cells).

C. Plates seeded with irradiated *try3^{+uv}* (30 viable cells) plus 1×10^8 ultraviolet-killed *try3* cells.

Time of Incubation before Respreading	Colony Counts			Increase Factor		
	A	B	C	A	B	C
0	34	37	39	—	—	—
	42	28	27			
	42	28	28			
4 hr.	57	69	66	×1.5	×2.4	×2.2
	61	72	79			
	63	81	60			
5½ hr.	259	229	358	×7.7	×9.9	×9.7
	319	361	324			
	318	332	224			

conclusion that none of the hypotheses advanced to explain the "delayed appearance" of induced mutants could account for the expression curve of the prototrophs obtained from *try3*. There did not seem to be any reasonable possibilities other than those already eliminated, nor did there appear to be any basis for suspecting a methodological artefact in the experimental procedure used to obtain data for expression curves, as has been suggested in the case of bacteriophage-resistant mutants (Newcombe, 1953). A route of escape from this impasse, however, became evident in the course of reappraising the assumptions made in handling the data from such experiments. There is no doubt about the fact that the number of residual divisions undergone by the irradiated population increases as a function of the amount of nutrient broth added to the minimal agar. There is no doubt, also, that the number of induced mutations increases as the enrichment level is raised (up to a certain point). It does not necessarily follow, however, that the number of induced mutations must be a function of the number of residual divisions. It is quite conceivable, *a priori*, that increased residual division and increased mutation frequency could be parallel, but independent, consequences of increased enrichment. The assumption of a causal relation between induced mutation frequency and residual division is implicit in the way expression curves have been plotted, and it is this logically vulnerable assumption that was the next object of experimental analysis.

The experiments to be described were based upon the likelihood that induced mutations are established irreversibly before the completion of the first postirradiation division (as suggested by the duration of susceptibility to posttreatments, by the prevalence of pure mutant clones in the cases analyzed, and by the results of the respreading experiments described above). If this assumption is correct, the expression curve for the induced prototrophs obtained from *try3* could be interpreted as demonstrating that high yields of induced mutations are favored by the presence of high levels of enrichment *during the first division*, and conversely, that low yields of mutations are obtained when the enrichment level of the medium upon which the *first postirradiation division* occurs is low. Changes in the enrichment level after the completion of the first division should affect only the final amount of residual growth, but should have little or no influence on the mutation frequency.

In the first series of experiments, irradiated *try3* cells were incubated for three hours (corresponding to the time required for an average of one division on nutrient agar) on minimal medium without added enrichment, or with a relatively low level of enrichment. The enrichment level was then raised by lifting the entire disc of agar out of the Petri dish with a sterile spatula (bacteria

TABLE 3. "TRANSPLANTATION" OF IRRADIATED *try3* FROM LOW TO HIGH LEVELS OF ENRICHMENT

4.1 × 10⁷ viable cells plated on low enrichment level, incubated three hours at 37°C, transplanted to high enrichment level, incubated 48 hours.

Initial Plating on	"Transplanted" to	Final No. Resid. Div.	No. Induced Prototrophs	No. Viable Cells at Time of Transplantation*
Unsupplemented "A"........	— (control)	0	0, 0	—
.05% n.b.....	— (control)	1.1	10, 12	—
2.5% n.b.....	— (control)	6.3	116, 108	—
Unsupplemented "A"........	5% n.b.	6.2	0, 0	3.9 × 10⁷
.05% n.b.....	5% n.b.	5.9	14, 11	4.3 × 10⁷

* Determined by plate-wash assays.

up) and placing it upon the surface of another minimal agar layer containing a relatively high amount of nutrient broth (5 per cent). Within one hour, the enrichment level is equilibrated by diffusion at the level of about 2.5 per cent, the maximum enrichment used in the previous experiments. The results of such a "transplantation" experiment are shown in Table 3. It will be seen that the number of prototrophs obtained on the transplant plates (incubated for three hours on unsupplemented "A" medium, or on "A" supplemented with 0.05 per cent broth, and then transplanted to "A" medium containing 5 per cent broth) is the same as the number of prototrophs obtained on the untransplanted low-enrichment controls. The residual division on the transplant plates, however, is the same as that on the high-enrichment controls having the maximal yield of induced prototrophs. Figure 2 shows the "expression" curve obtained by incubating irradiated *try3* on the standard series of enrichment levels for three hours, and then transplanting to agar enriched with 5 per cent broth. The curve obtained is indistinguishable from that obtained when the cells are incubated for the entire 48-hour incubation period on the initial enrichment levels, with no transplantation. The residual divisions ultimately achieved on the transplant plates was measured by platewashes after 24 hours of incubation, and was found to be uniform throughout, and to correspond to the level expected on the basis of the equilibrated enrichment level of about 2.5 per cent nutrient broth. It seems indisputable, therefore, that the number of induced mutations is determined by the enrichment level of the medium during the first three hours of postirradiation incubation (or some fraction of this period), and

128   EVELYN M. WITKIN

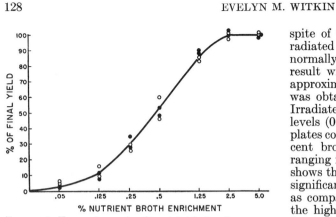

FIGURE 2. Expression of ultraviolet-induced proto-trophs obtained from *try3* on minimal medium enriched with different amounts of nutrient broth. Open circles, irradiated cells plated on enrichment levels indicated, and incubated 2 days. Closed circles, irradiated cells plated on enrichment levels indicated, incubated 3 hours, then "transplanted" to plates enriched with 5% nutrient broth and incubated two days.

is not critically related to the amount of residual division.

In the next series of experiments, an effort was made to determine more precisely the duration of the sensitive period during which the amount of nutrient broth added to the medium can still influence the mutation frequency. Ideally, this sort of experiment demands a method of instantaneous transfer of the irradiated cells from one enrichment level to another, since the "transplantation" method involves a period of about an hour before equilibration is completed. The use of ultrafine filter membranes (Millipore filters) was attempted, since bacteria growing on these membranes can be transferred quickly from the surface of one plate to another. This technique was found to be useless, however, since induced prototrophs were never obtained on the filter membranes, in

spite of the fact that residual growth of the irradiated *try3* cells took place, apparently quite normally. The probable reason for this unexpected result will be considered in the Discussion. An approximation of the timing of the sensitive period was obtained using the transplantation method. Irradiated *try3* cells were plated on low enrichment levels (0.05 per cent broth), and transplanted to plates containing minimal agar enriched with 5 per cent broth after various periods of incubation ranging from zero minutes to two hours. Table 4 shows the results. Immediate transfer results in a significant depression of the mutation frequency as compared with the control plated initially on the highly enriched medium, and the maximum depression (identical with the control plated initially on the low enrichment level) is obtained on plates transplanted after 30 minutes. Since equilibration is complete after one hour, the duration of the sensitive period can be set approximately at between one and one and a half hours of incubation.

In the experiments just described, the change in the enrichment level was that from low to high enrichment, and it was found that one to one and a half hours of incubation is sufficient to insure a low mutation yield, regardless of the subsequent increase in enrichment. Reciprocal experiments, in which the initial incubation is on high enrichment levels, followed by transfer to unsupplemented minimal medium, should result in high mutation yields, accompanied by low levels of final residual division. Obviously, the transplantation method could not be used here, since the equilibrated enrichment level after transplantation would differ too little from the initial high level. In these experiments, therefore, the changes in level of enrichment were made by washing the bacteria from the surface of highly enriched plates after one hour of incubation, concentrating and washing the cells by centrifugation, and replating on unsupplemented minimal medium. Low-to-high

TABLE 4. THE DURATION OF THE PERIOD OF SUSCEPTIBILITY OF INDUCED MUTATION FREQUENCY TO CHANGES IN LEVEL OF ENRICHMENT

3.4×10^7 viable irradiated *try3* cells plated on low enrichment level (0.05% n.b.), incubated for various lengths of time, then "transplanted" to high enrichment levels (5% n.b.). Equilibration of the enrichment level at 2.5% n.b. is complete in about one hour.

Time on Low Enrichment Level	No. Viable Cells per Plate at Time of Trans.	No. Induced Prototrophs per Plate (48 hr. inc.)	Final No. Residl Div.
0 (plated directly on 2.5% n.b.)........	—	126, 119	6.2
0 (immed. trans.)....	3.3×10^7	32, 27	5.9
30 min..............	3.6×10^7	0, 2	6.6
1 hr................	3.1×10^7	0, 1	5.7
2 hr................	3.7×10^7	0, 0	6.1
48 hr. (not trans.)....	—	1, 2	1.6

TABLE 5. TRANSFER OF IRRADIATED *try3* FROM HIGH TO LOW AND FROM LOW TO HIGH LEVELS OF ENRICHMENT AFTER ONE HOUR OF INCUBATION

4.3×10^7 survivors plated on initial enrichment level, incubated one hour at 37°C, washed from surface, concentrated by centrifugation, replated on final enrichment level, incubated 48 hours.

Low enrichment level: unsupplemented "A".
High enrichment level: 2.5% nutrient broth.

Enrichment Level		No. Viable Dells Replated	No. Induced Prototrophs per Plate	Final No. Resid. Div.
Initial	Final			
High	High	4.1×10^7	128, 127	6.5
High	Low	3.9×10^7	132, 122	1.5
Low	Low	4.3×10^7	2, 6	0.4
Low	High	4.0×10^7	4, 1	6.1

enrichment transfers were made by the same method. Table 5 shows the results, which provide further support for the conclusion that the frequency of induced prototrophs is determined by the amount of nutrient broth present in the medium during the first hour of postirradiation incubation, and is independent of the number of residual divisions.

The Generality of the Results. Since all of the experiments described above were done with a particular auxotrophic substrain of *S. typhimurium*, it seemed desirable to determine whether the behavior of other induced prototrophs, having expression curves previously interpreted as indicating extensive "delayed appearance", is similar to that of *try3*. Table 6 shows the results of transplantation from low to high enrichment levels after one hour of incubation for a number of different auxotrophic substrains of *S. typhimurium* and *E. coli*. In all cases, the results are in qualitative agreement with those obtained for *try3*. Table 7 shows the result of a more detailed analysis made with a tryptophane-requiring substrain of *E. coli* B/r, known as WP2. Reciprocal transfers from high to low, and from low to high levels of enrichment were made by the method of washing and replating. Again, it was found that the enrichment level present in the medium during the first hour after irradiation essentially determines the induced mutation yield, regardless of subsequent changes in the level of enrichment.

Thus, it appears to be generally true of ultraviolet-induced prototrophs in *E. coli* and *S. typhimurium* that the yield of mutations is independent of the number of residual divisions, but is determined by the amount of nutrient broth present in the minimal medium upon which the irradiated cells are incubated for approximately the first hour after irradiation, increasing as the amount

TABLE 6. "TRANSPLANTATION" OF IRRADIATED AUXO-TROPHS FROM LOW TO HIGH LEVELS OF ENRICHMENT AFTER ONE HOUR OF INCUBATION

Strain	% Nutrient Broth in Medium		No. Induced Prototrophs per Plate
	Initially plated on	Trans-planted to	
E. coli 12-33 (*meth-*)	0.05	—	77, 62
	2.5	—	562, 508
	0.05	5.0	59, 61
E. coli WP8 (*lys-*)	0.05	—	23, 25
	2.5	—	467, 441
	0.05	5.0	18, 29
E. coli 12-61 (*try-*)	0.05	—	34, 28
	2.5	—	1016, 884
	0.05	5.0	56, 37
S. typhimurium S1 (*pro-*)	0.05	—	0, 2
	2.5	—	76, 55
	0.05	5.0	1, 5

TABLE 7. TRANSFER OF IRRADIATED WP2 (*E. coli try-*) FROM HIGH TO LOW AND FROM LOW TO HIGH LEVELS OF ENRICHMENT AFTER ONE HOUR OF INCUBATION

4.1×10^7 survivors plated on initial enrichment level, incubated one hour at 37°C, washed from surface, concentrated by centrifugation, replated on final enrichment level, incubated 48 hours.

Low enrichment level: Unsupplemented "E".
High enrichment level: 2.5% nutrient broth.

Enrichment Level		No. Viable Cells Replated	No. Induced Prototrophs per Plate	Final No. Resid. Div.
Initial	Final			
High	High	3.6×10^7	736, 673	7.1
High	Low	3.5×10^7	585, 619	1.6
Low	Low	3.8×10^7	55, 64	1.1
Low	High	3.4×10^7	72, 58	6.8

of broth increases up to a certain level. Changes in the concentration of nutrient broth after this time have no further influence on the frequency of induced mutations, although the residual growth responds to increased enrichment at any time during the incubation period.

PART II. THE INFLUENCE OF THE RATE OF POSTIRRADIATION PROTEIN SYNTHESIS ON THE YIELD OF INDUCED PROTOTROPHS

In Part I, it was shown that the frequency of induced prototrophs increases as a function of the concentration of nutrient broth present in the minimal medium during the first hour of postirradiation incubation. The effect of nutrient broth can be considered another example of a posttreatment capable of modifying the mutagenic action of ultraviolet light. The nutrient broth effect, like that of temperature (Witkin, 1953), differs from most other known posttreatments in that it affects quite specifically the yield of mutations, having no effect upon survival. The goal of the next series of experiments was to determine, as far as possible, the nature of the metabolic pathway through which the concentration of nutrient broth exerts its modifying action on the induced mutation frequency.

The Active Component of Nutrient Broth. Since nutrient broth was used throughout as the source of enrichment in the experiments previously described, the concentration of the growth-controlling amino acid, tryptophane in the case of *try3*, and those of all other components of nutrient broth were necessarily varied proportionately as the enrichment level was changed. The first step in this analysis, therefore, was to determine whether the active component was the required growth factor itself. Table 8 shows results of an experiment in which irradiated *try3* was plated on minimal agar supplemented with increasing quantities of tryptophane, resulting in a range of residual divisions from 0.7 to 5.8. Essentially no

130   EVELYN M. WITKIN

TABLE 8. THE YIELD OF INDUCED PROTOTROPHS
OBTAINED FROM IRRADIATED *try3* ON MINIMAL
MEDIUM SUPPLEMENTED WITH TRYPTOPHANE

No. survivors plated: 3.9×10^7.

No. induced prototrophs on minimal agar enriched
with 2.5% nutrient broth: 122, 136, 111.

Enrichment	Number of Residual Divisions	Number of Induced Prototrophs/Plate
gamma tryptophane/ml		
0.01	0.7	0, 2, 0
0.05	2.3	0, 1, 1
0.1	3.3	0, 0, 0
0.25	4.3	1, 4, 0
0.5	5.8	2, 1, 0

induced prototrophs were obtained on plates
supplemented only with tryptophane, despite the
occurrence of extensive residual division. These
results demonstrate once again that the mutation
yield and the level of residual growth are independent responses to enrichment, and show also
that tryptophane, the factor that determines the
amount of residual division, is not the component
of nutrient broth that favors high yields of induced
mutations when present in the early postirradiation growth medium.

In the next experiments, minimal agar partially
enriched with tryptophane was further supplemented with metabolites known to be present in
nutrient broth. Table 9 shows the effects upon
the number of induced prototrophs of supple-
menting with a pool of purines and pyrimidines,
a pool of amino acids other than tryptophane, and
a pool of vitamins. Clearly, the amino acid fraction is the active component. It was further established that the presence of all other amino acids
in optimal concentrations, but without any tryptophane, results in failure to obtain induced prototrophs.

TABLE 9. THE YIELD OF INDUCED PROTOTROPHS OB-
TAINED FROM IRRADIATED *try3* ON MINIMAL MEDIUM
ENRICHED WITH VARIOUS METABOLITES PRESENT IN
NUTRIENT BROTH

No. survivors plated: 4.2×10^7.

Concentration of each vitamin: 2 γ/ml.

Concentration of each purine and pyrimidine: 20
γ/ml.

Concentration of each amino acid (except tryptophane): 20 γ/ml.

Minimal Medium Enriched with	No. Induced Proto-trophs per Plate
.5 γ tryptophane......................	0, 0, 1
.5 γ/ml. tryp. + vitamin pool.........	0, 3, 0
.5 γ/ml. tryp. + purine and pyrimidine pool..............................	1, 1, 0
.5 /ml. tryp. + amino acid pool.......	154, 163, 149
amino acid pool (minus tryp.)..........	0, 0, 0
2.5% nutrient broth..................	167, 138, 151

The "transplantation" experiments described
in Part I were repeated, using high and low concentrations of pooled amino acids instead of nutrient broth as the source of high or low enrichment
levels (always with a constant limiting amount of
tryptophane), and the results were duplicated
quite precisely. The nutrient broth effect is an
amino acid effect, and can be redescribed as follows: the yield of induced prototrophs is a function
of the concentration of amino acids (within a
certain range) present in the selective minimal
medium during the first hour of postirradiation
incubation.

Qualitatively comparable results were obtained
with a number of other auxotrophic substrains
of *S. typhimurium* and *E. coli*, including one strain
requiring adenine as a growth factor. In all strains
showing a pattern of "delayed appearance" according to former criteria, it was found that the
maximal yield of prototrophs was obtained when
a high concentration of amino acids (plus a limiting amount of the required growth factor) was
present during the early part of the postirradiation
lag phase.

The Specificity of the Amino Acid Effect. It has
been established that a pool of amino acids, present in the postirradiation growth medium during
the first hour of incubation, favors the production
of high yields of induced prototrophs. A preliminary study was made of the degree of specificity,
if any, of particular groups of amino acids. In
nutritional studies with microorganisms, amino
acids are often grouped for convenience into five
pools, the division being based upon biosynthetic
interrelationships. Table 10 shows the relative
effectiveness of each of these five pools individually, and in all possible combinations of two, three
and four pools, compared with the activity (already known) of all five pools. No single amino
acid pool approaches the effectiveness of the combination of all five. In general, the mutation frequency increases with the number of pools, although there is considerable variability, and pools
1, 2, 3 and 4, in combination, exceed the effectiveness of the combination of all five pools. The pattern is not one of great specificity, and a more
detailed investigation of this type was postponed
until the mechanism of the amino acid effect is
better understood.

The Role of the Preirradiation Culture Medium.
In all of the experiments previously described,
the auxotrophic cultures were grown initially in
nutrient broth. The finding that the presence of
amino acids during the early part of the postirradiation growth period favors the production of
high yields of induced prototrophs heightens the
possible significance of the previous history of the
irradiated cells with regard to amino acid metabolism. Specifically, it seemed important to determine whether *try3* cultures initially grown in
minimal medium supplemented with tryptophane,
and thereby adapted before irradiation to the

ULTRAVIOLET-INDUCED MUTATIONS IN BACTERIA 131

synthesis of other amino acids, would exhibit the same dependence upon an exogenous supply of amino acids in producing induced mutations. Table 11 shows the results of an experiment comparing the behavior of *try3* cultures grown initially in broth and in tryptophane-supplemented minimal medium. Whereas the broth-grown cells are unable to produce induced prototrophs on minimal plates supplemented only with tryptophane, the cells grown initially in tryptophane-supplemented minimal medium produce about half as many prototrophs on plates devoid of other amino acids as on plates supplemented with all amino acids. It is evident that the requirement for exogenous amino acids other than tryptophane for the production of induced prototrophs is less pronounced in bacteria initially "trained" to synthesize their own.

The Growth Rates of Irradiated Cells on Media of Various Enrichment Levels. The finding that the requirement for an exogenous supply of amino acids during the early postirradiation incubation period is less pronounced for bacteria previously adapted to the synthesis of amino acids suggests that the rate of protein synthesis during the criti-

TABLE 10. THE EFFECT OF ENRICHMENT WITH SINGLE AMINO ACID POOLS, AND WITH VARIOUS COMBINATIONS OF AMINO ACID POOLS, ON THE YIELD OF INDUCED PROTOTROPHS OBTAINED FROM IRRADIATED *try3*

No. of survivors plated: 3.8×10^7.

Pool 1 contains cysteine, methionine, argenine and lysine.

Pool 2 contains leucine, isoleucine and valine.

Pool 3 contains tyrosine and phenylalanine.

Pool 4 contains histidine, threonine, proline and glutamic acid.

Pool 5 contains serine, glycine, alanine and aspartic acid.

All plates contain 0.5 γ/ml tryptophane.

Final concentration of each other amino acid: 20 γ/ml.

Amino Acid Pools Added to Medium	No. Induced Prototrophs	Amino Acid Pools Added to Medium	No. Induced Prototrophs
1	5, 8	1 + 2 + 3	43, 52
2	14, 7	1 + 2 + 4	161, 189
3	8, 7	1 + 2 + 5	79, 127
4	16, 15	1 + 3 + 4	81, 94
5	11, 10	1 + 3 + 5	57, 40
1 + 2	31, 26	1 + 4 + 5	64, 58
1 + 3	14, 24	2 + 3 + 4	51, 67
1 + 4	23, 46	2 + 3 + 5	55, 35
1 + 5	38, 57	2 + 4 + 5	35, 49
2 + 3	23, 6	3 + 4 + 5	43, 26
2 + 4	65, 90	1 + 2 + 3 + 4	208, 199
2 + 5	41, 23	1 + 2 + 3 + 5	74, 119
3 + 4	21, 31	1 + 2 + 4 + 5	135, 117
3 + 5	31, 30	1 + 3 + 4 + 5	72, 72
4 + 5	24, 21	2 + 3 + 4 + 5	45, 54
		1 + 2 + 3 + 4 + 5	159, 143

TABLE 11. THE EFFECT OF THE PREIRRADIATION CULTURE MEDIUM ON THE YIELD OF INDUCED PROTOTROPHS OBTAINED FROM IRRADIATED *try3* ON MINIMAL AGAR WITH AND WITHOUT AMINO ACID ENRICHMENT

No. survivors plated: A, 2.3×10^7.

No. survivors plated: B, 2.6×10^7.

Preirradiation Culture Medium	No. Induced Prototrophs per Plate	
	On minimal agar + 0.5 γ/ml tryp.	On minimal agar + 0.5 γ/ml tryp. + amino acid pool*
Nutrient broth A......	0, 0, 2	102, 96, 116
Minimal medium + 20 γ/ml tryptophane B.................	122, 114, 121	240, 222, 231

* Each amino acid concentration: 20 γ/ml.

cal period may be the variable that is actually concerned in modifying the induced mutation frequency. This idea was given considerable support by measurements of the growth rates of irradiated *try3* cells on minimal agar enriched with various concentrations of nutrient broth. Figure 3 shows the growth curves obtained by incubating on various enrichment levels, and washing the cells from the surfaces of the plates after different periods of incubation. It is evident that the length of the lag phase and the slope during logarithmic growth, as well as the final level of residual growth, are functions of the amount of nutrient broth added to the minimal medium upon which the irradiated cells divide. Similar results were obtained when the plates were enriched with increasing concentrations of pooled amino acids (including tryptophane), using a pool containing 20 gamma/ml of each amino acid as the equivalent of undiluted nutrient broth. The fact that the growth rate under these conditions is limited by the concentration of amino acids in the medium suggests that the rate of protein synthesis is de-

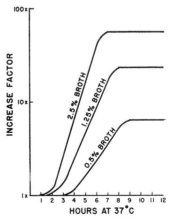

FIGURE 3. Growth curves of irradiated *try3* on minimal agar enriched with various amounts of nutrient broth.

132   EVELYN M. WITKIN

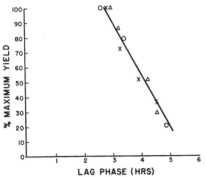

FIGURE 4. The relation between the length of the lag phase of irradiated auxotrophs (as determined by the enrichment level) and yield of induced prototrophs. Circles, *try3* (*S. typhimurium*); crosses, *WP2* (*E. coli*, tryptophaneless); triangles, MS31 (*E. coli*, leucineless.)

creased, and becomes growth-limiting when the concentration of exogenous amino acids drops below a certain level.

The striking parallel between the effect of amino acid concentration on the growth rate of the irradiated cells and on the yield of induced mutations is shown in Figure 4. This curve shows the relation between the length of the lag phase (presumably a measure of the rate of protein synthesis, since it is shortened by increasing the concentration of amino acids) and the percentage of the yield of induced prototrophs obtained on minimal medium having various levels of enrichment, for two different auxotrophic strains of *E. coli*, as well as for *try3*. The results for all three strains, each having a very distinctive pattern of "delayed appearance" as measured in expression curves, show the same linear relationship between the length of the lag phase and the proportion of induced prototrophs obtained. It appears quite reasonable to infer that the yield of induced prototrophs is proportional to the rate of protein synthesis during the first hour of postirradiation incubation.

Further support for this hypothesis was obtained by analysis of a group of *E. coli* auxotrophs that showed no influence of amino acid concentra-

TABLE 12. THE EFFECT OF POSTTREATMENT WITH CHLORAMPHENICOL ON THE YIELD OF INDUCED PROTO-TROPHS OBTAINED FROM IRRADIATED AUXOTROPHS

Strain	No. Viable Cells per Plate after One Hour		No. Induced Prototrophs per Plate	
	Chloram-phenicol treated	Control	Chloram-phenicol treated	Control
E. coli WP2 (*try-*)......	1.0×10^7	1.2×10^7	24, 20	216, 221
E. coli MS28 (*leu-*)......	5.5×10^6	6.0×10^6	6, 2	42, 39

tion on the yield of induced mutations. These auxotrophs had no "delayed appearance," or, in revised terms, yielded as many induced prototrophs on unsupplemented minimal medium as on minimal medium supplemented with nutrient broth or with amino acids. An analysis of the post-irradiation growth rates of these strains on unsupplemented minimal medium and on media supplemented with increasing amounts of amino acids showed no difference in the length of the lag phase nor in the generation time. In these strains, for reasons as yet not known, the rate of protein synthesis was not decreased in the absence of exogenous amino acids, nor was the yield of induced prototrophs.

The Effect of Posttreatment with Chloramphenicol. More direct support for the hypothesis that the frequency of induced prototrophs is determined by the rate of protein synthesis during the first hour of postirradiation incubation was obtained from experiments in which chloramphenicol, which is known to arrest bacterial division specifically by inhibiting the synthesis of proteins, without affecting the rate of RNA and DNA synthesis (Wisseman *et al.*, 1954; Hahn *et al.*, 1954) was used as a posttreatment. Irradiated cells of strains WP2 (try⁻) and MS28 (leu⁻) of *E. coli* were plated on optimally enriched plates containing 20 micrograms of chloramphenicol per ml, and were incubated on these plates for one hour. After this time, the cells were washed off the chloramphenicol plates and replated on optimally enriched medium without chloramphenicol. Controls were handled in the same way, except that chloramphenicol was omitted from the plates upon which the irradiated cells were incubated for the first hour. The results are shown on Table 12. It will be noted that one hour of incubation on plates containing chloramphenicol reduced the yield of induced prototrophs drastically with no appreciable effect upon the level of survival. The results with strain MS28 are of particular interest, since it is one of the exceptional strains described above that shows no response to amino acid concentration, with regard either to growth rate or to induced mutation frequency. Low concentrations of exogenous amino acids apparently do not decrease the rate of protein synthesis in this strain, hence there is no decrease in the yield of prototrophs. When an effective means of arresting protein synthesis, such as chloramphenicol, is used, this strain shows itself to be similar to the others, in that induced prototrophs are not obtained under these conditions.

According to the working hypothesis that has been adopted as a guide to further work, the effects of chloramphenicol and of low concentrations of amino acids have the same basis: their action in arresting or decreasing the rate of protein synthesis during the sensitive period. Since the period of susceptibility of induced mutation frequency to low concentrations of amino acids is known to

ULTRAVIOLET-INDUCED MUTATIONS IN BACTERIA 133

be about one hour, it can be predicted that the period of susceptibility to chloramphenicol should also be about one hour. Table 13 shows the results of an experiment in which irradiated cells were incubated for various periods of time on optimally enriched minimal medium containing chloramphenicol, and then washed from the surface and replated on chloramphenicol-free medium. The yield of induced mutations is found to be reduced to a minimum after one to one and a half hours of incubation in the presence of chloramphenicol. The periods of susceptibility to the antimutagenic effects of chloramphenicol and of low concentrations of exogenous amino acids are approximately coincidental.

Table 14 shows the results of experiments in which irradiated cells were incubated on optimally enriched minimal plates without chloramphenicol for one hour, then washed off and plated on chloramphenicol plates, incubated for one hour, and then again washed off and replated on chloramphenicol-free plates. It is clear that the yield of induced prototrophs is not affected by incubation in the presence of chloramphenicol *after* the first hour of incubation. These results are comparable to the "transplantation" experiments, in which it was found that decreasing the enrichment level after the first hour of incubation on high enrichment levels was not effective in reducing the mutation frequency.

The Effect of Incubation in Buffer. It has been shown that, when the rate of protein synthesis is decreased by limiting the concentration of exogenous amino acids, or by the presence of chloramphenicol, in a medium that is otherwise growth-supporting, the yield of induced prototrophs is irreversibly reduced within one hour after irradiation. It seemed important to determine whether the mutation yield would respond with the same timing to the inhibition of protein synthesis in a medium also suppressing other biosynthetic activities. Table 15 shows the results of an experiment in which irradiated cells were incubated at 37°C in buffer, and plated after various periods of time on optimally enriched medium. It is evident that the ability to produce induced prototrophs is

TABLE 13. THE EFFECT OF POSTTREATMENT WITH CHLORAMPHENICOL FOR VARIOUS PERIODS OF TIME ON THE YIELD OF INDUCED PROTOTROPHS OBTAINED FROM IRRADIATED WP2

Time on Chloramphenicol	No. Viable Cells per Plate after Chloramphenicol Treatment-	No. Induced Prototrophs per Plate (after Replating)
minutes		
0	9.5×10^6	110, 131
30	1.1×10^7	76, 92
60	1.2×10^7	34, 30
90	9.9×10^6	14, 12
120	9.9×10^6	16, 10

TABLE 14. THE EFFECT OF POSTTREATMENT WITH CHLORAMPHENICOL AFTER ONE HOUR OF INCUBATION ON CHLORAMPHENICOL-FREE MEDIUM ON THE YIELD OF INDUCED PROTOTROPHS OBTAINED FROM IRRADIATED WP2

Experiment No.	No. Viable Cells Plated on Final Plates		No. Induced Prototrophs per Plate	
	Chloramphenicol treated	Control	Chloramphenicol treated	Control
1	2.3×10^6	1.8×10^6	166	149
			188	155
			172	156
2	2.4×10^6	2.3×10^6	135	143
			135	134
			162	152

TABLE 15. THE EFFECT OF POSTIRRADIATION INCUBATION IN BUFFER ON THE YIELD OF INDUCED PROTOTROPHS OBTAINED FROM *try3*

Washed suspension of *try3* cells irradiated in phosphate buffer (pH 7); irradiated suspension incubated at 37°C as indicated; after each time interval, suspension assayed to determine no. viable cells per ml, and 0.1 ml aliquots plated on semienriched (2.5% nutrient broth) plates to determine yield of induced prototrophs.

Time of Incubation in Buffer at 37°C	No. Viable Cells per ml after Incubation in Buffer	No. Induced Prototrophs per Plate
minutes		
0	2.7×10^8	144, 121
30	3.1×10^8	143, 137
60	3.1×10^8	121, 137
120	3.3×10^8	158, 124
180	2.8×10^8	66, 94

preserved much longer in the absence of protein synthesis, if other metabolic activities are arrested as well. The sensitive period is not to be measured in absolute time, therefore, but in relation to the metabolic conditions under which the posttreatment is administered.

PART III. THE EFFECTS OF POSTIRRADIATION TEMPERATURE ON THE YIELD OF INDUCED PROTOTROPHS

The timing of the sensitive period during which the frequency of induced prototrophs can be influenced by the amino acid concentration of the medium and by chloramphenicol is strikingly similar to the temperature-sensitive period reported for a variety of induced mutations in *E. coli* (Witkin, 1953; Berrie, 1953), the first third of the lag phase (one hour at 37°C). A study of the effects of postirradiation temperature on the yield of ultraviolet-induced prototrophs obtained from *try3* was conducted, in the hope that the integra-

134 EVELYN M. WITKIN

TABLE 16. EFFECT OF POSTIRRADIATION TEMPERATURE
OF INCUBATION ON THE YIELD OF INDUCED
PROTOTROPHS OBTAINED FROM *try3*

Identical aliquots of irradiated suspension plated on minimal agar enriched with 2.5% nutrient broth; assays to determine survival made on same medium (*try3* colonies grow to visible size on this enrichment level when seeded sparsely).

Temperature of Incubation	Survival	No. Induced Prototrophs per Plate
	%	
45°C	7.1	0, 0
37°C	9.5	101, 121
24°C	10.5	256, 263
15°C	9.5	21, 25

tion of these effects and those of the nutritional factors described above might provide a more substantial basis for the interpretation of the complex events leading from irradiation to stable genetic change.

Table 16 shows the frequency of induced prototrophs obtained when irradiated cells are incubated on optimally enriched plates (2.5 per cent nutrient broth) at various temperatures. No important differences in the number of survivors, nor in the amount of residual division, were found, but the number of induced prototrophs was affected dramatically by the incubation temperature, ranging from none at 45°C to a number two to three times higher than the 37°C frequency at 24°C. Having established that postirradiation temperature influences the mutation frequency, experiments designed to determine the limits of the temperature-sensitive period were conducted. Plates seeded with irradiated *try3* were incubated for various periods of time at an initial temperature, and the plates were then transferred to a second temperature at which incubation was continued until the prototrophs could be counted. These "temperature-switch" experiments were done using all possible combinations of the four temperatures studied, two by two. In all of these

TABLE 17. THE DURATION OF THE PERIOD OF SENSITIVITY TO POSTIRRADIATION TEMPERATURE OF THE YIELD OF INDUCED PROTOTROPHS OBTAINED FROM *try3*

The temperature-sensitive period is the minimum time of incubation at the temperature indicated after which changes in temperature of incubation have no further influence on the yield of induced prototrophs.

Temperature	Length of Lag Phase	Duration of Temperature-Sensitive Period
37°C	3 hours	1–1½ hours
24°C	9 hours	3–4 hours
15°C	30 hours	10–12 hours

experiments, it was found that, if the length of time of incubation at the initial temperature before switching to the second temperature was as long as one third the length of the lag phase, the mutation frequency was the same as that in controls incubated throughout at the initial temperature. Changes in the temperature of incubation, as was found in the studies with *E. coli*, have no effect upon the yield of induced mutations after the first third of the lag period. Table 17 summarizes the results of some of these experiments, details of which will be presented elsewhere. The period of sensitivity to postirradiation temperature (when the initial incubation is set at 37°C) thus has about the same duration as the periods of sensitivity to the action of chloramphenicol and of low concentrations of amino acids.

The temperature-switch experiments revealed a particularly interesting phenomenon in the case of irradiated cells incubated for short periods of time at 37°C and then permitted to complete their growth at a lower temperature (24° or 15°C). The mutation yields obtained under these conditions are considerably higher than those obtained at any single temperature of incubation, the maximum frequency (about 5 times the 37° yield) resulting from a switch to a lower temperature after about 30 minutes at 37°C. A similar synergistic effect is obtained when short incubations at 45°C are followed by transfer to lower temperatures.

DISCUSSION

The basis of the "delayed appearance" of induced prototrophs can now be described as a response of the mutation frequency to increasing concentrations of amino acids present in the medium during the first hour of postirradiation incubation (at 37°C), or, by inference, as a response to increasing rates of protein synthesis during this time. The insight gained in these studies, although they appear to have general validity for induced prototrophy, cannot be applied to the interpretation of the expression curves for induced bacteriophage-resistance or streptomycin-resistance. In these cases, the measurement of delayed appearance involves the use of the same medium throughout, the control over residual growth residing in the time of application of the selective agent. The 10 to 12 generation delay described for phage-resistance, therefore, must have its basis elsewhere. The finding that the phenotypic lag for induced prototrophs is no more than one generation, however, does not exclude the possibility that other classes of mutations may exhibit much more prolonged phenotypic lag. Thus, the report by Jacob, Wollman and Hayes (this Symposium) that phage-resistance introduced into sensitive cells by recombination requires about four generations to achieve full phenotypic expression accounts, in part, for the delayed appearance of this class of mutants. Whatever additional factors may enter to contribute to the observed delay, the timing of the period of susceptibility to postirradiation temperature, as well as the prevalence of pure clones of induced phage-resistant mutants, makes it seem quite certain that this mutation, like that to prototrophy, is genetically established before the completion of the first postirradiation division.

Considerable evidence has been offered in support of the hypothesis that the yield of induced prototrophs is proportional to the rate of protein synthesis during the first hour of postirradiation incubation, under conditions in which protein synthesis is growth-limiting.[3] In considering the possible significance of this finding for the theory of mutagenesis, the nature of the sensitive period will be discussed as a starting point. It is known that, for an impressive variety of mutational types, the period of susceptibility to the influence of postirradiation temperature is a constant fraction (about ⅓) of the lag phase. It would seem likely, then, that the terminal event, the occurrence of which establishes the induced mutation yield irreversibly, is associated with the process of cell division. On the other hand, the sensitive period is *not* a constant fraction of the lag phase when growth is limited by the rate of protein synthesis rather than by temperature, but under these conditions, remains about one hour (at 37°C) regardless of the differences in growth rate imposed by the availability of exogenous amino acids. The terminal event marking the end of the sensitive period, therefore, must take place quite independently of protein synthesis. Both DNA synthesis and RNA synthesis are known to take place while protein synthesis is inhibited, as in the presence of chloramphenicol. Preliminary measurements made of the rate of DNA synthesis in irradiated *try3* cells, made by Dr. D. Kanazir, indicate that the DNA level is doubled in one hour at 37°C. It is tempting, therefore, to speculate that the modification of the mutation yield by posttreatments is possible only before the completion of DNA duplication. Whether or not the sensitive period proves to coincide with the period of DNA duplication, its termination is achieved through a process that is associated with cell division, that goes on independently of the rate of protein synthesis, and that is temperature-dependent with a Q_{10} of about 3.

In addition to the process leading to the termination of the period during which the mutation yield can be influenced, a second process, which we will in our ignorance call "X," must be invoked. The existence of X can be inferred from the temperature data alone, since their interpretation requires the postulation of at least two processes, having different temperature coefficients, the relative rates of which determine the yield of induced mutations. The studies involving the effects of the rate of protein synthesis also lead to the postulation of two processes, the first determining the duration of the sensitive period, which is independent of the rate of protein synthesis, and the second, most economically assumed to be the same X indicated by the temperature experiments, which is markedly affected by the rate of protein synthesis. Although much more information must be gathered before the construction of any detailed model would be fruitful, it is possible to speculate about the nature of X, and to map out some of the directions of future work that are most likely to lead to its identification.

One possibility is that X is the process of repair of genetic damage. According to this view, induced prototrophs arise primarily from a fraction of survivors that has undergone genetic damage (chromosome breakage?) as a result of irradiation. Protein synthesis favors repair of this damage (restitution of breaks?) and if repair is accomplished before the end of the sensitive period (DNA duplication?) the damaged cells survive, with a high probability of surviving as a mutant. If repair is not accomplished within the critical period, the damage is lethal. It is well known that the reunion of radiation-induced chromosome breaks is often delayed, and is influenced by postirradiation metabolism (see, for example, Wolff and Atwood, 1954; Wolff and Luippold, 1955). A more controversial question is the timing of the breakage process itself. Several investigators (Kaufmann, Hollaender and Gay, 1946; Swanson and Yost, 1951; Swanson, 1955) have suggested that breaks may occur after irradiation, and that they may or may not be realized depending upon postirradiation conditions, although the view that breakage is an immediate consequence of irradiation is still widely held. It is interesting in this connection that, within the framework of the hypothesis that X is a process of repair of genetic damage, the results of the temperature experiments can best be explained on the assumption that breakage, as well as reunion, is delayed, and is susceptible to postirradiation temperature. The synergistic effect of short exposures to high temperatures, followed by incubation at lower temperatures, is difficult to explain as an effect on reunion alone. All of the temperature-switch experiments, however, fit the assumption that a high frequency of breaks is favored by high temperatures during the first half-hour of the postirradiation growth period, while subsequent reunion is more efficient at low temperatures.

Another possibility is that X is the production or decay of an unstable mutagenic intermediate

---

[3] Preliminary work indicates that the failure to obtain induced prototrophs from irradiated *try3* incubated on Millipore filters (even when the latter are placed on the surface of optimally enriched plates) may be another example of the postirradiation effect of decreased rate of protein synthesis during the sensitive period. Irradiated *try3* cells have a longer lag phase on the filter membranes than on the surface of agar plates, when semi-enriched minimal medium is used.

resulting from ultraviolet irradiation. In this case, protein synthesis would favor the production of greater amounts of mutagen, or would enhance its stability. The yield of induced prototrophs would depend upon the amount of mutagen still present at the time of occurrence of the terminal event. The temperature results, including the synergistic effect of high and low temperatures in sequence, could be explained by the assumption that maximum amounts of mutagen are produced at high temperatures, and that low temperatures increase its stability. If a mutagenic intermediate is involved, it could not be the mutagenic cell poison postulated by Novick and Szilard (1949), since the survival level of the irradiated cells is not modified by the posttreatments affecting the induced mutation frequency.

A third hypothesis concerning the nature of X, the process affected by the rate of protein synthesis, is that it is the production or decay of an unstable condition within the genic material (Auerbach, 1946; McElroy and Swanson, 1951). The mutation yield would then depend upon the frequency of metastable configurations not yet dissipated at the critical time.

The evaluation of these hypotheses must await future developments. One differentiating prediction concerns the fate of potential mutants that fail to appear under antimutagenic postirradiation conditions. The breakage-reunion hypothesis leads to the prediction that such conditions result in the failure of these cells to survive, while the hypotheses concerning the decay of an extragenic mutagen or an intragenic state of metastability predict their survival as nonmutants. As yet, no experimental technique has been developed that could provide the answer.

It will be important to determine the extent to which a particular mutation responds to the posttreatments described after induction with different mutagenic agents, as well as to examine the behavior of different classes of mutations induced by the same mutagen. The changes generally grouped together under the heading of "induced mutations" may actually be the end results of very different kinds of primary events, and generalizations based on the study of one class of genetic changes is not justifiable. If, for example, it should be found that the rate of protein synthesis during the early postirradiation growth period has no influence upon the yield of induced auxotrophs, or phage-resistant mutants, but is limited in its influence to induced prototrophy, it may become necessary to seek the explanation of this effect in an entirely new direction. Since the change from auxotrophy to prototrophy appears to involve the rapid acquisition of a new enzymatically controlled synthetic capacity, the involvement of protein synthesis in the establishment of the new genotype could be related to the system of specificity transfer, mediated perhaps

through other channels in the case of phenotypes more remotely dependent upon protein structure.

Other important areas for future work are the implementation of the genetic studies with direct biochemical analysis of the metabolic events in the critical period, the integration of these findings with the extensive studies of other postirradiation effects, and their application to the work on spontaneous mutation.

SUMMARY AND CONCLUSIONS

Part I. A comparative study of the pattern of "delayed appearance" of ultraviolet-induced prototrophs and that of genetically identical prototrophs arising by transduction in a tryptophane-requiring substrain of *Salmonella typhimurium* LT2 was made. It was found that the maximal yield of prototrophs arising by transduction was obtained under conditions permitting only one residual division, while the maximal yield of induced prototrophs was obtained only on plates sufficiently enriched with nutrient broth to permit about six residual divisions. It was concluded that phenotypic lag, irregularity of the growth pattern of irradiated cells, and nuclear segregation were not primarily responsible for the "delayed appearance" of induced prototrophs.

The rate of increase in the number of newly induced prototrophs in clones developing from irradiated auxotrophs was compared, by respreading experiments, with the growth rate of a previously isolated stock of the induced prototroph under comparable conditions. The results ruled out the possibility that the induced mutations to prototrophy are typically delayed in their occurrence for several generations after irradiation, as well as the possibility that newly induced prototrophs are markedly delayed in their time of onset of division.

It was concluded that none of the hypotheses previously proposed to account for the "delayed appearance" of induced prototrophs was correct.

The validity of the assumption that the induced mutation frequency and the number of residual divisions, both of which are functions of the amount of nutrient broth added to the postirradiation culture medium, are causally related was questioned. Residual division and induced mutation frequency were shown, in fact, to be independent and separable functions of the enrichment level, in a series of experiments in which irradiated cells were transferred from low to high enrichment levels, and vice versa, after various periods of postirradiation incubation. These experiments showed that the yield of induced prototrophs is determined irreversibly by the amount of nutrient broth present in the culture medium during the *first hour* of postirradiation incubation, while the number of residual divisions responds to increasing amounts of nutrient broth added at any time during the postirradiation incubation period.

Part II. The active component of nutrient

ULTRAVIOLET-INDUCED MUTATIONS IN BACTERIA 137

broth was found to be the amino acid fraction. A pool of amino acids, including a limiting amount of the required growth factor, has the same effect as nutrient broth in promoting the maximal yield of induced prototrophs when present in the postirradiation culture medium during the first hour of incubation. The required growth factor alone was found to be inactive, as were pools of vitamins or purines and pyrimidines. In general, the yield of induced prototrophs increases as the number of different amino acids in the postirradiation medium increases.

The preirradiation culture medium was found to influence the degree of dependence upon postirradiation enrichment with amino acids. While irradiated *try3* initially grown in nutrient broth can give rise to induced mutants only if exogenous amino acids are supplied during the first hour of postirradiation growth, cells initially grown in minimal medium supplemented with tryptophane are relatively independent of this requirement, producing substantial numbers of induced prototrophs on minimal medium supplemented only with limiting amounts of tryptophane.

Studies of the growth rates of broth-grown irradiated *try3* on minimal agar supplemented with various amounts of nutrient broth, or with various concentrations of amino acids, revealed that the growth rate of the irradiated cells is proportional to the concentration of exogenous amino acids over a considerable range. The inference was made that the rate of protein synthesis is decreased, and becomes growth-limiting, as the concentration of exogenous amino acids is decreased within the critical range. It was found, for three different auxotrophic strains having very different patterns of "delayed appearance" as measured by expression curves, that the same linear relation was obtained when the percentage of the maximal yield of induced prototrophs obtained on various enrichment levels was plotted against the length of the lag phase (presumably a measure of the rate of protein synthesis) on these enrichment levels. *It was inferred that the yield of induced prototrophs is proportional to the rate of protein synthesis during the first hour of postirradiation incubation, when protein synthesis is growth-limiting.* Confirmatory evidence was obtained from experiments in which chloramphenicol was used as a posttreatment. It was found that incubation of irradiated auxotrophs in the presence of chloramphenicol for one hour (on an otherwise optimal medium for the production of the maximal yield of induced mutants) leads to drastic reduction of the number of induced prototrophs obtained, with no important effect upon the number of survivors. Exposure to chloramphenicol after one hour of incubation under optimal conditions without chloramphenicol has no effect upon the yield of induced prototrophs. The capacity to produce induced prototrophs is maintained for relatively long periods of time when other metabolic activities are inhibited along with protein synthesis (as in buffer).

Part III. The temperature of incubation during postirradiation growth was found to have a profound effect upon the yield of induced prototrophs, while having little influence upon the survival level. The period of susceptibility to the influence of postirradiation temperature was found to be about one-third of the lag phase. When the initial temperature is 37°C, the temperature-sensitive period coincides approximately with the periods of susceptibility to the postirradiation influence of chloramphenicol and low amino acid concentrations.

The yield of induced prototrophs is increased five-fold over the frequency obtained at 37°C when short incubations at 37°C are followed by incubation at lower temperatures. This synergistic effect is obtained also when brief incubations at 45°C are followed by lower incubation temperatures.

The implications of the results are discussed in relation to theories of mutagenesis.

ACKNOWLEDGMENTS

The author is indebted to Miss Ann Lacy, Miss Constance Thomas and Miss Miriam Schwartz for their invaluable assistance at various stages of these studies.

REFERENCES

ANDERSON, E. H., and BILLEN, D., 1955, The effect of temperature on X ray induced mutability in *Escherichia coli*. J. Bact. *70*: 35–43.

AUERBACH, C., 1946, Chemically induced mosaicism in *Drosophila melanogaster*. Proc. Roy. Soc. (Edinburgh) B*62*: 211–222.

BERRIE, A. M. M., 1953, The effects of temperature on ultraviolet-induced mutability in *Escherichia coli*. Proc. Nat. Acad. Sci. Wash. *39*: 1125–1133.

DAVIS, B. D., 1948, Isolation of biochemically deficient mutants of bacteria by penicillin. J. Amer. Chem. Soc. *70*: 4267.

1950, Studies on nutritionally deficient bacterial mutants isolated by means of penicillin. Experientia *6*: 41–50.

DEMEREC, M., 1946, Induced mutations and possible mechanisms of the transmission of heredity in *Escherichia coli*. Proc. Nat. Acad. Sci. Wash. *32*: 36–46.

DEMEREC, M., and CAHN, E., 1953, Studies of mutability in nutritionally deficient strains of *Escherichia coli*. J. Bact. *65*: 27–36.

HAHN, F. E., WISSEMAN, C. L., JR., and HOPPS, H. E., 1954, Mode of action of chloramphenicol II. J. Bact. *67*: 674–679.

HOLLAENDER, A., and STAPLETON, G. E., 1956, Studies on protection by treatment before and after exposure to X- and gamma radiation. Proc. Intern. Conf. Peacetime Uses Atomic Energy, Geneva, *11*: 311–314, New York, United Nations.

JACOB, F., WOLLMAN, E. L., and HAYES, W., 1956, Conjugation and genetic recombination in *Escherichia coli* K-12. Cold Spring Harbor Symp. Quant. Biol. *21*: 141–162.

KAPLAN, R. W., 1952, Auslösung von Farbsektor und anderen Mutationen bei *Bacterium prodigiosum* durch monochromatisches Ultraviolett verschiedener Wellenlängen. Z. Naturf. *7b*: 291–304.

138   EVELYN M. WITKIN

KAUFMANN, B. P., HOLLAENDER, A., and GAY, H., 1946, Modification of the frequency of chromosomal rearrangements induced by X-rays in *Drosophila*. I. Use of near infrared radiation. Genetics *31:* 349–367.

KELNER, A., 1949, Photoreactivation of ultraviolet-irradiated *Escherichia coli*, with special reference to the dose-reduction principle and to ultraviolet-induced mutation. J. Bact. *58:* 511–522.

LABRUM, E. L., 1953, A study of mutability in streptomycin-dependent strains of *Escherichia coli*. Proc. Nat. Acad. Sci. Wash. *39:* 280–288.

McELROY, W. D., and SWANSON, C. P., 1951, The theory of rate processes and gene mutation. Quart. Rev. Biol. *26:* 348–363.

NEWCOMBE, H. B., 1949, Origin of bacterial variants. Nature, Lond. *164:* 150–151.

1953, The delayed appearance of radiation-induced genetic change in bacteria. Genetics *38:* 134–151.

1955, The timing of induced mutations in *Streptomyces*. Brookhaven Symp. Biol. *8:* 88–102.

NEWCOMBE, H. B., and SCOTT, G. W., 1949, Factors responsible for the delayed appearance of radiation-induced mutants in *Escherichia coli*. Genetics *34:* 475–492.

NEWCOMBE, H. B., and WHITEHEAD, H. A., 1951, Photo-reversal of ultraviolet-induced mutagenic and lethal effects in *Escherichia coli*. J. Bact. *61:* 243–251.

NOVICK, A., and SZILARD, L., 1949, Experiments on light-reactivation of ultraviolet inactivated bacteria. Proc. Nat. Acad. Sci. Wash. *35:* 591–600.

RYAN, F. J., 1954, The delayed appearance of mutants in bacterial cultures. Proc. Nat. Acad. Sci. Wash. *40:* 178–186.

1955, Phenotypic (phenomic) lag in bacteria. Amer. Nat. *89:* 159–162.

RYAN, F. J., FRIED, P., and SCHWARTZ, M., 1954, Nuclear segregation and the growth of clones of bacterial mutants induced by ultraviolet light. J. Gen. Microbiol. *11:* 380–393.

SWANSON, C. P., 1955, In: Radiobiology Symposium. Ed. Z. M. Bacq, and P. Alexander. London, Butterworths, pp. 254.

SWANSON, C. P., and YOST, H. T., 1951, The induction of activated, stable states in the chromosomes of *Tradescantia* by infrared and X-rays. Proc. Nat. Acad. Sci. Wash. *37:* 796–801.

WAINWRIGHT, S. D., and NEVILL, A., 1955a, Modification of the biological effects of ultraviolet irradiation by post-radiation treatment with iodoacetate and peptone. J. Gen. Microbiol. *12:* 1–12.

1955b, Some effects of post-radiation treatment with metabolic inhibitors and nutrients upon ultraviolet irradiated spores of *Streptomyces* T12. Canad. J. Microbiol. *1:* 416–426.

WISSEMAN, C. L., JR., SMADEL, J. E., HAHN, F. E., and HOPPS, H. E., 1954, Mode of action of chloramphenicol I. J. Bact. *67:* 662–673.

WITKIN, E. M., 1951, Nuclear segregation and the delayed appearance of induced mutants in *Escherichia coli*. Cold Spring Harbor Symp. Quant. Biol. *16:* 357–372.

1953, Effects of temperature on spontaneous and induced mutations in *Escherichia coli*. Proc. Nat. Acad. Sci. Wash. *39:* 427–433.

WITKIN, E. M., and THOMAS, C., 1955, Bacterial Genetics II, Yearb. Carnegie Instn. *54:* 234–245.

WOLFF, S., and LUIPPOLD, H. E., 1955, Metabolism and chromosome-break rejoining. Science *122:* 231–232.

WOLFF, S., and ATWOOD, K. C., 1954, Independent X-ray effects on chromosome breakage and reunion. Proc. Nat. Acad. Sci. Wash. *40:* 187–192.

ZINDER, N. D., and LEDERBERG, J., 1952, Genetic exchange in *Salmonella*. J. Bact. *64:* 679–699.

DISCUSSION

D. LEWIS: Still at the cellular level but in a higher plant, *Oenothera organensis*, the pollen grains with their gene controlling incompatibility reactions give a different, and in some respects a more direct, means of detecting induced mutations which are unstable. The mutational event can be placed with some accuracy in the previous history of the nuclear divisions that gave rise to the pollen; and the phenotypic change produced by the mutation can be detected when as little as one nuclear division has occurred after the event. Apart from stable mutations of the type of S_1 to S_f, unstable mutations S_1 to S_{f1} and reversion to S_1 occur both spontaneously and with X rays (U.V. cannot be used effectively with this material). A higher frequency of unstable mutations of this type would have the effect of lowering the observed frequency of mutation in Dr. Witkin's experiments with *E. coli*.

There may also be a parallel in the pollen mutations with Dr. Witkin's findings on the effect of growth phase and nutrition. The number of unstable mutations depends upon the state of the nucleus at the time of the mutational event: in early prophase stable mutations are common but are much rarer or absent in the late stages of division.

An explanation advanced for the unstable mutations in pollen is similar to Dr. Witkin's repair hypothesis.

Only a fraction and not all of the strands of the chromosome could be damaged at the locus, and this fraction might be the outer strands which were the ones to become active in producing the immediate phenotype while the inner strands, which will not become phenotypically active for some time, are undamaged. A loss of the outer strands and a preservation of the inner ones during subsequent chromosomal duplication would then be an unstable mutation.

WITKIN: It is possible, undoubtedly, that some mutations induced in bacteria may be of the unstable type described by Dr. Lewis as occurring in *Oenothera*. If this were the case in our material, however, except as a rare exception, the occurrence of mutations to prototrophy after many cell divisions could be detected in the respreading experiments. These experiments, as well as the evidence based upon the duration of the period of susceptibility to modification by posttreatment, seem to me to establish beyond doubt that most of the mutations with which we are dealing occur before the completion of the first postirradiation cell division. Not only do these mutations occur early, but there is a striking fact that must be considered in any theory of mutagenesis: In the cases analyzed (including induced phage resistance, lactose nonfermentation and prototrophy) the colonies in which the induced mutants arise do not usually contain any cells of the nonmutated parent type. The irradiated cell that gives rise to an induced mutation typically produces *only* mutants, although in a small proportion of colonies some admixture of the original type is

ULTRAVIOLET-INDUCED MUTATIONS IN BACTERIA 139

found. By way of contrast, colonies in which prototrophs arise by transduction invariably contain also cells of the unchanged parental genotype. Any theory of mutagenesis, therefore, must incorporate the fact that the induced mutations usually affect the original locus in all its replicas, if such exist at the time of treatment. These facts would not support a copy-error concept of induced mutation, nor would they be easily reconciled with the hypothesis of Dr. Lewis involving damage to a fraction of multiple strands. This does not rule out the possibility that some bacterial mutations may be of the unstable variety, nor that a small fraction of the changes occurring in our material may be of this type. In the cases described, however, it seems to be most often true that the total population descended from a "promutant" exhibits the new genotype.

ALPER: The mechanisms by which ultraviolet light and ionizing radiation act on biological material are very different in many respects; Dr. Witkin's conclusions are therefore of great interest because it would appear that, for the genetic effects she describes, there are close parallels in the effects of the two types of radiation. There is evidence from several lines of investigation that the extent to which many types of X-ray damage are expressed greatly depends on the rate at which synthetic processes occur, immediately after irradiation. For example, Dr. Kimball has shown that in *Paramecium* lethal mutations are reduced in frequency if, after radiation, the animals are exposed to agents which slow their growth, such as cold, or high concentrations of hydrogen peroxide. On the other hand there is apparently a great difference in the lethal effects of U.V. and X rays on bacteria. Stapleton and his coworkers showed that bacteria "recovered" if they were held at temperatures below 37°, after they had been X-irradiated, and constructed curves which gave the optimal holding temperatures for recovery, for several strains of *E. coli*. Dr. Witkin has pointed out that after U.V. irradiation, prolonging the lag period reduces the number of biochemical mutants which appear, but not the number of lethal injuries. These results suggest that it may be more nearly correct to regard the X-ray injury as a lethal mutation than the U.V. effect which causes failure to multiply.

Dr. Witkin has told us that millipore membranes cannot be used in these studies, as they prolong the lag period and therefore interfere with the results. I have developed a technique for studying microcolonies and long forms after X-irradiation, and for this purpose have been using small pieces of cellophane laid over the surface of agar plates. From observations made on fixed and stained colonies, from periods of three quarters of an hour to four hours after seeding, it appears that lag period and generation time match those observed in liquid culture if the same nutrient medium is used. It seems possible therefore that cellophane might be useful where millipore membrane cannot be used.

WITKIN: Systematic studies comparing the responses to postirradiation metabolic conditions of a particular mutation induced by a variety of mutagenic agents would be of great value in interpreting the nature of these effects. Present indications are that the responses to particular posttreatments are not mutagen-specific, as, for example, the finding by Demerec and Cahn that the "pattern of delay," which in reality is a measure of the response to amino acid concentration in the postirradiation culture medium, is the same for a given mutation induced by either ultraviolet light or manganous chloride. Postirradiation temperature effects on mutation frequency, similar to those obtained with ultraviolet light, have been observed for X-ray induced prototrophs by Anderson and Billen, the maximum yield of mutants appearing with incubation at 24°, as was found for prototrophs induced with ultraviolet light in *try3*. If these indications are borne out, it seems to me that they tend to weaken the hypothesis of mutagenic intermediates, which might be expected to show some specificity related to the agent used. The breakage-repair hypothesis, and that of intragenic metastability, on the other hand, would predict a greater degree of uniformity of results with different mutagenic agents.

With regard to the effects of postirradiation temperature on the lethal action of X rays and ultraviolet light, I do not think that the comparison of our results with those of Dr. Kimball can be usefully made. Many kinds of posttreatments do affect the survival of ultraviolet-treated bacteria, and in some strains, such as the B strain of *E. coli*, there is marked heat-reactivation. The comparison of X rays and ultraviolet in this regard should be made with the same material.

GREER: I have evidence which supports the idea that increased protein synthesis favors repair of genetic damage. In experiments with *E. coli* it has been demonstrated that certain amino acid combinations depress the rate of mutation to purine independence and streptomycin resistance induced by 5 OH,7NO₂ benzimidazole. While the amino acids have no effect on the spontaneous mutant frequency, they depress the chemically induced mutant frequency twenty-fold.

LIEB: Several years ago I studied the delay in the appearance of U.V.-induced prototrophs in a histidine-requiring (h^-) culture of *E. coli* (Genetics *36:* 460–477). The bacteria were grown in synthetic medium with an optimal concentration of histidine, and after irradiation were allowed to grow for various lengths of time in the same medium. Growth for a period of time allowing one to two cell divisions was adequate for the expression of all h^+ mutants. It was also shown that growth of irradiated bacteria on plates with limited histidine enrichment did not result in the expression of mutants.

140   EVELYN M. WITKIN

Mutation from prototrophy to auxotrophy was studied under similar conditions. There was no growth-dependent delay in the expression of h^- mutants. This seems to favor Dr. Witkin's hypothesis of gene repair, if I understand it correctly. A process involving protein synthesis would appear to be necessary to restore a gene to its functional state, but would not be required in the case of mutation involving a loss of gene activity.

WITKIN: While the rate of postirradiation protein synthesis has a decisive influence on the yield of induced prototrophs, we have preliminary evidence that the frequency of certain other classes of induced mutations (phage resistance, for example) may be unaffected by the concentration of amino acids and by chloramphenicol posttreatment. It appears possible that protein synthesis in the early postirradiation period may be a critical factor only in the case of mutations involving the acquisition of new synthetic capacities, as Dr. Lieb suggests. If this should prove to be true, however, I do not see that it would support the repair hypothesis, at least without important revisions. This hypothesis, as well as those invoking the decay of unstable mutagenic intermediates, or of intragenic metastable configurations, presupposes a unitary mechanism of induced mutation, which, indeed, is supported by the qualitative uniformity of the postirradiation temperature effects. If protein synthesis is necessary for the fixation of genetic changes leading to new enzymatic activity, but not for other kinds of mutations, it seems to me that new parameters will have to be considered within the framework of any hypothesis. For example, if the effect of mutagenic treatment is considered to be the production of unstable configurations in genic DNA, it may be that proteins synthesized immediately after irradiation "capture" the transient configuration in stable form, and pass it back to the DNA at the time of gene duplication. In the absence of protein synthesis, the altered specificity is lost. Mutations in which postirradiation protein synthesis has no effect may utilize other routes of specificity transfer, via RNA perhaps. While it is certainly possible that the dichotomy, if it is proved to exist, between mutations that are affected by postirradiation protein synthesis and those that are not, may reflect entirely different pathways of mutagenesis, it seems more economical at first to seek the basis of the difference within one primary framework.

LIEB: The fact that several types of ultraviolet-induced mutations do not seem to require protein synthesis suggests that the basic mechanism, if there is only one, does not involve protein. However, some genes, while able to replicate, may not become functional in metabolism until certain intermediates or "primers" are synthesized. This synthesis may involve protein. Such intermediates might be closely associated with the genes, and might even be transduced. The destruction of such intermediates would account for non-hereditary growth factor requirements often observed in irradiated microorganisms.

MAALØE: Dr. Witkin's elegant analysis of the physiological conditions which favor the development of U.V.-induced mutants throws new light on experiments recently carried out in our laboratory. Exponentially growing cultures of *E. coli* and *S. typhimurium* were irradiated with moderate doses of U.V., and after various post-irradiation treatments the fractions of cells surviving to form colonies on broth plates were determined. It was found that *maximum killing* was obtained when the conditions maintained during the first 15 to 30 minutes after irradiation were such as, according to Witkin, will produce the *maximum number of mutants*. This parallelism extends to nearly all the different growth conditions, including the various combinations of nutrients and growth factors, described by Witkin.

This strongly suggests that the killing of bacteria by U.V. results from processes secondary to the absorption of radiation energy *and* that the same, or closely related processes may lead to the formation of non-lethal mutants. (A paper by Victor G. Bonce and myself, containing some of our observations on the effects of postirradiation treatment has been accepted for publication in Biochim. Biophys. Acta.)

WITKIN: In our experience, too, the survival of bacteria irradiated in the exponential phase of growth is exceedingly sensitive to postirradiation nutritional factors, and it is interesting to hear that these effects parallel those we have obtained for induced prototrophy. In our studies, we avoided the use of growing cultures, as well as the use of liquid culture media in the postirradiation growth period, specifically because of the fact that, under these conditions, survival is so greatly affected by altered conditions that the effects on induced mutation become very difficult to follow. In the strains used in this study, survival after ultraviolet treatment is remarkably uniform under the whole spectrum of posttreatments used, if stationary phase cultures and solid media are employed. While this facilitates the study of induced mutations independently of lethal effects, I do not wish to imply that the modifications of survival level are unimportant. The fact that effects on mutation and on survival are separable, however, supports the already well-documented idea that the postirradiation metabolic sequence is branched, as well as multi-step.

DNA and the T2 Phage Chromosome

Cairns J. 1961. **An Estimate of the Length of the DNA Molecule of T2 Bacteriophage by Autoradiography.** (Reprinted, with permission, from *J. Mol. Biol.* **3:** 756–761 [©Academic Press].)

John Cairns (right) with his son, William, and Matt Meselson by the water's edge at Cold Spring Harbor Laboratory, circa 1961.

Textbook accounts of molecular genetics give the impression that following the revelation of the double helix, discovery followed discovery as the confident pioneers of the new field almost instantly generated today's picture of molecular biology. In fact, "many people found it hard to imagine that biology could be reduced to something so simple" (Cairns 2000), and for some ten years after the double helix, much research was needed to link that elegant structure to genes and chromosomes. This was also a time when molecular geneticists were aspiring to the intellectual rigor of the physicists—nothing could be assumed and experiments had to be done to dot every *i* and cross every *t*. It was in this climate of a "...prevailing wish for precision..." (Cairns 1966a) that John Cairns set out to demonstrate that the chromosome of phage T2 is a single, double-stranded DNA molecule.

The first steps had been taken already in studies of phage chromosomes and by 1956, Hershey was able to conclude that these were indeed "...naked molecules of DNA..." (Hershey et al. 1957). The relationship of genetic maps to DNA molecules was more problematic, especially as genetic analysis indicated T2 had a single chromosome, whereas isolated T2 DNA came in several pieces. Thus, it was more than possible that the chromosome had a complicated structure, with, for example, fragments of DNA linked by proteins (Taylor 1959, 1997). However, when it was realized that even pipetting could break DNA molecules (Davison 1959), more careful extraction methods led to the isolation of much larger molecules. It was still not known whether a chromosome was a single DNA double helix, or, perhaps, following replication, it was made up of a pair of molecules attached side by side.

Cairns's first interaction with phage genetics came in 1957 on his first visit to the United States when he stayed in a house whose tenants included Matthew Meselson, John Drake, and Howard Temin. He had come to Renato Dulbecco's laboratory at the California Institute of Technology to learn the tissue culture techniques being used for studying animal viruses. After four months, Cairns returned to Australia but came back in 1960 to spend a year in Alfred Hershey's laboratory at Cold Spring Harbor. Hershey was developing techniques to isolate unbroken phage DNA, so the material was at hand for determining the size of intact double helices. Cairns's goal was to measure the length of a T2 DNA molecule, and to compare the mass of such molecules, calculated from the average mass of a T2 DNA nucleotide and the number of nucleotides in a molecule of the observed length, with the mass as determined by other means (in particular, from the direct and unambiguous measurement of the amount of phosphorus per T2 virus particle).

Two techniques were available—electron microscopy and autoradiography. Beer and Zobell (1961) had used electron microscopy to determine the sizes of quarter-length T2 DNA fragments produced by physically shearing T2 chromosomes. Cairns rejected the high resolution of the electron microscope approach because the intact DNA molecules would not fit on a specimen grid. Instead, he turned to autoradiography, a technique that has lower resolution but was one he had already used for studying the replication of vaccinia virus (Cairns 1960). The advantage of this technique over electron microscopy was that, by adding an excess of unlabeled carrier DNA, it was possible to know that long labeled molecules could not be due to artificial aggregation of smaller molecules. This was something that could not be excluded by electron microscopy because its resolution, although high, was not high enough in those days to distinguish one double helix from two double helices lying side by side.

The first problem faced by Cairns was to find a method to attach DNA molecules to glass slides. These would then be covered with "stripping film"—a gelatin film with a photographic emulsion that would record the images of the DNA molecules after sufficiently long exposure (about 2 months). As Cairns candidly admitted, the successful method was discovered by chance, but even here, only a small proportion of the labeled DNA molecules were fully extended. Nevertheless, he found a sufficient number to indicate that the maximum fully extended length was a little over 50 μM. After reviewing the possible sources of error in the measurements, Cairns assumed that the molecules were 52 μM long. Taking that there was one nucleotide per 3.4 Å and that the average molecular weight of a nucleotide was 357, he calculated that the mass of his T2 DNA molecules was 110×10^6. This was slightly lower than estimates using other methods, but "... probably not enough to warrant, at this stage, postulating anything other than an uncomplicated double helix as the form of the T2 DNA molecule".

This was not the last occasion on which Cairns used autoradiography to study chromosomes. Watson and Crick had recognized that a double helical structure posed the problem of how the DNA strands could unwind and separate for replication "...without everything getting tangled." Their solution was to assume that because the cell could actually do it, there must be a mechanism waiting to be discovered (Watson and Crick 1953). Theoretical considerations of unwinding and replication continued to exercise some of the best minds (Delbrück and Stent 1957) so Cairns wanted to *see* what was going on during replication. Using replicating DNA molecules from *E. coli*, he showed that the chromosome was probably a circle, 700 to 900 microns long, and that DNA replication produced a fork as the helix separated and each strand was duplicated (Cairns 1963, 1966b).

Later still, he carried out similar experiments on eukaryotic chromosomes. Cairns had done the experiment using HeLa cells while in Australia but had not examined the autoradiographs before, in 1963, he returned to Cold Spring Harbor as director of the Biological Laboratory. Then, other pressing matters at the Laboratory preoccupied him. Fortunately, he was reminded of the experiments three years later when Matt Meselson phoned him to suggest that they collaborate in examining *Drosophila* chromosomes. These would be labeled by growing larvae in vast quantities of tritiated thymidine; Meselson would isolate the fourth chromosome and Cairns would do the autoradiography. At this point, Cairns recalled that he had the HeLa cell autoradiographs in his desk drawer awaiting analysis. He did the analysis and saw the short replicons characteristic of eukaryotic DNA replication. In contrast to the three years between experiment and analysis, Cairns's note was sent to the *Journal of Molecular Biology* within two weeks (Cairns 1966c)!

John Cairns was born on November 21, 1922, in Oxford. He graduated in medicine from the University of Oxford in 1943. He was appointed Clinical Pathologist at the Radcliffe Infirmary in 1947 before moving to Australia in 1950. He remained there until 1963 except for two years (1952–1954) at the Virus Research Institute, Entebbe, and Research Fellowships at the Rockefeller Institute (1957) and Cold Spring Harbor (1960–1961). Cairns was the Director of Cold Spring Harbor Laboratory (1963–1968) and oversaw its creation by amalgamation of the Biological Laboratory with the Carnegie Institution's Department of Genetics. (Hershey and McClintock then constituted the Carnegie's Unit of Genetics, with the former as its director.) He remained a staff member at the Laboratory until 1972 when he was appointed Head of the Mill Hill Laboratories of the Imperial Cancer Research Fund, London. After eight years there he moved to the Harvard School of Public Health, from which he retired in 1991. Cairns was made a Fellow of the Royal Society in 1974.

Beer M. and Zobell C.R. 1961. Electron stains 2: Electron microscopic studies on visibility of stained DNA molecules. *J. Mol. Biol.* **3:** 717–726.

Cairns J. 1960. The initiation of vaccinia infection. *Virology* **11:** 603–623.

———. 1961. An estimate of the lengths of the DNA molecule of T2 bacteriophage by autoradiography. *J. Mol. Biol.* **3:** 756–761.

———. 1963. The bacterial chromosome and its manner of replication as seen by autoradiography. *J. Mol. Biol.* **6:** 208–213.

———. 1966a. The autoradiography of DNA. In *Phage and the origins of molecular biology* (ed. J. Cairns et al.), pp. 252–257. Cold Spring Harbor Laboratory, Cold Spring Harbor, New York.

———. 1966b. The bacterial chromosome. *Sci. Am.* **214:** 36–44.

———. 1966c. Autoradiography of HeLa cell DNA. *J. Mol. Biol.* **15:** 372–373.

———. 2000. The size of the units of heredity. In *We can sleep later: Alfred D. Hershey and the origins of molecular biology* (ed. F.W. Stahl). Cold Spring Harbor Laboratory Press, Cold Spring Harbor, New York. (In press.)

Davison P.F. 1959. The effect of hydrodynamic shear on the deoxyribonucleic acid from T2 and T4 bacteriophages. *Proc. Natl. Acad. Sci.* **45:** 1560–1568.

Delbrück M. and Stent G.S. 1957. On the mechanism of DNA replication. In *The chemical basis of heredity* (ed. W.D. McElroy and B. Glass), pp. 699–736. Johns Hopkins Press, Baltimore, Maryland.

Hershey A.D., Burgi E., Mandell J.D., and Tomizawa J. 1957. Growth and inheritance in bacteriophage. *Carnegie Inst. Wash. Year Book* **56:** 362–364.

Taylor J.H. 1959. The organization and duplication of genetic material. *Proc. Int. Congr. Genet.* **1:** 63–78.

———. 1997. Tritium-labeled thymidine and early insights into DNA replication and chromosome structure. *Trends Biochem. Sci.* **22:** 447–450.

Watson J.D. and Crick F.H.C. 1953. Genetical implications of the structure of deoxyribonucleic acid. *Nature* **171:** 964–967.

J. Mol. Biol. (1961) **3,** 756–761

An Estimate of the Length of the DNA Molecule of T2 Bacteriophage by Autoradiography

JOHN CAIRNS†

*Department of Genetics, Carnegie Institution of Washington,
Cold Spring Harbor, N.Y., U.S.A.*

(*Received 29 June 1961*)

T2 bacteriophage, labelled with [^{3}H]thymidine or [^{3}H]thymine, is subject to suicide on storage. The efficiency of suicide from ^{3}H-decay is apparently the same as that from ^{32}P-decay.

Autoradiography of T2 DNA, labelled with [^{3}H]thymine and extracted in the presence of 1000-fold excess of cold T2, shows that the molecule can assume the form of an unbranched rod about 52 μ long. If the molecule is throughout its length a double helix in the *B* configuration, this indicates a molecular weight of 110×10^6.

1. Introduction

The decay of tritium gives rise to electrons whose mean range in autoradiographic emulsion is less than one micron (Fitzgerald, Eidinoff, Knoll & Simmel, 1951). It should therefore be possible to obtain a high resolution image of individual molecules of ^{3}H-labelled DNA by autoradiography, using the very highly labelled [^{3}H]thymine that is now available; thus DNA containing [^{3}H]thymine of specific activity 10 c/m-mole will have roughly one disintegration per micron of double helix per week and should produce a near-continuous line of grains along its length after a few weeks' exposure.

Bacteriophage T2 seemed in most respects the best material with which to launch such a procedure. Most of the precursors for T-even thymine synthesis come from the medium after infection (Weed & Cohen, 1951; Kozloff, 1953) so there should be extensive incorporation of labelled thymine given at the time of infection; T2 DNA can be extracted in a pure and homogeneous state with phenol (Mandell & Hershey, 1960), each particle providing a single molecule with a molecular weight of over 100×10^6 (Rubinstein, Thomas & Hershey, 1961); lastly a molecule of such great size is ideal for testing a method of measuring molecular length the accuracy of which is, in theory at least, independent of length.

2. Materials and Methods

Phage. T2, strain T2H (Hershey), was used throughout.

Bacteria. Escherichia coli strain S was used for the production of stocks of unlabelled phage and for phage assays. Labelled phage was prepared in the thymineless strain B3 (Brenner).

† Present address: The Australian National University, Canberra, Australia.

Media. Stocks of unlabelled phage were prepared in M9 (Adams, 1959) supplemented with 0·5 g/l. NaCl. All experiments on the production of phage in the presence of limited thymine or thymidine were carried out using the glucose-ammonium medium described by Hershey (1955).

E. coli strain B3 was grown in the presence of 5 μg/ml. of thymine or thymidine. Phage assays were performed using the standard methods (Adams, 1959). Dilution of phage stocks was made in 10^{-3} M-MgCl$_2$, 0·05% NaCl, 0·001% gelatin, buffered with 0·01 M-tris (2-amino-2-hydroxymethylpropane-1:3-diol) pH 7·4.

[^{3}H]*Thymine and* [^{3}H]*thymidine.* These were obtained from the New England Nuclear Corp. and from Schwartz Inc. In the case of the former source these materials are prepared by reducing 5-hydroxymethyl-uracil with tritium so that the label is confined to one hydrogen atom in the methyl group; for this reason they can be specifically designated as 5-[^{3}H]methyl-uracil and 5-[^{3}H]methyldeoxyuridine.

Preparation of labelled phage. In order to ensure extensive incorporation of [^{3}H]thymine (or thymidine) into phage it was necessary to engineer the situation so that phage would only be made if thymine was present and would then be made in an amount which was proportional to the amount of thymine present. The thymineless B3 strain of *E. coli* was used as host since this strain readily incorporates thymidine even at low concentrations, whereas the prototroph does not (Crawford, 1958). Since T2 infection causes the formation of thymidylate synthetase even in thymineless bacteria (Barner & Cohen 1959), it was necessary to block this enzyme by the addition of 5-fluorodeoxyuridine (FUDR) (Cohen, Flaks, Barner, Loeb & Lichtenstein, 1958). At the same time, uridine (UR) was added to ensure that FUDR derivatives were not incorporated into RNA. Thus the final procedure was as follows:

E. coli B3 was grown to 2×10^8 cells/ml. in glucose-ammonium medium with 5 μg/ml. thymidine, and then centrifuged and resuspended in one third volume of fresh medium without thymidine. FUDR and UR were added to give final concentrations of 10^{-5} and 10^{-4} M. Five minutes later the bacteria were infected with T2 at a multiplicity of 4. After 4 min, 0·02 ml. of these infected bacteria was added to 0·02 ml. of double strength medium (with 2×10^{-5} M-FUDR and 2×10^{-4} M-UR) and 0·02 ml. of [^{3}H]thymidine in water. The final concentrations in this growth tube were 2×10^8 B3/ml. 10^{-5} M-FUDR, 10^{-4} M-UR, 2 to 8 μg/ml. thymidine (40 to 160 μC/ml.). (When thymine was the label, the growth tube was supplemented with 10^{-3} M-deoxyadenosine.) After aeration for a further 60 min, the bacteria were lysed with chloroform and the contents of the growth tube were made up to 10 ml. with diluting fluid.

The yield of phage from such a system was 1×10^{10} phage/μg thymidine and 6 to 8×10^9 phage/μg thymine; it was slightly less with [^{3}H]thymidine and [^{3}H]thymine. These yields are 2 to 4 times less than would be expected on the basis of the known thymine content of T2 (Hershey, Dixon & Chase, 1953), but they were not lowered further by raising the concentration of FUDR. In the absence of thymine or thymidine the yield was about 5 phage per bacterium. Purified phage prepared from cold thymidine in this way had the normal optical density per infective particle at 260 mμ. Phage prepared from hot or cold thymine or thymidine showed no rise in the frequency of *r* mutants.

Extraction of phage DNA. Phage was extracted with phenol according to the method of Mandell & Hershey (1960). When labelled phage was extracted, enough cold carrier phage was added to bring the concentration up to the requisite 3×10^{12} phage/ml.; the mixture was then packed in the centrifuge, resuspended and extracted with phenol.

Autoradiography. Once the DNA had been extracted it was diluted to a suitable concentration in various salt solutions and spread in various ways upon glass microscope slides which had previously been coated with various materials. These slides were, on occasion, then coated with chrome-gelatin (0·5% gelatin, 0·05% chrome alum). They were overlaid in the usual manner with Kodak autoradiographic stripping film, AR 10, and stored at 4°C over silica gel in an atmosphere of CO$_2$ to prevent latent image fading (Herz, 1959). After exposure the film was developed with Kodak D19b for 20 min at 16°C.

J. CAIRNS

3. Results

(a) *Suicide of* ³H-*labelled phage*

Two lots of [³H]thymidine and one of [³H]thymine were used at various times for making labelled phage. In each case the resulting phage was diluted 10² to 10⁴-fold, stored at 4°C and repeatedly assayed for surviving phage. Excess cold phage mixed with the hot phage and stored under the same conditions proved to be stable, as did phage prepared by the same procedure but with cold thymidine. Thus the observed

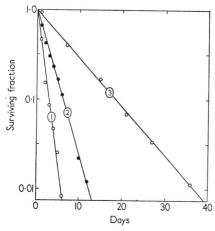

FIG. 1. The suicide of T2 labelled with (1) [³H]thymine 11·2 c/m-mole. (2) [³H]thymidine 5·2 c/m-mole and (3) [³H]thymidine 1·8 c/m-mole.

inactivation of labelled phage was due neither to indirect effects of radiation within the storage tube nor to any innate instability of phage prepared, for example, in the presence of FUDR. Since tritium has a half-life of 12 years, no correction has to be made for decline in radioactivity during the period of storage.

TABLE 1

The suicide of T2 phage labelled with [³H]thymine and [³H]thymidine.

Label	sp. A (c/m-mole)	k (lethals/phage. day)	$\dfrac{k.\text{m-mole}}{c}$	αN† ("lethal" methyl H atoms/phage)
(1) 5-(³H-methyl)-uracil	11·2	0·78	0·069	$3·9 \times 10^4$
(2) 5-(³H-methyl)-deoxyuridine	5·2	0·37	0·071	$4·1 \times 10^4$
(3) ³H-thymidine	1·8	0·13	0·072	$4·1 \times 10^4$

$$† \; \alpha N = \frac{\text{lethals}}{\text{disintegration}} \times \frac{\text{methyl H atoms}}{\text{phage}}$$

$$= \frac{\text{lethals}}{\text{phage . day}} \times \frac{\text{m-mole}}{c} \times \frac{\text{molecules}}{\text{m-mole}} \times \frac{\text{methyl H atoms}}{\text{molecule}} \times \frac{\text{Curie. day}}{\text{disintegration}}$$

$$= k \times \frac{\text{m-mole}}{c} \times 6·02 \times 10^{20} \times 3 \times \frac{1}{3·20 \times 10^{15}}$$

$$= 5·64 \times 10^5 \times \frac{k.\text{m-mole}}{c}$$

The results are shown in Fig. 1 and Table 1. The suicide of phage is seen to be a first order process with a rate constant (k) which is directly proportional to the specific activity (c/m-mole) of the thymine or thymidine. From this the number, αN, of "lethal" thymine methyl H atoms per phage may be calculated to be 4×10^4. Since the burst size in these experiments was around 120, the contribution of cold thymine from the pool of bacterial DNA (about 24 phage equivalents of thymine) (Hershey *et al.*, 1953; Hershey & Melechen, 1957) will have lowered the specific activity of the incorporated thymine by 20%. Correction for this raises αN to 5×10^4.

This value is not significantly different from the number of "lethal" P atoms per phage, determined from the rate of ^{32}P-suicide (Hershey, Kamen, Kennedy & Gest, 1950; Stent & Fuerst, 1955). Since one third of the bases in T2 DNA are thymine, the total number of thymine methyl H atoms equals the total number of P atoms. It follows therefore that the efficiency of inactivation by decay of ^{32}P and ^{3}H are the same. This is an unexpected result. First, ^{32}P and ^{3}H differ greatly in the energy of the electrons they emit (max. energy 1700 and 17 kev respectively). Second, the sites of their incorporation into DNA seemingly could scarcely differ more; the decay of ^{32}P, in the sugar-phosphate chain, and its conversion to sulfur must necessarily break that chain; the decay of ^{3}H, in the methyl group of thymine, and its conversion to helium need not necessarily cause chain breakage nor perhaps any lasting local alteration in the DNA at all.

Practically, these results indicate that at least 99% of the phage is fairly uniformly labelled.

(b) *Autoradiography of ^{3}H-labelled T2 DNA*

Although the production and extraction of highly-labelled DNA presented no problem, there was little prior information on how best to fix this DNA in a sufficiently extended state so that its contours could be followed by autoradiography. Electron microscopy has shown that DNA can be adsorbed from phosphate-buffered solutions of pH 5 to 6 to a variety of surfaces as straight rods many microns long (Hall & Litt, 1958; Beer, 1961) and various methods have been used to ensure that at the time of adsorption the molecules are subject to sufficient shear to align them.

In the course of several months many combinations of DNA concentrations, suspending fluids, varieties of shearing force and adsorbent surfaces were tested. Interestingly, fibres drawn from DNA at high concentration show very poor extension of the minority of molecules that are labelled. The most satisfactory method was found by accident. On testing the appearance of labelled DNA adsorbed to slides partly coated with a co-polymer of polyvinylpyridine and styrene (generously supplied by Dr. Michael Beer), numerous straight molecules were seen adsorbed to the glass on either side of the area coated with polymer; this glass had been cleaned with chromic acid, coated with DNA by drawing the slide across the surface of a solution of DNA in M/15 phosphate buffer pH 5·6, drained and rinsed with distilled water, and then coated with chrome-gelatin (to ensure that the autoradiographic film remained stuck to the slide on drying). Even here, however, there were only localized regions where the DNA was suitably extended. In most regions, the individual molecules were apparently folded back on themselves several times to form a short "rod" of densely packed grains. It seems therefore that the best method for displaying DNA molecules —at least those as long as T2 DNA—has not yet been found.

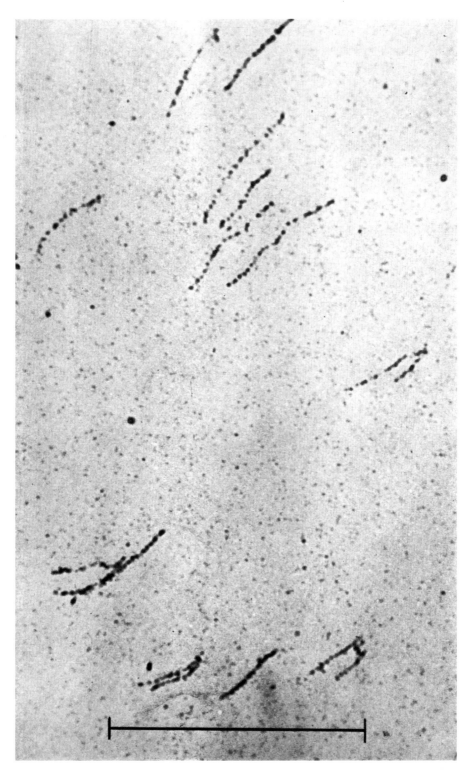

PLATE I. T2 DNA, labelled with [³H]thymine (11·2 c/m-mole), extracted with phenol in the presence of 1000-fold excess of cold T2, and adsorbed to glass at a total DNA concentration of 25 μg/ml. in M/15 phosphate buffer pH 5·6. The autoradiographic exposure was 63 days. The scale shows 100 μ.

The appearance of the extended DNA is shown in Plate I. Of the 13 labelled molecules (or fragments of molecules) shown, 7 have a length between 49 and 53 μ; of the remaining 6, one (immediately over the center of the scale) seems from its grain density and length to be folded about its center. Other samples of DNA prepared on different occasions likewise showed that the maximum length, when adsorbed on to glass, was slightly more than 50 μ.

As an estimate of length and hence of molecular weight this is subject to certain errors and variables. For example:

(i) There are several configurations which the molecules might assume, giving values of 2·55 to 3·46 Å per base pair (Langridge, Wilson, Hooper, Wilkins & Hamilton, 1960). Of these the most likely, particularly in the case of T2 DNA (Hamilton *et al.*, 1959), is the *B* configuration with 3·4 Å per base pair.

(ii) Since it is the autoradiographic image and not the molecule itself which is seen, any stretching of the film between exposure and measuring will produce an apparent lengthening. This, however, seems to be a rare occurrence with stripping film.

(iii) Since the molecule is indicated as a series of grains which one may assume to be randomly placed along its length, it is simple to show for this case, where the mean number of grains (M) per molecule of length L is more than about 10, that the mode, mean and variance of observed lengths (that is, between the centres of the outermost grains) will be approximately $L(1 - 1/M)$, $L(1 - 2/M)$ and $L^2(2/M^2)$ respectively. One would therefore expect for molecules such as these, marked with 50 to 100 grains, that the length of the average molecule would be underestimated by 2 to 4%.

(iv) The resolution of the technique can best be judged by the fact that the grains appear to deviate little to either side of the apparent line of each labelled molecule. It is therefore unlikely that the length of any molecule is overestimated by more than 1μ for the reason of poor resolution.

Thus if any single length has to be selected as the most likely for T2 DNA, that length is probably 52 μ. Taking a value of 3·4 Å per base pair and 357 as the average molecular weight of a base in the sodium salt of T2 DNA, this indicates a molecular weight of 110×10^6 and a phosphorus content of $3\cdot0 \times 10^5$. These are slightly below the accepted values though probably not by enough to warrant, at this stage, postulating anything other than an uncomplicated double helix as the form of the T2 DNA molecule.

I am greatly indebted to Dr. A. D. Hershey for his advice and encouragement and for the hospitality of his laboratory, to Dr. Michael Beer for information and advice on spreading DNA, and to the National Institutes of Health (U.S.A.) for a post-doctoral fellowship during which this work was done.

REFERENCES

Adams, M. H. (1959). *The Bacteriophages.* New York: Interscience Publishers.
Barner, H. D. & Cohen, S. S. (1959). *J. Biol. Chem.* **234**, 2987.
Beer, M. (1961). *J. Mol. Biol.* **3**, 263.
Crawford, L. V. (1958). *Biochim. biophys. Acta*, **30**, 428.
Cohen, S. S., Flaks, J. G., Barner, H. D., Loeb, M. R. & Lichtenstein. J. (1958). *Proc. Nat. Acad. Sci.*, *Wash.* **44**, 1004.
Fitzgerald, P. J., Eidinoff, M. L., Knoll, J. E. & Simmel, E. B. (1951). *Science*, **114**, 494.
Hall, C. E. & Litt, M. (1958). *J. Biophys. Biochem. Cytol.* **4**, 1.

AUTORADIOGRAPHY OF T2 DNA   761

Hamilton, L. D., Barclay, R. K., Wilkins, M. H. F., Brown, G. L., Wilson, H. R., Marvin, D. A., Ephrussi-Taylor, H. & Simmons, N. S. (1959). *J. Biophys. Biochem. Cytol.* **5,** 397.

Hershey, A. D. (1955). *Virology,* **1,** 108.

Hershey, A. D., Dixon, J. & Chase, M. (1953). *J. Gen. Physiol.* **36,** 777.

Hershey, A. D., Kamen, M. D., Kennedy, J. W. & Gest, H. (1950). *J. Gen. Physiol.* **34,** 305.

Hershey, A. D. & Melechen, N. E. (1957). *Virology,* **3,** 207.

Herz, R. H. (1959). *Lab. Investigation,* **8,** 71.

Kozloff, L. M. (1953). *Cold Spr. Harb. Symp. Quant. Biol.* **18,** 209.

Langridge, R., Wilson, H. R., Hooper, C. W., Wilkins, M. H. F. & Hamilton, L. D. (1960). *J. Mol. Biol.* **2,** 19.

Mandell, J. D. & Hershey, A. D. (1960). *Analyt. Biochem.* **1,** 66.

Rubinstein, I., Thomas, C. A. & Hershey, A. D. (1961). *Proc. Nat. Acad. Sci., Wash.* **47,** 1113.

Stent, G. S. & Fuerst, C. R. (1955). *J. Gen. Physiol.* **38,** 441.

Weed, L. L. & Cohen, S. S. (1951). *J. Biol. Chem.* **192,** 693.

The Sticky Ends of Phage λ and Their Uses

Hershey A.D., Burgi E., and Ingraham L. 1963. **Cohesion of DNA Molecules Isolated from Phage Lambda.** (Reprinted, with permission, from *Proc. Natl. Acad. Sci.* **49:** 748–755.)

Al Hershey settles down in 1971 to read his copy of the Cold Spring Harbor Monograph "Lambda Phage" that he edited.

Hershey continued his studies of phage infection—principally by T2—through the 1950s. He published papers on protein and nucleic acid synthesis following infection, as well as ones examining recombination between phage chromosomes. In the late 1950s, however, his interests began to change and he became interested in the biophysical properties of phage chromosomes, especially in determining their size and the effects of shear and other conditions on them. This change was reflected in the titles of his reports in the Department of Genetics' section of the Carnegie Institution of Washington's Annual Report. These had been entitled "Growth and Inheritance in Bacteria" but were changed for 1962–1964 to "Some Idiosyncrasies of Phage DNA Structure." Two years later, however, biology reappeared in "Structure and Function of Phage DNAs."

As early as 1958, Hershey was interested in the paradox that although physical studies suggested that T2 contained between 5 and 20 molecules of DNA, genetic data were unambiguous in showing that there was a single chromosome. Hershey and his colleagues, especially Elizabeth Burgi who was his collaborator and co-author on many experiments throughout the 1960s, began studying this paradox. By 1961, Hershey believed that he had shown that earlier physical experiments were wrong—there was a single DNA molecule in T2 and its molecular weight was at least 130 million (Rubenstein et al. 1961). He concluded that the isolation procedures were breaking the molecules, accounting for the plethora of T2 DNA molecules detected previously. Burgi and Hershey pursued this, using simple tools—changes in temperature and salt concentration, and mechanical shearing to break chromosomes—and analyzing the resulting fragments by chromatography. Later, they turned to density centrifugation using sucrose and cesium chloride gradients to separate and determine the sedimentation coefficients of the molecules. These are measured in S units (named for Svedberg, the Swedish Nobel Laureate who developed high-speed centrifugation).

By 1962, Hershey and Burgi were working also on T5 and phage λ, but the results with phage λ were unusual and suggested that phage λ DNA existed in "... a diversity of molecular shapes or aggregation products..." (Hershey 1962). Phage λ had been discovered by Esther Lederberg as a phage that infects the bacterium *Escherchia coli* strain K12 (Lederberg and Lederberg 1953). Normally, DNA from the infecting phage particles begins to replicate, phage proteins are synthesized, and new virus particles are assembled and burst from the cell. The phage DNA can also become integrated into the bacterial chromosome, where its lytic functions remain latent, until, many bacterial generations later, the phage reappears. This phenomenon of *lysogeny* had been observed as early as 1921, but its mysterious nature was used in the controversy over the nature of bacteriophage (Summers 1999). Opponents of Felix d'Hérelle's assertion that bacteriophage was an intracellular bacterial virus argued that viruses killed cells and could not coexist with the infected cell as lysogeny implied. It was not until many years later that Lwoff's classic work demonstrated that lysogenic bacteria do contain phage (Lwoff and Gutmann 1950). Phage λ has several desirable properties as an experimental organism (Hershey and Dove 1971), and it became a staple of phage research despite not being one of the phage approved by Delbrück (Brock 1990). (Yarmolinsky, in his brief tongue-in-cheek "psychohistory" of phage research, has suggested that Delbrück, a physicist, was attracted to the no-nonsense virulent T phages rather than the more complex phages like λ; Yarmolinsky 1981.)

Hershey and Burgi decided to pursue their unusual findings because of the interesting biological properties of phage λ, as well as technical advantages of using a virus with a small chromosome (50 kb) that could be easily isolated. In the paper reprinted here, they showed that DNA molecules isolated from phage λ aggregated at high concentration and that these aggregates could be broken by spinning a flat blade through the solution. First, 32S molecules were produced at low speeds and then smaller 25.2S molecules were formed at higher speeds. The former molecules would reaggregate while the latter did not. The behavior of the 32S molecules was similar to that of linear T2 DNA molecules, and Hershey concluded that the 32S molecule was the λ linear chromosome. A third form of phage λ DNA—37S molecules—was obtained when dilute solutions of DNA were heated and then cooled slowly. Hershey suggested that these were folded molecules and these, like the broken molecules, did not aggregate. Finally, aggregation and folding was specific to λ DNA—T5 DNA did not behave in the same way, nor could it form aggregates with, or affect the folding of, λ DNA. Hershey synthesized these findings by proposing that each phage λ chromosome had two cohesive sites that are far apart on the chromosome and small in relation to the overall length of the chromosome. Aggregates were formed by end-to-end joining of molecules through these cohesive sites whereas folded molecules arose when the two sites on a single molecule joined together.

These speculations were confirmed when Hershey and Burgi broke λ DNA molecules in half, separated the two halves, and showed that only left and right halves would rejoin (Hershey and Burgi 1965). In 1971, Wu and Taylor, in what was then a technical tour-de-force, completed sequencing the cohesive ends and showed that they were only 12 nucleotides long out of the 50,000 nucleotides of the complete molecule (Wu and Taylor 1971; Wu 1994). As to the biological functions of the cohesive ends, they are involved in replication of the phage λ chromosome and its packaging into the protein shell of the virus. λ DNA is replicated as a "rolling circle" that produces a chain of λ chromosomes joined through their cohesive ends. An enzyme, terminase, cuts the chain at each cohesive site and the individual chromosomes are packed into the virus shells. A second role for the cohesive ends is in integration of λ DNA into the bacterial chromosome during lysogeny. By first making a ring, the λ chromosome can insert itself into the bacterial chromosome in a single recombinational event (the Campbell model; Campbell 1962).

Finally, cos sites (as the cohesive ends came to be called in the 1970s) assumed a much wider significance with the advent of recombinant DNA techniques. Phage λ was rapidly adopted as a cloning vehicle, in large part because extensive genetic analysis, carried out over the previous 20 years, showed that 40% of the λ could be removed to make room for cloned genes (Murray 1983). In addition, the cos sites themselves were key to an important vector, the cosmid (Collins and Hohn 1978; Sambrook et al. 1989). This combined the advantages of a plasmid that can be grown in *E. coli* with the large size and other advantages of a phage λ vector. Their utility was shown by the isolation of a clone containing a 30-kb stretch of the chicken ovalbumin gene, the largest single DNA fragment cloned up to that time (Royal et al. 1979). Cosmids played a major role in the assembly of large stretches of cloned DNA, although they have now been largely superseded by bacterial, yeast, and P1 phage artificial chromosomes (BACs, YACs, and PACs).

Biographical information on Al Hershey can be found on page 202.

Brock T.D. 1990. *The emergence of bacterial genetics*, p. 127–128. Cold Spring Harbor Laboratory Press, Cold Spring Harbor, New York.

Campbell A. 1962. The episomes. *Adv. Genet.* **11:** 101–137.

Collins J. and Hohn B. 1978. Cosmids: A type of plasmid gene-cloning vector that is packageable *in vitro* in bacteriophage λ heads. *Proc. Natl. Acad. Sci.* **75:** 4242–4246.

Hershey A.D. 1962. Growth and inheritance in bacteriophage. *Carnegie Inst. Wash. Year Book* **6:** 443–448.

Hershey A.D. and Burgi E. 1965. Complementary structure of interacting sites at the ends of lambda DNA molecules. *Proc. Natl. Acad. Sci.* **53:** 325–328.

Hershey A.D. and Dove B. 1971. Introduction to Lambda. In *The bacteriophage Lambda* (ed. A.D. Hershey), pp. 3–11. Cold Spring Harbor Laboratory, Cold Spring Harbor, New York.

Hershey A.D., Burgi E., and Ingraham L. 1963. Cohesion of DNA molecules isolated from phage lambda. *Proc. Natl. Acad. Sci.* **49:** 748–755.

Lederberg E.M. and Lederberg J. 1953. Genetic studies of lysogenicity in *Escherichia coli. Genetics* **38:** 51–64.

Lwoff A. and Gutmann A. 1950. Recherches sur un *Bacillus megatherium* lysogène. *Ann. Inst. Pasteur* **78:** 711–739.

Murray N. 1983. Phage lambda and molecular cloning. In *Lambda II* (ed. R.W. Hendrix et al.), pp. 395–432. Cold Spring Harbor Laboratory, Cold Spring Harbor, New York.

Royal A., Garapin A., Cami B., Perrin F., Mandel J.L., LeMur M., Brégégègre F., Gannon F., LePennec J.P., Chambon P., and Kourilsky P. 1979. The ovalbumin gene region: Common features in the organisation of three genes expressed in chicken oviduct under hormonal control. *Nature* **279:** 125–132.

Rubenstein I., Thomas C.A. and Hershey A.D. 1961. The molecular weights of T2 bacteriophage DNA and its first and second breakage products. *Proc. Natl. Acad. Sci.* **47:** 1113–1122.

Sambrook J., Fritsch E.F., and Maniatis T. 1989. Cosmid vectors. In *Molecular cloning: A laboratory manual*, pp. 3.5–3.58. Cold Spring Harbor Laboratory Press, Cold Spring Harbor, New York.

Summers W.C. 1999. *Felix D'Herelle and the origins of molecular biology*, chapter 6, The nature of phage: Microbe or enzyme. Yale University Press, New Haven, Connecticut.

Wu R. 1994. Development of the primer-extension approach: A key role in DNA sequencing. *Trends Biochem. Sci.* **19:** 429–433.

Wu R. and Taylor E. 1971. Nucleotide sequence analysis of DNA. II. Complete nucleotide sequence of the cohesive ends of bacteriophage DNA. *J. Mol. Biol.* **57:** 491–511.

Yarmolinsky M.B. 1981. Summary. *Cold Spring Harbor Symp. Quant. Biol.* **45:** 1009–1015.

Reprinted from the PROCEEDINGS OF THE NATIONAL ACADEMY OF SCIENCES
Vol. 49, No. 5, pp. 748–755. May, 1963.

COHESION OF DNA MOLECULES ISOLATED
FROM PHAGE LAMBDA

BY A. D. HERSHEY, ELIZABETH BURGI, AND LAURA INGRAHAM

GENETICS RESEARCH UNIT, CARNEGIE INSTITUTION OF WASHINGTON, COLD SPRING HARBOR,
LONG ISLAND, NEW YORK

Communicated March 25, 1963

Aggregation of DNA is often suspected but seldom studied.   In phage lambda we found a DNA that can form characteristic and stable complexes.   A first account of them is given here.

Materials and Methods.—DNA was extracted from a clear-plaque mutant (genotype cb^+) of phage lambda[1] by rotation[2] or shaking[3] with phenol.   Sodium dodecylsulfate, ethylenediaminetetraacetate, citrate, or trichloroacetate was sometimes included in the extraction mixture without effect on the properties of the DNA.   Phenol was removed by dialysis, with or without preliminary extraction with ether, against 0.1 or 0.6 M NaCl.

Sedimentation coefficients were measured[4] at 10 μg DNA/ml in 0.1 and 0.6 M NaCl in aluminum cells at 35,600 rpm with consistent results, and are reported as $S_{20,w}$.

Zone sedimentation[5] of labeled DNA's[6] was observed in 0.1 M NaCl immobilized by a density gradient of sucrose.   A sample, usually containing less than 0.5 μg of DNA in 0.15 ml of 0.1 M NaCl, was placed on 4.8 ml of sucrose solution, and the tube was spun for 5 or 6 hr at 28,000 rpm in an SW39L rotor of a Spinco Model L centrifuge at 10°C.

Solutions containing 5–40 μg DNA/ml in 0.1 or 0.6 M NaCl were stirred on occasion for 30 min at 5°C with a thin steel blade turning in a horizontal plane.[7] Since we used two stirrers of different capacities, stirring speeds given in this paper are comparable only within a context.

Salt solutions were buffered at pH 6.7 with 0.05 M phosphate.

Results.—Disaggregation and breakage: Solutions containing 0.5 mg/ml of lambda DNA in 0.1 M NaCl acquire an almost gel-like character on standing for some hours in a refrigerator.   Diluted to 10 μg/ml, the DNA exhibits in the optical centrifuge an exceedingly diffuse boundary sedimenting at 40–60 s (Fig. 1*A*).   If the diluted solution is aged for several days, the sedimentation rate may fall somewhat (not below 40 s), but the boundary remains diffuse and often appears double.

Stirring the diluted solution at 1,300 to 1,700 rpm yields a single component sedimenting at 32 s (Fig. 1*B*).   The product so obtained is stable for a week or more in the cold in 0.1 M NaCl.   We call this process disaggregation by stirring.

If samples of the diluted solution are stirred at increasing speeds between 1,800 rpm and 2,100 rpm, one sees a stepwise transition from 32 s to 25.2 s components, each by itself exhibiting a sharply sedimenting boundary (Figs. 1*C* and 1*D*).   We call this phenomenon breakage.   Broken DNA can form aggregates, but the characteristic 32 s species cannot be regained.

Aggregation: Disaggregation, in contrast to breakage, is reversible, as shown by the following experiment.   Lambda DNA at 40 μg/ml in 0.6 M NaCl was

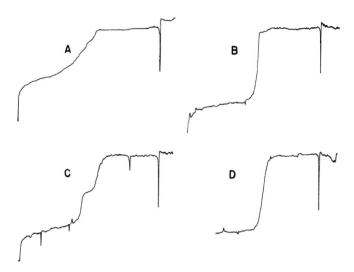

Fig. 1.—Sedimentation pattern of initially aggregated DNA after stirring at several speeds. *A*, unstirred; *B*, 1,600 rpm; *C*, 1,900 rpm; *D*, 2,000 rpm. The meniscus shows at the right.

disaggregated by stirring at 1,700 rpm, and samples at either 40 μg/ml or 10 μg/ml in the same solvent were warmed to 45°C. After measured time intervals, the tubes were chilled and their contents diluted to 10 μg/ml, if necessary, with cold 0.6 M NaCl. Sedimentation coefficients were measured over the course of some hours. Unheated samples showed the same sedimentation rate at the beginning and end of the series of measurements. The heated samples were analyzed in random order, so that the results reflect mainly the duration of heating, not the duration of subsequent storage.

The results, presented in Figure 2, show that the sedimentation rate of the DNA increases rapidly on heating at 40 μg/ml, and less rapidly on heating at 10 μg/ml. The reversibility of disaggregation, and the dependence of rate of aggregation on concentration of DNA, justify our choice of language.

Similar experiments showed that heating in 0.1 M NaCl under the same conditions does not cause appreciable aggregation. Aggregation occurs in that solvent at higher DNA concentrations, however. Thus, aggregation is accelerated by high DNA concentrations, high temperatures, and high salt concentrations.

Linear molecules: According to the description given above, aggregated lambda DNA can be reduced under shear to a uniform 32 s product, which is evidently the structure subject to breakage at higher rates of shear. The maximum stirring speed withstood by 32 s lambda DNA is 1,800 rpm at 10 μg/ml. When

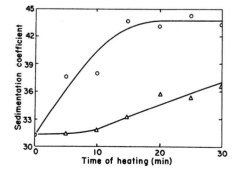

Fig. 2.—Aggregation at 45°C. Circles, 40 μg DNA/ml in 0.6 M NaCl; triangles, 10 μg/ml. The scale on the ordinate refers to observed sedimentation coefficients.

T2 DNA is stirred under the same conditions, it is reduced to fragments sedimenting at 31 s.   Thus, lambda DNA exhibits a fragility under shear that is appropriate
to linear molecules[7] sedimenting at about 32 s.   We therefore conclude that 32 s
lambda DNA consists of linear molecules.   These and other DNA structures are best
identified by zone centrifugation, as illustrated below.

Linear molecules can also be prepared (irrespective of the initial state of aggregation of the DNA) by heating a solution in 0.1 or 0.6 M NaCl to 75°C for 10 min
and cooling the tube in ice water (Fig. 3A, solid line).   This procedure is effective
at concentrations up to 10 μg/ml at least.

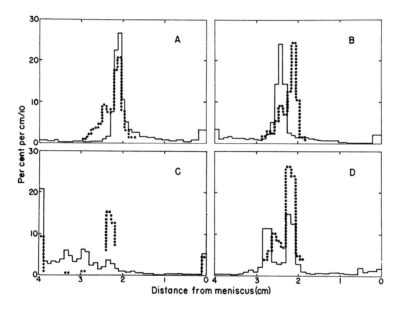

FIG. 3.—Zone sedimentation of several molecular forms.   Solid lines: P[32]-
labeled linear molecules (A), folded molecules (B), aggregates (C), dimers and
linear monomers (D).   Broken lines, a preparation of H[3]-labeled marker DNA
containing linear and some folded molecules.

Linear molecules are obtained directly by extracting the DNA (by rotating, not
shaking, the tubes)[6] at 2 μg/ml into 0.1 M NaCl (Fig. 3A, dotted line).   Control
experiments showed that the mechanical operations involved in the extraction do
not destroy previously formed complexes in solutions diluted to 2 μg/ml.

We conclude that the 32 s form of lambda DNA is analogous to more conventional phage DNA's and is a typical double-helical molecule.

Folded molecules: Another form of lambda DNA we usually prepare by heating
a dilute solution (5 μg/ml or less) in 0.6 M NaCl to 75°C for 1 min, and allowing
the container to cool slowly (0.4° per min at 65°) in the heating bath with the
heater disconnected.   The resulting product sediments as a narrow band moving
1.13 times faster than linear molecules in zone centrifugation (Fig. 3B).   The
expected sedimentation coefficient is 32 × 1.13 = 36.2 s.   Material prepared as
described and then concentrated by dialysis against dry sucrose followed by 0.6
M NaCl shows in the optical centrifuge a sharp boundary at 37 s.

The formation of 37 s material is equally efficient at several DNA concentrations between 5 μg/ml and 0.1 μg/ml (at higher concentrations it is obscured by simultaneous aggregation). The 37 s product, therefore, is composed of monomers that we shall call folded molecules.

When a dilute solution containing either linear or folded molecules in 0.6 M NaCl is heated to 75°C, one gets only linear molecules by rapid cooling and only folded molecules by slow cooling. Partial conversion of linear to folded molecules occurs on heating to 45°C for 30 min followed by rapid cooling, and nearly complete conversion at 60°C. Thus, at 75°C linear molecules are the stable form of lambda DNA. At low temperatures, folded molecules are more stable but the conversion is slow. The slow cooling from 75°C serves to find a temperature near 60°C at which the conversion to folded molecules is rapid and the product is stable.

Folded molecules are formed on heating and slow cooling in 0.1 M NaCl as well as in 0.6 M NaCl, but the conversion is not complete at the lower salt concentration. Some molecular folding also occurs when linear molecules are stored at low concentration and low temperature for a few weeks in 0.1 M NaCl or a few days in 0.6 M NaCl. This is the origin of the faster-sedimenting component of the tritium-labeled marker DNA whose sedimentation pattern appears in Figure 3.

Folded molecules can be converted back into linear molecules by stirring as well as by heating, though the margin between the stirring speed required to accomplish this and the speed sufficient to break linear molecules is rather narrow.

It should be added that heating DNA at 10 μg/ml and 45°C in 0.6 M NaCl produces many folded molecules whose formation competes with the simultaneous aggregation. For this reason the dependence of rate of aggregation on DNA concentration is not truly represented in Figure 2.

Folded molecules themselves do not aggregate. Solutions concentrated for analytical centrifugation continue to yield sharp boundaries after aging in 0.6 M NaCl. Neither do folded molecules form complexes with linear molecules. This was shown by mixing P^{32}-labeled folded molecules with unlabeled linear molecules (20 μg/ml) and aging the mixture in 0.6 M NaCl for 4 days at 5°C. A similar mixture containing labeled linear molecules served as control. Zone centrifugation of each mixture with added H^{3}-labeled marker DNA showed that the labeled linear molecules but not the folded molecules had formed complexes with the unlabeled DNA.

The similarity between the conditions, other than DNA concentration, controlling formation and destruction of folded molecules, and formation and destruction of aggregates, suggests that similar cohesive forces are involved in both phenomena. The folding implies that each molecule carries at least two mutually interacting cohesive sites, which join to form a closed structure. The uniformity of structure of folded molecules, indicated by the narrow zone in which they sediment, suggests that there are not more than two cohesive sites, and that these are identically situated on each molecule.

Dimers and trimers: Aggregated DNA often shows multiple boundaries in the optical centrifuge and always shows multiple components in zone centrifugation. An example, prepared by heating linear molecules for 30 min at 45°C and 40 μg/ml in 0.6 M NaCl, is shown in Figure 3C. Since the characteristic folding seen in monomers is incompatible with aggregation, as already described, it is likely that some of the differently sedimenting products of aggregation are polymers differing in mass rather than configuration.

One form of aggregate can be obtained in moderately pure state by allowing aggregation to occur during a day or so in the cold at 100 μg/ml in 0.1 M NaCl (Fig. 3*D*). Such material contains a fraction of the molecules in linear form, and presumably contains in addition mainly the smaller and more stable aggregates. One of these, as shown in the figure, always predominates, and we assume that it is a dimer. It sediments 1.25 times faster than linear molecules.

In a study of zone centrifugation to be reported separately, we found a relation

$$\frac{D_2}{D_1} = \left(\frac{L_2}{L_1}\right)^{0.35} \tag{1}$$

between molecular lengths (L) and distances sedimented (D) of two DNA's, which is valid for linear molecules. According to this relation, dimers are about twice as long as linear molecules of lambda DNA. The only alternative compatible with the sedimentation rate is a second form of folded monomer, which is ruled out by the requirement for high DNA concentrations during formation. Therefore, dimers are tandem or otherwise open structures. (For definitions of "open" and "closed," see hereafter.)

In more completely aggregated material (Fig. 3*C*) one sees few or no linear molecules, a very few folded molecules (fewer the more concentrated the solution in which aggregation occurred), a considerable fraction of dimers sedimenting 1.25 times faster than linear molecules, and another characteristic component sedimenting 1.43 times faster than linear molecules. According to its sedimentation rate, the last component could be a tandem trimer or a folded or side-by-side dimer. We believe that it is an open trimer for the following reasons.

A folded dimeric structure is ruled out because material sedimenting at rate 1.43 does not form when a dilute solution containing dimers (similar to that shown in Fig. 3*D*) is aged for two weeks in the cold in 0.1 or 0.6 M NaCl, or is heated in 0.6 or 1.0 M NaCl at 45°C. At high DNA concentrations, trimers do form under these conditions. At low DNA concentrations, dimers and trimers are stable and one sees only the conversion of linear to folded monomers.

A side-by-side dimeric structure can be ruled out on the basis of susceptibility to hydrodynamic shear. Figure 4 shows the result when samples of a mixture of trimers, dimers, and folded and linear monomers are stirred at increasing speeds. Trimers disappear first, being converted to dimers or linear molecules or both. Next to go are dimers. Folded monomers are much more resistant, but can be reduced to linear monomers at stirring speeds just insufficient to break the molecules. Thus, trimers, as expected if they are open structures, are more fragile than open dimers, whereas closed dimers should be more stable. We note, however, that a small amount of the material sedimenting at the rate of trimers is relatively resistant to stirring and could signify a minority of closed dimers.

We note also that destruction of dimers and trimers does not liberate any folded monomers, a result consistent with the evidence from sedimentation rates for an open polymeric structure, and with our finding that folded monomers do not form complexes. The fact that aggregation and folding are mutually exclusive processes implies that both utilize the same limited number of cohesive sites, which must be small in size to account for the open polymeric structure. As already suggested by the unique configuration of folded monomers, there may be only two sites per molecule.

Specificity of aggregation: If tracer amounts of P[32]-labeled lambda DNA are mixed with unlabeled lambda DNA at 25 μg/ml in 0.6 M NaCl, and the mixture is brought to 75°C for 1 min and allowed to cool slowly, subsequent zone sedimentation with added H[3]-labeled marker shows that most of the P[32]-labeled DNA has been converted to aggregates and a small remainder to folded molecules. When the same procedure is followed with H[3]-labeled or unlabeled T5 DNA substituted for the unlabeled lambda DNA, the T5 DNA sediments (at its normal rate) 1.20 times faster than the P[32]-labeled lambda DNA, which now consists entirely of folded molecules. Thus, lambda DNA shows no tendency to form complexes with T5 DNA, T5 DNA itself does not form stable aggregates, and T5 DNA does not inhibit molecular folding in lambda DNA. The cohesive sites in lambda DNA are therefore mutually specific, as our model requires.

Role of divalent cations: Divalent cations probably do not play any specific role in the phenomena described in this paper. In NaCl solutions, molecular folding and aggregation are not inhibited by added citrate or ethylenediaminetetraacetate. Neither are these processes appreciably accelerated, in the presence of NaCl, by added calcium or magnesium ions. In a solution of 0.01 M MgCl₂, 0.01 M CaCl₂, and 0.01 M tris (hydroxymethyl) aminomethane, pH 7.2,[8] linear monomers at 10 μg/ml are about as stable as they are in NaCl solutions.

Interpretation of sedimentation rates: Equation (1) shows that if two identical DNA molecules were joined end to end their sedimentation rate would increase by the factor 1.27, evidently owing to the loss of independent mobility. Perhaps the result would be about the same whether they were joined end to end or to form a V, a T, or an X. Thus, we are led to the definition of an *open dimeric structure* as one formed by the joining of two linear molecules at a single point, recognizable by a 1.27-fold increase in sedimentation rate. The principle of independent mobility of parts suggests that, as the structure departed from the tandem arrangement, its sedimentation rate

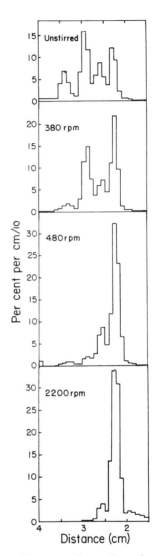

FIG. 4.—Successive destruction by stirring of trimers, dimers, and folded molecules, as seen by zone centrifugation. Linear molecules began to break at 2,400 rpm in this series. The starting material was prepared by heating a sample of DNA, already containing some spontaneously formed folded molecules and dimers, for 30 min at 45°C and 40 μg/ml in 1.0 M NaCl, and diluting to 5 μg/ml in buffered water.

could only increase, not decrease, and in the order V, T, X.

Our results also show that molecules of lambda DNA undergo some sort of folding, apparently as the result of bonding between two cohesive sites lying at some distance from each other on each molecule. If that interpretation is correct, it would appear that when the molecule (regarded as two halves joined end to end) forms an additional point of attachment between its parts, the sedimentation rate increases by an additional factor of 1.13, evidently owing to a further loss of independent mobility of parts. Thus, we are led to the definition of a *closed dimeric structure* as one formed by joining two linear molecules at two points. Such a structure ought to sediment 1.27×1.13 or 1.43 times faster than the linear monomer. We have not found closed dimers, but the question remains how the factor 1.13 would depend on the point of closure of a threadlike molecule. The principle of independent mobility of parts suggests that the sedimentation rate would approach or pass through a maximum as the fraction of the molecular length contained in the loop increased. In some measure it may be possible to answer such questions empirically by determining the locations of cohesive sites on the molecules.

Discussion.—Lambda DNA can exist in at least four characteristic forms that we call linear monomers, folded monomers, open dimers, and open trimers, which sediment respectively at the rates 1.0, 1.13, 1.25, and 1.43, expressed in arbitrary units. These structures are interconvertible with certain restrictions according to the scheme

$$\text{open polymers} \leftrightharpoons \text{linear monomers} \leftrightharpoons \text{folded monomers}$$

As the scheme indicates, linear monomers are subject to two distinct processes: aggregation, seen at DNA concentrations exceeding 10 μg/ml, and folding, seen at any concentration but forced to compete with aggregation at high DNA concentrations. Both processes are accelerated as the temperature is raised to about 60°C, beyond which only linear monomers are stable, and as the salt concentration is raised from 0.1 to 1.0 M. Both processes are rapidly reversed at 75°C or by hydrodynamic shear. All four structures are stable at low temperatures, low DNA concentrations, and low salt concentrations, except for a slow conversion of linear to folded molecules.

Since folded molecules exist in only one stable configuration, and since molecular folding and aggregation are mutually exclusive processes, we postulate that each molecule carries two cohesive sites in prescribed locations, and that these are responsible for both processes. To account for the considerable effect of molecular folding on sedimentation rate, the sites must lie rather far apart along the molecular length. To account for the moderate effect of dimerization on sedimentation rate, the cohesive sites must be small compared to the total molecular length.

According to the proposed model, one might anticipate two dimeric forms, open (that is, joined by one pair of cohesive sites) and closed (joined by two). We find only open polymers, though a minority with closed structures is not excluded. Failure to detect closed polymers may be explained, at least in part, by the fact that the rate of folding must decrease as the length of the linear structure increases.[9]

Whether all details of our model are correct or not, it is clear that lambda DNA forms a limited number of characteristic complexes, not the continuously variable series that might be expected if the molecules could cohere at random. The

VOL. 49, 1963   *BIOCHEMISTRY: HERSHEY ET AL.*   755

limited number of mutually specific cohesive sites implied thereby suggests a specialized biological function, one that remains to be identified.

Summary.—The DNA of phage lambda undergoes reversible transitions from linear to characteristically folded molecules, and from linear monomers to open polymers. Some conditions favoring one state or another have been defined. It may be surmised that each molecule carries two specifically interacting cohesive sites.

This work was aided by grant CA-02158 from the National Cancer Institute, National Institutes of Health, U.S. Public Health Service. Its direction was determined in part in conversation with Dr. M. Demerec about heterochromatin, synapsis, deletions, and speculations to be pursued. Professor Bruno Zimm contributed useful suggestions about the manuscript.

[1] Kellenberger, G., M. L. Zichichi, and J. Weigle, these PROCEEDINGS, **47**, 869 (1961).

[2] Frankel, F. R., these PROCEEDINGS, **49**, 366 (1963).

[3] Mandell, J. D., and A. D. Hershey, *Anal. Biochem.*, **1**, 66 (1960).

[4] Burgi, E., and A. D. Hershey, *J. Mol. Biol.*, **3**, 458 (1961).

[5] Hershey, A. D., E. Goldberg, E. Burgi, and L. Ingraham, *J. Mol. Biol.*, **6**, 230 (1963).

[6] Burgi, E., these PROCEEDINGS, **49**, 151 (1963).

[7] Hershey, A. D., E. Burgi, and L. Ingraham, *Biophys. J.*, **2**, 423 (1962).

[8] Kaiser, A. D., *J. Mol. Biol.*, **4**, 275 (1962).

[9] Jacobson, H., and Stockmayer, W. H., *J. Chem. Phys.*, **18**, 1600 (1950).

DNA Synthesis and the Case of the Missing Enzyme

De Lucia P. and Cairns J. 1969. **Isolation of an *E. coli* Strain with a Mutation Affecting DNA Polymerase.** (Reprinted, with permission, from *Nature* **224**: 1164–1166 [©Macmillan Magazines Ltd.].)

Paula De Lucia in John Cairns's laboratory in 1968.

ALTHOUGH IT HAD NOT ESCAPED WATSON AND Crick's notice that the double helix suggested "...a possible copying mechanism for the genetic material" (Watson and Crick 1953a), how that came about was quite another matter. In their second paper, for example, they entertained the possibility that a specific enzyme might not be needed to join the nucleotides that would become the newly synthesized DNA strand (Watson and Crick 1953b). One person who can have had no doubt that enzymes were involved in DNA synthesis was Arthur Kornberg (1989). Beginning with an analysis of the synthesis of nucleotides, Kornberg and his associates developed an in vitro system based on extracts of *E. coli* in which there was net synthesis of DNA (Kornberg et al. 1956).

This work caused something of a sensation. Crick, in a letter to Sydney Brenner, described Kornberg's 1956 presentation at the "Chemical Basis of Heredity" Symposium (McElroy and Glass 1957) as "by far the most exciting story" (Judson 1996). It was especially exciting because of the findings that the reaction required DNA and all four deoxynucleotide triphosphates, as expected on the basis of the Watson–Crick model. Later, Kornberg was able to show that the nucleotide composition of the newly synthesized DNA matched that of the original DNA (Lehman et al. 1958). Finally, biologically active DNA was prepared by two rounds of replication of a bacterial gene in vitro (Litman and Szybalski 1963). This established once and for all that a single enzyme was, in principle, enough to duplicate a limitless number of genes including, of course, the gene that coded for the enzyme doing the synthesis.

Nevertheless, while Kornberg and his colleagues continued to examine the properties and functions of DNA polymerase, during the late 1960s evidence was accumulating that led John Cairns to think that, perhaps, the cell was not restricted to this one polymerase (Cairns 1972; Friedberg 1997). Kornberg himself discovered a curious property of his polymerase—it possessed a 5′ to 3′ exonuclease activity that cut nucleotides from the end of a DNA strand. It was hard to see why an enzyme that made DNA should have this activity. At the Cold Spring Harbor Symposium in 1968, Cairns heard that Roy Curtiss III had isolated an *E. coli* mutant that produced minicells that had no DNA but did have Kornberg's polymerase, suggesting that most of the polymerase in *E. coli* is, at any moment, not associated with DNA (Cohen et al. 1968). Finally, Cairns did not think that the cell would use the same polymerase to carry out the very different processes of DNA replication and DNA repair. These considerations led him to design an experiment to isolate *E. coli* mutants that had no Kornberg polymerase.

Cairns could not, however, be certain that *E. coli* could survive without the Kornberg enzyme, so he searched for a strain that could grow at low temperature but contained a version of the Kornberg enzyme that was inactive at high temperature. This required an assay for polymerase that had to be very simple to perform, because it seemed likely that the desired mutation would be rare and that very many colonies would therefore have to be tested. In addition, the assay had to be specific for the in vitro activity of the Kornberg enzyme; that is, it must not give a positive result for normal in vivo synthesis. It took Cairns six months to develop a suitable assay and then his technician, Paula De Lucia, and he began the experiment. *E. coli* cultures were treated with a powerful mutagen; the cells were plated out; individual colonies were picked and grown overnight, and extracts were made and assayed for polymerase activity. After testing 3477 colonies, the cells of colony p3478 were found to have no measurable activity, containing, at the most, perhaps 5–10 molecules of functioning enzyme. Despite this, the mutant had the same growth properties as normal cells and it supported growth of infecting viruses to the same extent.

However, p3478 was much more sensitive to UV light, and only very few colonies grew in low concentrations of methylmethanesulfonate (MMS), a carcinogen that did not affect normal cells. Interestingly, those cells that were MMS-resistant had become also UV-resistant and regained their DNA polymerase, showing that the three phenotypes were linked. The genetic analysis of p3478 was worked out in the next two weeks (with Cairns acting as laboratory technician) by Julian and Marilyn Gross, who happened at the time to have just finished teaching the Bacterial Genetics Course. In an accompanying paper, the Grosses described the map position of the mutation and reported that it was an amber mutation that was recessive in diploid strains (Gross and Gross 1969). When Cairns asked what the gene should be called, Julian Gross instantly replied that it should be called *polA* and would be pronounced *Paula* by those in the know.

De Lucia and Cairns could not rule out the possibility that although the polymerase, present at 5–10 molecules per cell, could not manage repair processes, it was still capable of replicating DNA. Nevertheless, they felt that as the mutant strain lacked the irrelevant part of in vitro DNA synthesis (repair), it would allow people to assay for the enzyme that does normal synthesis. Within a couple of years, two other DNA polymerases were found (Cairns 1972). The Kornberg polymerase became pol I and, not surprisingly, the new enzymes were named pol II and pol III. The former is involved in DNA repair while the latter, in a complex with many other proteins, is the polymerase responsible for DNA replication in *E. coli* (Kornberg and Baker 1992).

For many years there was a tension between biochemistry and the newly developing field of molecular biology, the former regarding the latter as an arrogant upstart. It is exemplified in the exchanges between Chargaff and Watson and Crick (Watson 1968), and in Chargaff's remark that molecular biology is the practice of biochemistry without a license (Chargaff 1963). But biochemical studies of a soluble enzyme system tell us how that simple system works in the test tube and not necessarily about its role in the life of the cell. As Zubay and Marmur put it, "One cannot necessarily assign a function to an enzyme merely on the basis of its *in vitro* properties or its abundance. A genetic approach using mutants makes meaningful *in vivo* correlates possible" (Zubay and Marmur 1973).

Biographical information on John Cairns can be found on page 280. Paula De Lucia came to John Cairn's laboratory as a technician in 1964. She left Cold Spring Harbor Laboratory in 1971.

Cairns J. 1972. DNA synthesis. *Harvey Lect.* **66:** 1–18.

Chargaff E. 1963. *Essays on nucleic acids*, p. 176. Elsevier, Amsterdam, The Netherlands.

Cohen A., Fisher W.D., Curtiss III, R., and Adler H.I. 1968. The properties of DNA transferred to minicells during conjugation. *Cold Spring Harbor Symp. Quant. Biol.* **33:** 635–641.

De Lucia P. and Cairns J. 1969. Isolation of an *E. coli* strain with a mutation affecting DNA polymerase. *Nature* **224:** 1164–1166.

Friedberg E.C. 1997. *Correcting the blueprint of life: An historical account of the discovery of DNA repair mechanisms*, pp. 129–131. Cold Spring Harbor Laboratory Press, Cold Spring Harbor, New York.

Gross J. and Gross M. 1969. Genetic analysis of an *E. coli* strain with a mutation affecting DNA polymerase. *Nature* **224:** 1166–1168.

Judson H.F. 1996. *The eighth day of creation: Makers of the revolution in biology*, p. 320. Cold Spring Harbor Laboratory Press, Cold Spring Harbor, New York.

Kornberg A. 1989. *For the love of enzymes.* Harvard University Press, Cambridge, Massachusetts.

Kornberg A. and Baker T.A. 1992. *DNA replication*, ch. 5. W.H. Freeman, New York.

Kornberg A., Lehman I.R., Bessman M.J., and Simms E.S. 1956. Enzymic synthesis of deoxyribonucleic acid. *Biochim. Biophys. Acta* **21:** 197–198.

Lehman I.R., Zimmerman S.B., Adler J., Bessman M.J., Simms E.S., and Kornberg A. 1958. Enzymatic synthesis of deoxyribonucleic acid. V. Chemical composition of enzymatically synthesized deoxyribonucleic acid. *Proc. Natl. Acad. Sci.* **44:** 1191–1196.

Litman R.M. and Szybalski W. 1963. Enzymatic synthesis of transforming DNA. *Biochem. Biophys. Res. Commun.* **10:** 473–481.

McElroy W.D. and Glass B., eds. 1957. *The chemical basis of heredity.* Johns Hopkins University Press, Baltimore, Maryland.

Watson J.D. 1968. *The double helix: A personal account of the discovery of the structure of DNA.* Atheneum, New York.

Watson J.D. and Crick F.H.C. 1953a. Molecular structure of nucleic acids. *Nature* **171:** 737–738.

————. 1953b. Genetical implications of the structure of deoxyribonucleic acid. *Nature* **171:** 964–967.

Zubay G.L. and Marmur J. 1973. DNA synthesis. In *Papers in biochemical genetics,* 2nd edition (ed. G.L. Zubay and J. Marmur), pp. 1–13; quote on p. 5. Holt, Rinehart and Winston, New York.

1164

NATURE VOL. 224 DECEMBER 20 1969

Isolation of an *E. coli* Strain with a Mutation affecting DNA Polymerase

by

PAULA DE LUCIA
JOHN CAIRNS
Cold Spring Harbor Laboratory,
Cold Spring Harbor,
New York 11724

By testing indiscriminately several thousand colonies of mutagenized *E. coli*, a mutant has been isolated that on extraction proves to have less than 1 per cent of the normal level of DNA polymerase. The mutant multiplies normally but has acquired an increased sensitivity to ultraviolet light.

KORNBERG's discovery of an enzyme that could faithfully copy DNA *in vitro*[1] was a crucial step in the history of molecular biology because it firmly established the fact that only a small part of a cell's DNA is needed to code for a mechanism that can duplicate the whole. Whether this is the enzyme responsible for DNA duplication *in vivo* was rightly thought, at that time, to be of secondary importance. Since then, however, circumstantial evidence has accumulated suggesting that, at least in bacteria, this particular enzyme is used for the repair of DNA rather than for its duplication. The various mutants of *Escherichia coli* and *Bacillus subtilis* that are unable to duplicate their DNA at high temperature have all been shown to contain normal polymerase and normal deoxyribonucleoside triphosphate pools at the non-permissive temperature[2-6], and at least one of them has been shown to carry out repair synthesis at high temperature[7]. Repair replication and the process of DNA duplication apparently differ in the extent to which they discriminate against 5-bromouracil as an acceptable substitute for thymine, suggesting that the two reactions involve different polymerases[8]. Finally, the 5'-exonucleolytic activity, recently shown to be an intrinsic property of the *E. coli* polymerase[9], is clearly a desirable attribute for an enzyme responsible for excision and repair but is of no obvious advantage for an enzyme carrying out semiconservative replication.

These and other less persuasive arguments prompted us to look for mutants of the polymerase, in the hope that they would either establish a role for the polymerase in DNA duplication or exclude it and, at the same time, provide convenient strains in which to search for the right enzyme. Although we have not succeeded in these more distant objectives, we have isolated such a mutant and here describe the method of isolation and some of its properties. The accompanying article describes a genetic study of the mutation.

The Selective Procedure

The successful isolation of mutants of *E. coli* lacking ribonuclease I[10] demonstrated that it is possible to find the mutant one wants simply by testing individually several hundred colonies grown from a heavily mutagenized stock. Because we wished to avoid having to guess what symptoms, if any, would result from a lack of DNA polymerase, we decided to follow that example and assay the polymerase in clones of a mutagenized stock until we found what we were looking for. We had to allow for the possibility that the mutation we sought might be a conditional lethal, so we began by assaying at 45° extracts made from clones grown at 25° or 30°; later we tested clones grown at 37°, thinking that temperature-sensitive mutants of the polymerase might be more readily detectable if the enzyme had been assembled at a higher temperature. As it turned out, the mutant we eventually isolated would have been found whatever approach had been adopted, and we shall therefore simply give the history of the mutant when we describe its isolation and properties.

Extraction of Polymerase

Because we expected to have to test many hundred colonies, we required a very simple method for preparing extracts. In addition, we needed a procedure which made the bacteria incapable of incorporating deoxyribonucleosides, to ensure that labelled triphosphates could not enter DNA by way of breakdown to nucleosides and incorporation by those few cells that might have survived the extraction procedure. These two requirements were satisfied by the slight modification of a method devised for extracting polysomes, using the non-ionic detergent Brij-58 (ref. 11). *E. coli* is suspended at a concentration of about 3×10^9/ml. in ice cold 10 per cent sucrose 0.1 M Tris (pH 8.5); lysozyme and EDTA are added to final concentrations of 50 μg/ml. and 0.005 M, respectively, and the mixture is kept on ice for 30 min; addition of a mixture of Brij and $MgSO_4$ (at room temperature) to give final concentrations of 5 per cent and 0.05 M, respectively, results in partial clearing; following centrifugation (1,500g for 30 min), the deposit contains 99.9 per cent of the DNA and the supernatant contains the polymerase, which may then be assayed simply by adding sonicated calf thymus DNA (to 50 μg/ml.) and the four deoxyribonucleoside triphosphates (to a final concentration of 4 nmoles/ml. dATP, dGTP, dCTP and 2 nmoles/ml. ³H-TTP).

This extraction procedure demonstrates one point of interest: any method of lysis that liberates fragmented DNA will automatically create sites for the attachment of polymerase and therefore cannot give a true picture of the location of the polymerase *in vivo*[12]. Extraction with Brij yields cells which still contain their DNA but, on resuspension, have little if any ability to incorporate deoxyribonucleoside triphosphates. Because Brij apparently does not dissociate polymerase from its template (the polymerase being assayable in the presence of Brij), we can conclude that most of the polymerase in *E. coli* is normally not attached to DNA but lies free within the cell—as might befit an enzyme awaiting the summons to repair synthesis. This conclusion is supported by the observation that when *E. coli* segregates daughter cells which lack DNA these cells nevertheless retain their full quota of DNA polymerase[13,14].

NATURE VOL. 224 DECEMBER 20 1969

Isolation of the Mutant

E. coli W3110 *thy⁻*, growing in minimal medium, was washed and suspended in 0·15 M acetate (*p*H 5·5), treated with N-methyl-N′-nitro-N-nitrosoguanidine (1 mg/ml.) for 30 min, and then centrifuged and suspended in Penassay broth[15]. Following growth at 25° C for 18 h, the culture was plated; after incubation overnight at 37° C, the colonies were picked into 1 ml. lots of Penassay broth which were incubated overnight at 37° C and then centrifuged and extracted with lysozyme and Brij.

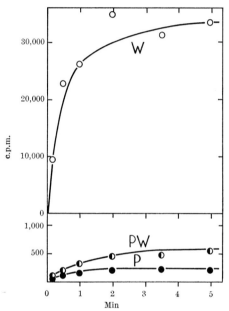

Fig. 1. Triphosphate incorporation by extracts of the parent strain (W), the mutant strain (P) and a mixture of 99 per cent mutant and 1 per cent parent (PW). *E. coli* W3110 *thy⁻* and the mutant derivative, p3478, were grown with aeration in Penassay broth at 37° to about 5×10^8/ml. Each culture was then chilled, centrifuged and suspended in 0·1 M Tris-0·01 M $MgSO_4$ (*p*H 7·4) at a concentration of 6×10^9/ml. A mixture of 1 per cent parent, 99 per cent mutant strain was prepared. This and the two unmixed suspensions were centrifuged and suspended in Tris-Mg²⁺ at a concentration of 1×10^{10}/ml. The three suspensions were disrupted by sonic vibration and mildly centrifuged (1,000*g* for 10 min). To 0·9 ml. of each supernatant at 25° 0·1 ml. sonicated calf thymus DNA was added (final concentration 20 μg/ml.) and, 5 min later, 0·3 ml. triphosphate solution (final concentrations 100 nmoles/ml. dGTP, dATP, dCTP, and 0·6 nmoles/ml., 2·5 μCi/ml., ³H-TTP). Samples of 0·2 ml. were taken from each reaction mixture into 5 ml. 5 per cent trichloroacetic acid–1 per cent sodium pyrophosphate[16]. These samples were then washed on Whatman GFA filters with 5 per cent trichloroacetic acid and with 5 per cent acetic acid, dried and counted in a scintillation counter.

After testing a few thousand colonies in this way we found a clone, p3478, that appeared to lack polymerase activity. It was therefore tested again using a more conventional method for extracting the enzyme. According to this test (Fig. 1), extracts of the mutant have 0·5–1·0 per cent of the normal activity. This decrease in activity does not seem to arise from the presence of an inhibitor.

Some Properties of the Mutant

As far as we can determine, the mutant multiplies at the same rate as the parent strain, in minimal and complete media, and at temperatures from 25° to 42° C. On plating, it forms slightly smaller colonies than those of the parent strain, and occasionally it seems to have difficulty in getting out of stationary phase, but we have not investigated further either of these phenomena.

Parent and mutant are equally susceptible to infection with T4, T5, T7 and λ bacteriophages. When converted to spheroplasts, they are equally susceptible to φX174 DNA and produce equal yields of phage (personal communication from David Dressler). This finding was somewhat surprising, but it should be remembered that all stages in the replication of φX174 DNA are temperature sensitive in a mutant that is temperature sensitive for normal DNA replication[17] but not for repair synthesis[7].

With regard to host cell reactivation, there is no detectable increase in the rate of inactivation of T7 by ultraviolet light, when the survivors are assayed on the mutant rather than the parent. Thus the mutant is *hcr⁺*.

The mutant has a marked increase in sensitivity to ultraviolet light. For convenience, this effect will be documented in the following article[18], where the sensitivities of various derivative strains are compared.

The parent strain will form colonies normally in the presence of 0·04 per cent methylmethanesulphonate, whereas the mutant plates with an efficiency of about 10⁻⁷. We assume that these rare methylmethanesulphonate-resistant cells are revertants that have either arisen spontaneously or been created by the methylmethanesulphonate. Because every one of twenty such independently arising revertants exhibited normal sensitivity to ultraviolet light and had normal or near-normal levels of polymerase, it is clear that the three basic properties of the mutant (UVs, MMSs and lack of polymerase) are the result of a single mutational step.

Repair or Replication

The accompanying article[18] demonstrates that we are dealing with an amber mutation which is recessive in partial diploids. We assume that it is in the gene coding for DNA polymerase, although proof will require the demonstration that it—or other similar mutations—results in changes in the polymerase protein. Because the mutation produces an increased sensitivity to ultraviolet light, it seems likely that recovery from the effects of ultraviolet light is partly the responsibility of this polymerase.

Unfortunately, it is not going to be easy, by a study of this or other such mutants, to show that this polymerase plays no part in normal DNA duplication. Because *E. coli* contains several hundred polymerase molecules per bacterium[19], the residual activity found in extracts of our mutant could represent perhaps 5–10 molecules per cell—a number that could well be sufficient for normal duplication. Even if we could somehow prove that the residual activity were entirely that of another enzyme (in other words, that this amber mutation is not measurably leaky), we should still not have proved that duplication is carried out by some other enzyme, for it could readily be argued that those few polymerase molecules concerned with duplication are necessarily incorporated into some larger enzyme complex the activity of which is not assayable *in vitro*. It could even be argued that more of the polymerase molecule must be intact for it to serve as a repair enzyme (and, incidentally, to survive extraction) than for it to act when part of the replicating machinery. We therefore believe that the question will be resolved either by engineering a total deletion of the polymerase gene or by determining, in some direct manner, which enzymes and what precursors are used for normal DNA duplication. It is our hope that each of these exercises will have been made easier now that the polymerase gene has probably been located[18] and a mutant is generally available.

We thank Dr Raymond Gesteland (who pioneered this kind of mutant hunt) for encouragement; Dr David Dressler for testing our mutant with φX174 and for permission to cite his results; and Drs Julian and Marilyn Gross for arranging to stay on at Cold Spring Harbor to conduct most of the experiments reported in the next article.

1166

NATURE VOL. 224 DECEMBER 20 1969

The work was supported by a grant from the US National Science Foundation.

Received November 26, 1969.

[1] Lehman, I. R., Bessman, M. J., Simms, E. S., and Kornberg, A., *J. Biol. Chem.*, **233**, 163 (1958).

[2] Bonhoeffer, F., *Z. Vererbungslehre*, **98**, 141 (1966).

[3] Buttin, G., and Wright, M., *Cold Spring Harbor Symp. Quant. Biol.*, **33**, 259 (1968).

[4] Fangman, W. L., and Novick, A., *Genetics*, **60**, 1 (1968).

[5] Gross, J. D., Karamata, D., and Hempstead, P. G., *Cold Spring Harbor Symp. Quant. Biol.*, **33**, 307 (1968).

[6] Hirota, Y., Ryter, A., and Jacob, F., *Cold Spring Harbor Symp. Quant. Biol.*, **33**, 677 (1968).

[7] Couch, J.. and Hanawalt, P. C., *Biochem. Biophys. Res. Commun.*, **29**, 779 (1967).

[8] Kanner, L., and Hanawalt, P. C., *Biochim. Biophys. Acta*, **157**, 532 (1968).

[9] Kornberg, A., *Science*, **163**, 1410 (1969).

[10] Gesteland, R. F., *J. Mol. Biol.*, **16**, 67 (1966).

[11] Godson, G. N., and Sinsheimer, R. L., *Biochim. Biophys. Acta*, **149**, 476 (1967).

[12] Billen, D., *Biochim. Biophys. Acta*, **68**, 342 (1963).

[13] Cohen, A., Fisher, W. D., Curtiss, R., and Adler, H. I., *Cold Spring Harbor Symp. Quant. Biol.*, **33**, 635 (1968).

[14] Hirota, Y., Jacob, F., Ryter, A., Buttin, G., and Nakai, T., *J. Mol. Biol.*, **35**, 175 (1968).

[15] Adelberg, E. A., Mandel, M., and Chien Ching Chen, G., *Biochem. Biophys. Res. Commun.*, **18**, 788 (1965).

[16] Hurwitz, J.. Gold, M., and Anders, M., *J. Biol. Chem.*, **239**, 3462 (1964).

[17] Steinberg, R. A., and Denhardt, D. T., *J. Mol. Biol.*, **37**, 525 (1968).

[18] Gross, J. D., and Gross, M., *Nature*, **224**, 1166 (1969).

[19] Richardson, C. C., Schildkraut, C. L., Aposhian, H. V., and Kornberg, A., *J. Biol. Chem.*, **239**, 222 (1964).

Appendix 1

Charles Benedict Davenport, 1866-1944

A Study of Conflicting Influences

by

E. Carleton MacDowell
Carnegie Institution of Washington

(Reprinted from BIOS XVII: No. 1, March, 1946)

Bios   3

Charles Benedict Davenport, 1866-1944
A Study of Conflicting Influences

*by E. Carleton MacDowell**

Carnegie Institution of Washington

ONE of the most influential personalities in the biology of his day was Charles B. Davenport. Few in this field have wielded the power he won, primarily by virtue of the infectious quality of his enthusiasm. Flashes of this shot out in all directions, started uncountable new activities and organizations and gave him control of large funds. Power usually brings admirers and enemies; he had both, but it is probable that a large proportion of those who knew him have a feeling of not understanding, of not knowing just how to regard him. And small wonder; although he himself in his most candid self-analysis failed to recognize it, his life was dominated by a turmoil of conflicting loyalties, and his actions gave the impression of a paradoxical mixture of qualities.

Such complexity breeds legends. We so insist upon a simple classification of people, that we tend to modify our picture of a person to fit some familiar category. Different people shift the picture in different directions. Legends grow, until the real man is lost from view and the difficulties in the way of any true understanding are greatly increased. By praise we ease the sorrow of personal loss; and few there are without praiseworthy qualities. To treat a man of prominence as we would a personal loss is to belittle him.

Objective evaluation of a man's contributions and influence awaits the test of time. But Davenport's life as a whole has a significance for mankind quite apart from that of his various activities. His great problem in life was the adjustment to diverse influences, a problem that concerns everyone. In his case, such influences were so definite and the documentation is so extensive, that his life may be approached as an experiment, in which relationships between given conditions and results may be recognized. As in any experiment, the present purpose is to interpret the results, rather than to estimate their ultimate importance.

Such an intimate interpretation of a man so recently in our midst is usually not attempted. Yet even though subject to emendations and modifications of emphasis, this study may have greater interest while memories of him are still clear in many minds.

---

* The writer lived in continuous personal contact with Doctor Davenport during most of the second half of his life and has had access, during the past year, to official files and an extensive collection of personal letters, autobiographical sketches, diaries and other contemporary documents going back to childhood.

4   Bios

Background

The pioneer American eugenist and father of the Eugenics Record Office was deeply interested in biography, for he considered that what a man does is an important index of heredity. In supporting this thesis he wrote innumerable biographical sketches, emphasizing the interests and traits of the subject that also appeared among the relatives. It would be appropriate to follow this example in preparing his own biography, but the material now collected gives evidence that his life and personality, as well as those of his Puritan forebears, were so conspicuously influenced by doctrine and experience that to identify inherited capacities and trace them to their sources is virtually impossible. Repressions not only conceal capacities, they occasion reactions that belie native traits: affection may become spitefulness, humility reappear as ambition, and a weakness assume the role of a talent.

However undefinable the parental influence transmitted by genes, that conveyed by personal contact was specific and highly potent. Moreover, Davenport's father and mother were so different that their influence was often conflicting. Outwardly this diversity raised no serious problems during his youth, because the actions of the young Charles were unquestioningly regulated by his father. Internally, a conflict of loyalties must have been set up at a very early age. On one side, a loyalty of his mind justifying his actions, on the other, a loyalty of his heart in defense of his dreams. He was twenty-one before the paternal domination over his actions was broken and his mother's influence finally triumphed. This, however, did not bring a resolution of the inward conflict. The father's domination had lasted so long and been accepted so completely that its effects were never entirely outgrown.

This father, Amzi Benedict Davenport (1817-1894) would have interested a psychiatrist. There were strong signs of frustration: in his face bitter unhappiness, in his attitude toward his children, harsh masterfulness. He grew up as a farmer's boy at Davenport Ridge, near Stamford, Conn., property that had been owned by four generations of Davenport farmers and carpenter-farmers. He went to school in New Canaan and early became a school teacher. Before he was 20, he established a private academy in Brooklyn in which he taught for 16 years. In 1853 he set up a real estate office, dealt in insurance and managed estates, eventually acquiring a reputation for great honesty and reliability. The business obviously thrived. He continued farming as a recreation and every summer he took his family from the home at 11 Garden Place, Brooklyn Heights, to Davenport Ridge, where he built a large and imposing dwelling and surrounded it with extensive formal gardens. Deep down he must have had a true love for the country and especially for the rolling meadows and brooks and woods that his ancestors knew. He was well versed in family history and proud of his ancestry. He published in two editions (1851 and 1876) an elaborate genealogy of the Davenport Family that went back continuously to 1086. Before the farming ancestors, the direct line came down through a series of ministers among whom Rev. John Davenport 1st, from England, was one of the founders of New Haven, and Rev. John Davenport 3rd, of Boston, then of Stamford, was granted Davenport Ridge. Short of being a minister, Amzi B. carried on the family's traditional church connection with great emphasis. Besides attending Sunday services and prayer meetings unfailingly, he conducted family prayers every morning in the parlor before breakfast (and woe to him who was late!). He aided in the establishment of two churches in Brooklyn and held the office of Ruling Elder and, four times, that of deacon. He read extensively on religious matters and collected a large general library that was not accessible to his children. The stern repressive teachings of his Puritan ancestors were accepted with reactionary enthusiasm. He was so strictly righteous that he never owed a dollar and therefore had no standing in the mercantile registers. Pleasure, being tainted with evil, had to be denied or disguised. He married twice and sired a total of eleven children, who appeared in most cases at two-year intervals. His attitude toward these children would appear to indicate that they were the source of great displeasure. He was probably a man who wanted to be loved so much more than he could permit himself to be, that he concealed lovable qualities. His letters to Charles's mother are said to have been charming, tender and loving.

But the only tender love that the nine children of the second marriage knew came from their mother, Jane Joralemon Dimon. Her affection was freely given; her rational philosophy, guided by the wholesome human attitudes of her Dutch ancestors, permitted her to be her natural self and to find satisfaction where she could. She maintained an emotional balance and with dignity and tact filled her difficult role under the domineering master of the home. In religious matters she was inclined to be skeptical, but in the results of science she had a lively interest. She studied natural history and collected specimens, cultivated her flower garden, and found a continuing satisfaction in reading French. Her father, John Dimon, a farmer's son and carpenter from East Hampton, Long Island, became an active citizen of Brooklyn, serving as Commissioner of the Alms House and, for several years, Alderman. Her maternal grandfather, Teunis Joralemon, son of a Dutch farmer in New Jersey, acquired a farm on Brooklyn Heights and became a trustee of the village of Brooklyn and, later, a judge. He amassed a considerable fortune, some of which gave financial independence to his granddaughter after she, too, married a farmer's boy (1850).

Davenport's immediate forebears were responding to the call of the city and rapidly changing their modes of life. Farmer's boys, with whatever education the country could give, from Connecticut, Eastern Long Island and New Jersey, went to town and found wives who had already become urbanized. Their sons, growing up in town, were meeting experiences the fathers as boys had not known.

6   Bios

Expression and Repression (*1866-1887*)

One of these sons was Charles Benedict Davenport, the 8th child and last son of Amzi and Jane. Since the date of his birth was June 1st (1866), the place was Davenport Ridge, for, as always, the family had gone to the country early in the spring. Little can be said of his early childhood, save that he was extremely active and read the entire New Testament between the ages of seven and eight. The only contemporary record at hand is a book of daily school marks for the winter of 1874-75, which cover spelling, reading, penmanship, arithmetic, grammar, parsing, and verses. In writing of his childhood, years later, he did not remember this early year at school or how he learned to read; the impression that persisted was of the later years when his father was his teacher until he was 13.

With the beginning of his diaries in 1878, the picture suddenly becomes detailed. In winter, he was office boy and janitor of his father's real estate office, which he opened, swept and dusted every morning; there he spent most of the day doing sums or "stuiding" (*sic,* and so for two years) Smith's "Grammer," except for errands to change "For Rent" signs, to collect frequently sizable rent bills or to the tax office. His father secured all these services for the sum of 25c per week, and of course, the lessons which he gave when he had the time. Frequently this was in the evening while he clipped papers. If Charles failed, he had to go to bed at once. The same penalty was paid for not knowing his Sunday School lesson, which was heard by Amzi even more regularly than other lessons. Sunday School sometimes included black-board exercises that Charlie liked and there he could get a book to read, such as the Rollo books. He attended midweek prayer meeting and Sunday service regularly; the text of the sermon was almost always noted in the diary and the sermon was almost always "very good."

At home he was handyman, blacking his father's boots, carrying coal and ashes and shovelling snow. He was stable boy too, when "Pet" was in town and, when his infant nephew was living with them, willingly and happily "played with baby after breakfast" to relieve the nurse. He had often accomplished a great deal by the time he reached the office about 8 A.M.

All these services he seems to have rendered with meekness. But boxed up in a real estate office, away from his contemporaries, it is small wonder that he occasionally burst out in his diary with "O! I want to go to school." "I hate be in the office." ". . . . that Prison House, as I call it." His father's departures from town and his returns were regularly noted. "Pa gone to the country (Good!?) Did not sweep out the office." "Pa was not at home so we regaled ourselves with an extra long sleep." But on another morning when he overslept, " . . . I descended to the parlor, where sat Pa with a frown on his brow consequently my breakfast did not consist of much." "Got up a little late and after breakfast departed for the office. 'Swept out' and Pa came in and I sat in his office stuiding S.S.

lesson from 8:30 A.M. to 9:30 P.M. All day! *Very* interesting time of it. Did a few examples in arithmetic, made up a little Grammer, but no S.S. lesson. Such is father." And one sad Monday, Dec. 26th, a legal holiday: "Woke at 6:30 and was late for prayers. After breakfast father sent me to bed for that reason for two hours."

Life was serious and mostly too busy for play. He told jokes and considered himself in those days something of a humorist, but the sense of humor that mellows life by quick recognition of distorted values, never developed. Spontaneous or organized group games or athletics seem to have been unknown at any age. Parlor games of an evening were "revels," or "had fun" and this was pathetically seldom. Even to talk to someone was an event. But he was free to write and he did so voluminously. Writing became and remained throughout life his major and more fluent mode of communication. Thoughts that most people would share in conversation, went down on paper; every lecture, and even informal talks, were prepared in full, and much of this writing was published. At the age of eleven he originated and edited a little four-page monthly paper, THE TWINKLING STAR, an excellently printed sheet, complete with editorials, news items, continued stories by "Carlos," jokes, advertisements and foreign correspondence. His father's letters during his grand tour of England and Europe ran through many of the 24 regular numbers that appeared and even required several Extra editions. This trip may have had a very serious genealogical excuse, but the itinerary shows that in fact it was mostly pure sight-seeing. The ancestral neighborhoods called out rhapsodic paragraphs and considerable early Davenport family history, but for the most part the comments specialized on dates, money values and the exact size of buildings, as well as other data to be found in Baedeker.

At this period Charles was buying for himself St. NICHOLAS, THE YOUNG FOLKS and THE YOUNG SCIENTIST, and two library memberships.

He had a few young friends, mostly relatives, but the only group activity was a Natural History Society. At first it was the Excelsior, very much of a family affair, that held meetings and supported an attic museum. "Today begins the third epoch. 1st) April, 1874, organization; 2nd) Jan. '77, adoption of Constitution; 3rd) March 8th, 1878, Revival." But by November of that same year, ". . . . decided to reorganize Excelsior, if others are willing."

Excelsior eventually faded out entirely and in its place came membership (1881) in the Agassiz Association, a more formal natural history society with several chapters, about which Charlie had heard and had written to the permanent secretary of the home chapter in Lennox, Mass. But Charlie found that the B Chapter of Brooklyn, which had elected him, was in a bad way. It consisted of only five boys and at his second meeting he was made vice-president. The next meeting was an attempt to unite chapters A and B, but Charles was the only member of Chapter B to appear.

"I took the president's chair and passed two laws. Made agreement to meet on Saturday to make up constitution and by-laws." As a result of this reorganization he became secretary and member of the executive committee and a few "good Meetings" are recorded, but following one "short and dull" meeting, just three months after his election to the association, he resigned. "Saw Hadden about a new Society. He and I are to form one by ourselves." According to an autobiographical note written many years later, such a society was formed and prospered; during the later school years, papers were written for the meetings. At this time Charles was establishing contacts with more serious scientific organizations by corresponding with the Smithsonian Institution, sending bird lists to the American Ornithological Union and weather records to the Signal Corps. Whenever he made an entry in a diary throughout life, weather was noted.

This interest in natural history was clearly guided by his mother, although in the diaries there is no reference to her in this connection. Indeed she is mentioned only rarely in any connection; whereas "Pa" appears almost daily. The evidence appears in the dedication of Davenport's first book, EXPERIMENTAL MORPHOLOGY ('97) "To the memory of the first and most important of my teachers of Natural History—my mother."

Through all his boyhood, Saturday afternoons were usually given over to an exploring trip, with a brother or sister, and later with a neighbor. There was usually much walking; notes were made or specimens collected —bird lists, colors of winter leaves, pockets full of minerals. Frequently the place was Prospect Park, the Parade Grounds, or Central Park, or one of the museums in New York, and one memorable day it was Trinity Church steeple. From this superior position in those days the panorama was unobstructed. Its commanding sweep and the incredibly small men and horses on the street below made such a deep impression that Charlie filled the odd pages at the end of that diary with the details. At the very top of the steeple he carved his initials. To carry his name to the top was to become a dominating urge, but this urge bore less the marks of an ordinary ambition for rewards, than of an endless struggle to live down the feeling of subservience instilled by his father.

When, at last, he was permitted to go to school (November 26th, 1879), he plunged eagerly into the new life, which he recorded at first in great detail: floor plans of the Brooklyn Collegiate and Polytechnic Institute, the daily routines and procedures in passing from room to room, ". . . after all was quiet, a stamping sound was heard as Dr. Cochran walked up the isle (R) and took position before the desk (B) and gave out a hymn."

From the beginning he was set apart; his isolation and early responsibilities gave him premature seriousness and independence. And to the end of his days he remained a lone man, living a life of his own in the midst of others, and feeling out of place in almost any crowd. His informal preparation was a handicap at first and intensified the spirit of competition called out by contact with his equals. To surpass them grew into an all-absorbing purpose. The habit of competition, thus early established, became fixed. Day by day in his diary he recorded the degree of his success in each recitation; examination and term marks were inevitably put down, often with comments. Over one especially good record he exulted, "Clap me, why don't you. I'm getting to know more than my teachers, haint me. Grammar, O." The determination to succeed drove him from good, in the first years, to excellent, and to the top of the class in the last year. And he did it by hard work. Spelling and mathematics were not easy for him, but he mastered them. In history and composition he excelled, while whatever Natural History was offered "gave me a thrill." Some forty school essays have been preserved, of which about half are on nature subjects, and he gave a commencement oration on "Woodland Lore."

But going to school did not excuse him from working at the real estate office. Saturday mornings and odd times were spent there and every morning he continued doing the cleaning. This, added to the home chores and family prayers, crowded the early morning hours, to which he often added one or more by getting up extra early to study; the period of oversleeping had passed.

His first eighteen summers were spent on his father's farm, Davenport Ridge, and they were long ones of six months. He early became a regular farm hand, tended stock, worked in the fields and drove to and from the station with his father who made frequent trips to Brooklyn. One summer his father was injured in an accident in the city and Charlie was the one he wanted to attend him. So the next month was spent in Brooklyn as a nurse. Although he was usually kept busy with farm work, and lessons were continued, these summers gave splendid opportunity to watch bird migrations and collect insects and profit by his mother's knowledge of natural history. Thus was established the interest that determined his life work.

During the summer of his eighteenth birthday, he spent two months as field reporter on the WHITE MOUNTAIN ECHO, of Bethlehem, N. H. He skipped around from resort to resort gathering items, reporting meetings and generally enjoying the mountains, although his earnings barely covered his expenses. For the next summer he prepared a remarkable program leading to the life work on which he had just decided. This was in March 1885, while in the Polytechnic class corresponding to the fourth year of high school. He had given his father less trouble than any of his brothers, but he knew that it would require diplomacy to win his father's approval of the program he had in mind. He realized that his father was a highly practical man, but that he liked admiration and appreciation. Although he was seeing his father every day, he trusted his pen more than his tongue and composed a letter of twenty pages which began :—

> "My dear Father:
>
> As the summer is now rapidly drawing upon us, I have begun to look about to find how I may best employ myself during the vacation and it seems to me that some employment by which I could put in practice these facts I have learned in the Polytechnic and which would have some bearing upon my future intended occupation, would best and most fully carry out the ideas of educating me, which your kindness in sending me to such a good Institute indicate. For the opportunity I cannot thank you too much, and hope I may be able to repay you by a life spent in the service of humanity.
>
> "The course of studies in the Polytechnic has taken me through geology, zoology and to some extent farm surveying. In addition to what has been taught in school I have become profoundly interested in the science of meteorology, the migration of birds, the growth of plants, the value of manures. An extended course in chemistry, blowpipe analysis and qualitative analysis, has prepared me for the investigation of soils and minerals.
>
> "All of the above studies have a direct bearing upon the Science of Agriculture which I should like to study as my life work, if you approve of it."

The proposal for the summer's work that follows is divided into 10 headings: survey of farm, meteorology, progress of seasons (a, bird migrations, b, blooming and falling of flowers, c, leafing and falling of leaves of trees, d, weather changes), examination of mammals and birds, examination and collection of insects, observations on the growth of garden vegetables and grains, experiments with manures, experiments with colored glass, determination of soils, geology of region. The importance of each topic in turn was discussed, its practical applications emphasized and the proposed work outlined with estimates of the necessary time and expense. It is a masterpiece of under-estimation.

> "I should like to spend the remaining time in trying to compensate directly to you for my support. . . . Three and a half hours then will be yours every day, . . . and I hope you will regard the whole idea not as a selfish scheme to get rid of work, but as a proposition to aid science, in which I am particularly interested. . . . I have much more to say but hold up this already long communication. I hope for an early response either by letter or word of mouth.
>
> Very reverently, Your son,
> Chas. B. Davenport."

For seven weeks there was no response of any sort. Then came a letter from his father, speaking with appreciation of the display of mental development shown by the wide range of theories considered. "But you have failed somewhat in meeting my view of the *practical* parts of the subject. . . . you are too theoretical." And therewith science was disposed of "It had been my intention to have you make a survey of my farm," etc. But he was quite well aware of the nature of the soil and of the manures required, etc., etc. "In fine, the question of prime importance is how much money can you make for yourself and for me . . ."

So in the summer of 1885, Charles surveyed the farm, made some des-
ultory bird notes and dutifully went back to the Polytechnic Institute for
another year in Civil Engineering and the B.S. degree. During this last
year in school he was editor-in-chief of the full-fledged school monthly,
The Polytechnic, and his literary capacity was further displayed in another commencement oration. This time the subject, more appropriate for
a civil engineer, was "The Father of the Republic."

But in spite of top grades and excellent testimonials from teachers, jobs
were hard to find. For a month in the middle of summer he did odd jobs
for a Capt. Palmer of the U. S. Barge Office and earned enough to cover
his expenses and meals for another month, which he spent in town, presumably on the job hunt. By September he was desperate and sent his first
communication to Science (Vol. 8, p. 236, 1886) as follows:

Science for a Livelihood

"Some time ago I read in your journal a stirring editorial, calling for
young men to devote their life to the cause of science, and deploring the
lack of persons who were willing to encounter hard work and poor pay because of love for investigation and study.

"Early this summer after graduating from a first class scientific school
I made application to four agricultural stations in this and other states
for some position, pay no consideration whatever. Having been brought up
on a farm and having a first-rate scientific education, a love of natural
sciences (in which I have done a little practical work), an excellent
physique, I thought myself fitted for investigation in scientific fields, particularly as I love it above all else.

"In every case I received answer, 'Places all full.' I have begun to
doubt whether investigators and workers are needed in the natural or experimental sciences, and think that a poor young man who cannot afford to give
money to the work has no call in this field. Am I right?
Brooklyn, N. Y., Sept. 4.   C. B."

This letter brought him a sympathetic response from the Museum at
Albany, but no job. In Science there followed a general discussion of the
difficulties of getting started in scientific work that may very well have
smoothed the way for those who followed. In his later years, Davenport
never forgot his own experience and never lost his warm sympathy for the
young fellow trying to get started, as scores of biologists can testify. It
is clear that his own enthusiasm and willingness were boundless, but the
only specialized training he could offer was in civil engineering. So it was
that in November he went off to a railway surveyor's camp in Michigan.

Emergence (1887-1899)

The rough life in the winter woods brought very new experiences, and
there were compensations in observing the animals and trees and in collecting minerals, all of which were fully reported to his mother. Perhaps it was
necessary for him to have this job to make him realize his mistake and turn
against his father's wishes. After nine months as a rodman, he withdrew

from engineering. Several reasons have been given for this, but actually it was a real crisis—the first and final revolt against his father's domination. He had perseveringly learned methods of civil engineering, but he was not an engineer. The interest in nature, which his mother had so long been quietly fostering, had been growing steadily; at this point it was to be followed openly. The prolonged physical separation from his father may have helped, and so may the fact of reaching his legal maturity. He had not expected to go to college; his engineering course did not meet college entrance requirements. And so, after reaching his decision, he began to study the Latin necessary for college entrance, for to give up engineering meant going to college and studying zoology. His classmate and close friend, Herbert H. Field, later founder of the Concilium Bibliographicum in Zürich, having taken the liberal course, was prepared to enter Harvard the fall after they finished school. Undoubtedly the fact that Field was at Harvard was a primary reason for selecting that college, and Field's persuasion may very well have eased the decision by providing an excuse. Indeed, this excuse was often given as *the* reason, as in a letter to his mother the following summer (at this time he had a job cou intngroganisms in water supplies and earning just enough, on an hourly basis, to meet his living expenses):

"Don't you think this is my grandest opportunity for reading and study? More money than I can use assured, my work thrillingly interesting, and yet ½ to ⅓ of my time free to myself. And I owe this in largest measure to Field, who insisted in my coming out of the woods to Harvard."

Davenport gave up the engineering job on the first of September and by the end of the month he was registered, as a junior, in practically all the courses in Natural History Harvard was giving. (By the end of the college year he had completed seven courses with an A in each.) He was welcomed by a notable group of young men who were galvanized by Prof. E. L. Mark's inspiring teaching. But to Davenport, who had just broken away from his father, coming to Doctor Mark meant something much deeper. The devotion he felt as a student was just as warm at the end of his life. There was a spirit of camaraderie in the group; much borrowing back and forth of money, in which even Doctor Mark was at times involved; mutual aid in collecting material and in other ways; symphony concerts as a major diversion (Davenport regularly subscribed, although the Seashore tests showed him tone deaf); and all of them working as hard as youthful zeal could drive them. It is doubtful if the unalloyed satisfaction and good fellowship of those student years at Harvard were ever equalled. The summers were variously spent at Woods Hole (1888, the year M.B.L. opened), at Agassiz' private Newport laboratory, or in Cambridge; and in 1890, the year the Biological Laboratory was started, Cold Spring Harbor was briefly visited.

With his mother's backing, tutoring and a job in the Division of Water Work, Massachusetts State Board of Health, he scraped through the frugal student years. The university granted him various offices: monitorship, proctorship, scholarship, assistantship, and finally an instructorship (1891-1899).

Undergraduate work was completed with an A. B. in 1889; graduate work with the Ph.D. in 1892. The exquisite plates in his thesis, OBSERVATIONS ON BUDDING IN PALUDICELLA AND SOME OTHER BRYOZOA, as well as in two other morphological papers on Bryozoa, indicate a rare skill in drawing that was not further utilized. The Harvard instructorship was renewed from year to year, until 1899. During this period he taught, at different times, the various elementary and intermediate courses offered by the Department of Zoology, and in 1893 he succeeded in introducing an entirely new course of his own, called Experimental Morphology. The first description of this course in the catalog includes the phrases: growth and development, movement of living substance, processes of nutrition, laws of growth and form, conservation and development of form, modification and control of development, regeneration. His interest had become strongly physiological and had branched out in numerous directions. Answers to great problems seemed almost within reach. Early in 1895 he wrote in a letter to his mother:

> "I have begun to work on movements of Amoeba. Amoeba is the simplest, least differentiated or specialized animal I know of, but nevertheless is in constant motion under certain conditions. I am trying to unravel the cause of its movements in any direction, indeed to see if I can fix things so that motion in a definite direction will not occur. When I have obtained a knowledge of all the causes which produce its motion and determine the direction of its movements, I shall be in a position to understand how the more complex organisms have come to have the movements they possess. This, it seems to me, is the most fundamental study in animal (or plant) physiology and morphology."

The catalog description of the Experimental Morphology course for 1895-96 begins with the phrase, "Ontogenetic and phylogenetic development of organisms." The next year the course became two, given in alternate years. One dealt with "Ontogenesis studied as a process" the other with "Phylogenesis studied as a process," and included: physiological processes of the development of the race, statistical and experimental study of variation, individual variation and sports, normal inheritance and crossing, selection, and different theories of phylogenesis (given 1896-97). The following year the ontogenetic course included the effects of chemical and physical agents on general functions of protoplasm, on growth, on cell division and on differentiation. Each year the courses continued to evolve and led to research studies which Davenport published with his students as co-authors: acclimitization to high temperatures (with W. E. Castle) and to poisonous chemicals (with H. V. Neal); heliotaxis (with W. B. Cannon); geotaxis (with H. Perkins); comparative variability (with C. Bullard); the role of water in growth (a series of 8 determinations of water content at different stages of developing frog's eggs, to which Davenport was always very fond of referring). The adjective *experimental* had exposed morphology to the general swing towards physiology.

In 1897 and 1899 appeared parts I and II of his first book, which took its title and most of its material from the course, EXPERIMENTAL MORPHOLOGY. As is true of a large proportion of Davenport's writing (over 400 titles, including 18 books) this book was forward-looking and important rather in its current influence upon other workers than in its lasting contribution to knowledge. In effect, this book was a plea for the use in biology of the exact quantitative methods of physics and chemistry. In twelve years the plea of the farmer's boy had acquired the terminology of the laboratory.

This emphasis on quantitative methods, combined with the mathematics of his engineering course, prepared Davenport to adopt enthusiastically the statistical methods of Galton and Pearson, and for several years he and his students, with whatever material they could obtain in sufficient quantities, were measuring, counting and calculating, as an approach to problems of evolution. His biometric papers, largely based on Pectens, his lectures and general discussions, and especially his manual, STATISTICAL METHODS WITH SPECIAL REFERENCE TO BIOLOGICAL VARIATION, a compilation of procedures, formulae, literature and useful tables, which appeared in four editions (1899, 1904, 1914, 1936), had unquestionable importance in introducing such methods to this country. At the time *Biometrika* was started (1901) the relations between Davenport and Pearson were most cordial, and Davenport was made one of the three editors. But Pearson could tolerate no inaccuracy or loose reasoning and a controversy arose in which Davenport demonstrated, first in private correspondence and later in print, that he could not meet criticism objectively.

Besides the Harvard courses, he was teaching in the "Annex," which was the everyday name of the "Society for the Collegiate Instruction of Women, by Professors and other Instructors of Harvard College," later called Radcliffe College. Here he met a graduate student who had been an instructor in Zoology at the University of Kansas, Gertrude Crotty. He claimed to be girl-shy, but she was small, sympathetic, capable, had an excellent and realistic mind and a gay laugh, and did not have the least fear of men who attracted her. The prolonged domination of his father and his infatuation for his work left a need for a level-headed person to hold him steady and take charge of mundane affairs. According to the wedding certificate and the engraved announcements of the wedding, they were married in June 1894, although he repeatedly gave the year as 1893 and, in the schedule of Family History filed in the Eugenics Record Office, put down the birth of each daughter one year early. However outwardly reserved and undemonstrative, underneath he was warmly affectionate and in marrying he gave his heart and soul; his wife became his reliance and guide. After the infancy of their two daughters she took an active part in his work, was co-author of several papers and for 26 years shared with him the management of the Biological Laboratory at Cold Spring Harbor. His loyalty and complete devotion to her never wavered; he never gave up his practice of writing to her virtually every day they were separated. She, and she alone, was his confidante and chief counselor. As the mother of a growing family she was keenly aware of the necessity of a growing income, which the prolonged instructorship did not provide. During these years week by week she would turn to the death notices in SCIENCE to see what position might be open. He became highly money-conscious. Advancement was imperative; his ambition soared. His letters home usually included money matters and reported almost apologetically the exact amounts of incidental expenditures. But, however sensitive on the subject, he never seemed able to put at rest the financial concerns of his spouse, or overcome a penuriousness in his use of funds, which seemed to prefer a good bargain to good quality.

Advancement (1898-1904)

The turn of the century was the critical period in Davenport's career. Not the death of a biologist, but the rapid growth of biology at this period brought the opportunities. The entering wedge into the Harvard stalemate was his appointment as Director of the Summer School of the Biological Laboratory at Cold Spring Harbor (1898-1923). The year 1899 brought an assistant professorship at the University of Chicago. This was advanced to an associate professorship in 1901, which in 1904 he resigned in favor of the directorship of the Station for Experimental Evolution, then established by the Carnegie Institution of Washington.

With his birthplace, Davenport Ridge, in the view across Long Island Sound, and his boyhood winter home on the same island 30 miles to the west, going to Cold Spring Harbor was like the return of a native. Davenport's love of planning was given full scope in organizing the Summer School, engaging instructors, inviting special lecturers, arranging details for dining hall and lodgings. With Mrs. Davenport taking charge of the domestic affairs, as well as teaching histological methods, the couple made a strong team and their devotion to the Biological Laboratory grew as they gave more and more of themselves to it. Each season closed gloriously with a clambake of legendary proportions, at which Dr. H. H. Wilder's "The Sad Fate of a Youthful Sponge" was always sung, and, after belts had reached their limits, the menu was reviewed in a special version of "These Dead Bones Shall Rise Again."

The Field Zoology course became famous for its on-the-spot study of the many diverse habitats close at hand, with most of the day spent in bathing suits and lectures outdoors. When Davenport lectured, those who huddled close could hear, as he communed with his notes. He was at a disadvantage before a class, and yet he was impressive. The halting speech and the uneasy manner masked, but could not entirely conceal, the radiance of inner enthusiasm, which spoke for itself in a language of its own, delighting those who understood. But the popularity of the course was due in no small measure to Herbert E. Walter, who combined a native talent for teaching with genial good fellowship, human understanding and great admiration for Doctor Davenport. As at Harvard, into this course went materials Davenport himself was investigating, and these were determined by the nature of the locality. Using modern quantitative methods as the key to the relation between specific form and geographical distribution, he hoped (as he wrote Whitman in '99) to throw light on the origin of species out-of-doors. But preliminary descriptions of the local fauna and the different assortments of organisms living under the various contrasted conditions and in the intermediate zones led almost inevitably to ecology. While he published only one strictly ecological paper, THE ANIMAL ECOLOGY OF THE COLD SPRING HARBOR SAND SPIT, his interest and influence was so effective at that time that animal ecologists hail him as one of their pioneers. At Chicago he found the opportunity to advance the new subject with the co-operation of Whitman and Coulter. Several of his graduate students took problems in ecology, and to Charles C. Adams, one of three students who had followed him from Harvard, he gave the opportunity to present what was probably the first course in animal ecology ever to be given in a university. Looking back to that time, Adams wrote to Davenport in 1941: ". . . . I believe that your departure from Harvard was fortunate. No eastern university had an atmosphere favorable for the development of ecology and not even today have they caught up! . . . You were at Chicago at the critical period and made a very substantial contribution to the movement . . . ."

The University of Chicago was not only peculiarly receptive to an interest that Cold Spring Harbor had developed in Davenport, but this University was also peculiarly influential in developing an interest that brought great changes to Cold Spring Harbor. An experimental approach to organic evolution was being widely discussed. The Royal Society of London had appointed its Evolution Committee, and Bateson and Saunders had started their breeding experiments in 1897. In this country, Prof. Charles O. Whitman of the University of Chicago, director of the Marine Biological Laboratory at Woods Hole since its beginning, had long been developing plans for a new kind of institution—a biological farm, equipped to carry on prolonged and uninterrupted studies of heredity, variation and related subjects. He formulated such plans in detail for a meeting of the American Naturalists in December 1897, and emphasized the many mutual advantages that would accrue if this biological farm could be located near the Marine Biological Laboratory at Woods Hole. At this same meeting Davenport also spoke of the needs of a farm, or zooligical preserve, for the study of phylogenetic problems. Whitman knew very well Davenport's interest in evolution and, before the end of the first season at Cold Spring Harbor, Davenport was tendered an invitation to take charge of the department of beginning investigation at Woods Hole the next year. This was declined on the basis of the inexpensiveness and pleasures of the Cold

Spring Harbor summer; his "mercenary motives," he wrote. Cooperation, even on a national scale, had no attraction for the man who was having his first experience as first in command. Whitman, however, persisted; if not Woods Hole, then an assistant professorship in his department at Chicago. The offer was received in September 1899 and was accepted. And so Davenport was brought into intimate contact with the most active discussion in the country of plans for a new institution to study evolution. To be sure, the details were all fitted to Woods Hole.

Cold Spring Harbor, however, had a summer school and highly varied habitats and the proximity to New York might weigh heavily against Woods Hole's greater isolation. Moreover, at the end of 1901, the future of the Laboratory at Cold Spring Harbor was uncertain. There was talk of transferring its control from the Brooklyn Institute of Arts and Sciences to Columbia University, with an inevitable change of director. If the Brooklyn Institute should expand the Laboratory by the establishment of a permanent resident staff to study evolution, this change could be forestalled. The associate professorship was an advance in status, but financially it was still inadequate. The Directorship of a new laboratory would mean a real advance and give tremendous opportunities to organize and map plans; no more classes or university routines, no more worry about recognition or advancement; to do as he pleased and shoot ahead as fast as he desired—nothing could be more alluring.

The Carnegie Institution of Washington was incorporated on January 4th, 1902. Davenport's first communication to this Institution was dated January 16th, 1902, and was delivered by hand to the secretary, Dr. Charles D. Walcott, by Mr. Charles L. Hutchinson, vice-president of the Corn Exchange Bank of Chicago, who went to Washington to attend the first meeting of the Institution's Board of Trustees. This was the opening move of a two-year campaign whose final success gave Davenport a position of extraordinary influence and power, and gave his name a lasting place in the history of science.

That there was such a campaign may seem surprising, but the lengths to which it was carried would be unimaginable without the original documents in the archives of the Carnegie Institution, in Washington. For once, the major influences and urges of his life worked in the same direction and their unified pressure became excessive. Davenport's procedures in this critical period have the greatest importance for an understanding of the man. The pressure was possibly never again so great; but on subsequent occasions, though less well documented, it is possible to recognize repeated use of the methods seemingly approved by success. In the present period the incandescence of his enthusiasm distorted his judgment and permitted exaggeration that bordered closely upon misrepresentation.

The application was repeatedly submitted to the Institution in different restatements; the financial requirements were progressively reduced, as by the Biological Laboratory's offer of free land and, finally, by Davenport's proposal to raise funds for a building from other sources. One of these applications, dated May 5th, 1902, was sent directly to the Carnegie Trustees and published in the first year book of the Institution. On July 26th, 1902, Doctor Walcott received another application forwarded by Prof. Franklin W. Hooper, Director of the Brooklyn Institute of Arts and Sciences, with a covering letter stating the approval of the Board of Managers of the Biological Laboratory. Doctor Davenport's special qualifications were reviewed and so were the advantages of establishing the new laboratory at Cold Spring Harbor, with partial support from the Brooklyn Institute. The members of the Institution's Zoological Advisory Committee (H. F. Osborn, Alexander Agassiz, W. K. Brooks, C. Hart Merriam, E. B. Wilson) were individually appealed to. To Osborn Davenport wrote, among other things:

> "I think the work ought to be limited pretty closely to experimentation and a careful, usually quantitative, study of results. There are a number of subjects of great importance for evolution, such as the statistical study of variation and geographic distribution which must be distinctly avoided in order that energy shall not be directed away from the main point. . . . I trust that the single-mindedness of this proposal will not be called in question. . . . Finally, I may say that I urge the establishment of this Station just as heartily if it appears to the committee that a better director is available."

This last sentence drew the response: "Your feeling that you desire the best Director available is exactly the right one. I myself know of no one better qualified to do this work than you; but the matter will certainly have to come before the committee to be considered from every point of view."

To Alexander Agassiz, the anti-evolutionist, President of the National Academy of Sciences and *ex-officio* Trustee of Carnegie Institution, he urged the careful consideration of the plan to establish a station for the study of certain zoological problems that require continuous attention over a long series of years, such as the influence of light, temperature and pressure upon specific characteristics, and so on, outlining numerous experiments, including the study of normal variation under different natural environments. The dangerous word was not once used. This was intended, of course, to be diplomatic, but it was in fact more objective than his other statements that dealt so freely with the term *evolution*.

Dr. John S. Billings, Director of the N. Y. Public Library, was the key man in the Carnegie Institution, the trustee assigned special responsibility for zoological matters. Davenport repeatedly visited him and wrote him numerous long letters. One of these (Aug. 18th, 1902) included three pages on the desirability of Cold Spring Harbor as the location for the new station and three pages on the desirability of himself as its director, which began:

"Concerning my fitness for the position of director of the Station, I am embarassed to speak freely. Professor Osborn has said in his letters that there is no one better equipped for the work. I feel that I can advance science in the direction proposed by the Station because of:

"a. My training. I am the author of the only book devoted to the statistical study of Evolution. Have also written two volumes on Experimental Morphology. . . . I believe I am stating merely the cold truth when I say that in training and in work accomplished in the study of heredity and variation—the elements of evolution—no one in the country is as well prepared for experimental and quantitative studies in evolution as I am. . . .

"b. Evolution studies require time. My age is 36. The chances are that I shall have 25 years to work upon the problems for which the Station stands, and to assist others who work at the laboratory. I propose to give the rest of my life unreservedly to this work. . . .

"c. I am about to spend four months in Europe investigating all Experimental Evolution Stations there: I shall do this to better fit myself for the work of directing the Station for Experiments on Evolution, whenever the Carnegie Institution establishes it."

Two days after writing this letter, he and Mrs. Davenport sailed for Liverpool. The actual purpose of the trip was to collect *Pecten* shells for statistical studies of geographical variation; so they visited most of the European Marine Biological Laboratories from Bergen to Naples, where they spent three weeks. From Paris he wrote to W. H. Hays of the

1900

Gertrude Crotty Davenport
Charles B. Davenport
Triest, 1902

Plant and Animal Breeders Association, suggesting that the Association bring its influence to bear upon the Carnegie Institution in behalf of the Cold Spring Harbor project.

Contact with Francis Galton had been established in 1897 by sending him reprints. The courteous acknowledgment and discussion of these had started a correspondence, in which Davenport's glowing praise was graciously accepted and Karl Pearson was recommended as the best authority on correlation statistics. The personal acquaintance of both Galton and Pearson was first made in London during this trip. Almost at once both of these gentlemen wrote letters on the proposed Cold Spring Harbor Laboratory. Pearson sent a characteristically clear-headed statement to the Carnegie Institution on the desirability and requirements of a laboratory for the study of evolution; concerning Davenport's qualifications he wrote: "Personally he seems to me stronger than his published work. . . . I should say he would not be wanting in energy and keenness of interest and would keep himself in touch with European workers and methods." Galton, however, wrote only to Davenport, a letter of such manifold importance that it is quoted in full:

"42 Rutland Gate, London SW
Oct. 16, 1902

"Dear Mr. Davenport:

"It gives me great pleasure to learn that there is a chance of an experimental station being established in America for the study of evolution. The question of how such a station might be managed so as to secure abundance of valuable results is by no means new to me. A few years ago I had to consider it seriously owing to a letter to me from Mr. Herbert Spencer who was then in communication with Mr. Carnegie. I also at that time was chairman of the committee of the Royal Society on Evolution, so I had plenty of opportunity of discussion with very competent men. I mention this to show you that my conclusions were not lightly formed.

"They were, that any such establishment must be a *costly* affair, both at first and in maintenance. That there are alternative ways of administering it, the best of which in each case would depend on *personalities,* I mean the number of available men who would, between them, be capable of devising appropriate experiments, and be of such ballast, so to speak, as to stick steadily to their work for many consecutive years, and gradually to create a tradition for good persistent work. On the other hand any attempt at showy work, for the sake of mere show, would lead to vast waste and to error.

"There must be of course a resident Director for administrative purposes, but it is an open question whether he ought himself to be a first class man in respect to original research, who should be given every opportunity of carrying on his own experiments. There is much to be said against this, which may be summed up by a paradoxical saying current among the librarians of great collections—that the librarian who *reads* is lost. He ought to know the outside of books, and to have a general notion of their merits and contents, but if he specializes and studies, it is so much energy diverted from the work he is engaged to do, and gives him a bent towards *disproportion.* The resident Director certainly ought to give by far the larger part of his time to see that the experiments ordered by the committee are conscientiously carried out, and to help outside experimenters to whom the committee may have allotted plots of garden or cages.

"There are, therefore, so many unknown conditions that I do not see my way to answer your wish that I should write some sort of testimony in your behalf. But I am sure of this, that your published works combined with that personal knowledge of yourself and of your administrative powers, which the managing committee are sure to have, would be far more trustworthy than any opinion of my own founded on partial knowledge.

"The idea of a station for the study of Evolution is, when taken in its entirety, a very great one. To carry it out properly needs an abundance of money, many-sided brains, earnest and honest effort and much preliminary planning. But I must stop.

Sincerely yours,
"Francis Galton"

But, in spite of all the pressure, the year (1902) ended without action by the Carnegie Institution on Davenport's proposal.

Biology was outgrowing its support in many places and the Laboratory at Woods Hole*was experiencing even greater difficulties than the Laboratory at Cold Spring Harbor. The sudden availability of Carnegie funds, which were deemed to be "beyond the dreams of avarice," offered a possible solution, and during the greater part of this year negotiations between the Marine Biological Laboratory and the Carnegie Institution were taking place. The hopes and wishful thinking of the Marine Biological Laboratory rose to such a pitch that in August its Corporation voted, 60 to 3, to execute and deliver to the Carnegie Institution of Washington a deed conveying all the properties of the Marine Biological Laboratory to the Institution in return for permanent support of the Laboratory. This action was not actually completed, for the policies subsequently established by the Institution brought sharp dissention among the trustees of the Laboratory and by the end of the year this project was entirely abandoned in favor of substantial annual grants from the Carnegie Institution for three years, to give time to organize permanent support. Before this solution had been reached, a special subcommittee of the Carnegie Institution's executive committee had recommended that action on Professor Davenport's proposition be deferred until final action was taken by the Woods Hole Laboratory, since his object and proposed methods appeared to be much the same as those proposed by the Trustees of the Marine Biological Laboratory; to acquire and support two separate institutions for substantially the same purpose was probably not admissible. "It is possible that Professor Davenport may be able to carry out his proposed investigations at Woods Hole."

This last would hardly have satisfied Davenport; indeed, he had made up his mind to start a new laboratory at Cold Spring Harbor even if the Institution turned him down. While he was still abroad he wrote at length

---

*Frank R. Lillie, *The Woods Hole Marine Biological Laboratory,* Univ. of Chicago Press, 1944.

22   BIOS

to a wealthy Cold Spring Harbor neighbor, paving the way for a local
campaign for funds, should this become necessary. As Whitman's associate
at Chicago, Davenport could not have failed to know of the many indications during the first part of the year (1902) that Carnegie Institution
would probably accept responsibility for the support of the Woods Hole
Laboratory. He had ample reason to doubt the success of his own plans
and did not need a committee's report to tell him that if Carnegie Institution supported Woods Hole it would not support Cold Spring Harbor. He
had even gone to Woods Hole the week before he sailed for Europe, in order
to attend the August meeting of the Corporation and record his vote (one
of the three) against the transfer of Marine Biological Laboratory property
to the Carnegie Institution. Whitman, seeing no other course, had gone
along with his trustees in the discussions and voted for the transfer, but
he had serious doubts as to the wisdom of this course and, later in the year,
fought strenuously against it.

The failure to establish the Biological Farm at Woods Hole did not
bring with it the decision to establish a similar project at Cold Spring Harbor. Another year of uncertainty intervened. Every stone had been turned. Idling was one thing Davenport could not tolerate—whether waiting
for a job or a green light! Early in 1903 he started working on plans for
another project—to carry out Whitman's biological farm idea near Chicago. He located a 700-acre tract that seemed suitable, at Porter, Indiana,
and mapped it. He called it "The Field," and drew up a far-reaching program that called for experimental breeding as well as experimental activities not only in other biological fields but also in nine other departments
of the University of Chicago. In presenting this project to President
Harper, Davenport included an essay, "A Summary of Progress in Experimental Evolution," which emphasized the widespread recognition
of the need for an approach to evolution and the inadequacy of the experimental work that already had been undertaken.

This same essay was also sent to the Carnegie Institution, again with
a covering letter from Doctor Hooper of the Brooklyn Institute and a re-
application excused by the increasing needs revealed by his recent European
contacts. The amount applied for was $7500 a year for 25 years. Eight
specific experiments were proposed:

Experiments for which assistance is asked.   (Condensed)

1. *Helix nemoralis;* breeding experiments with 1000 pairs. Are the
    varieties incipient species? How are the characters inherited in crosses?
    Can distinct species be produced from the variations by selection or
    isolation?

2. Chinch bug; How are long *vs.* short wings inherited? Is this variation
    the beginning of a new species?

3. *Limneae*; Can we induce the various "species" at will by altering con-
    ditions?

4. Spiders; How do offspring of dimorphic males vary?

5. Goats; Is the *sport,* neck pendants, prepotent in crosses with normals? Similar experiments with polydactyl cats and abnormal rodents.

6. Guinea pigs, mice and moths; Test Mendel's law of dichotomy in hybridization for animals.

7. Lady-bird beetles; Test effect of isolation in permitting differentiation. Breed two entirely isolated groups of one species for 8-10 generations and compare the last generation. Experiments to be started at once and continued for many years. Other series with other species should be started simultaneously.

8. Beach fleas, snails (*Littorina, Melampus*); Can partially aquatic animals be transformed into complete terrestrial ones by gradual adaptation?

But the fever of his enthusiasm had faded and his hopes for Carnegie support were low; he wrote to his wife that he would give up Cold Spring Harbor entirely if the University of Chicago would put him in charge of *The Field,* and offer an adequate salary. The property was inspected by a special committee and the response was favorable. He selected the building site and the acres he would buy for his own home. ("I have money enough in my desk now to buy a five acre farm with a ten-mile view and in excellent soil—cool in summer.") Although *The Field* was one of the many projects that did not materialize, plans went so far that, even after the Carnegie Institution had appointed him Director of the new Laboratory at Cold Spring Harbor and Davenport had resigned his professorship, he proposed that his connection with Chicago be maintained by making him Director of *The Field*—to ensure integration with the work at Cold Spring Harbor.

In October 1903 President Hooper was attempting to arrange for Doctor Billings to visit Cold Spring Harbor, but in spite of the offer of a special car on the train it seemed impossible to arrange a date. However, these two gentlemen met in New York, and President Hooper reported the conference in full to Davenport, with the news that Billings was disposed to recommend the adoption of the Cold Spring Harbor plan to the Carnegie Executive Committee, and that the Executive Committee's action would probably be approved by the trustees in December.

But even this seems to have raised little hope, for Davenport proceeded with plans to go to Tucson, Ariz., in December, to work at the new Carnegie Desert Laboratory during a three months' leave of absence from the University. His whole family was going too. The train passes had arrived and the trunks and camping equipment had started on ahead (tenting seemed the only way to avoid exposure to past or present tubercular patients) when word came that Doctor Billings wanted to see him in New York. The Trustees had authorized the organization of a Department of Experimental Biology with laboratories at Dry Tortugas, Fla., and at Cold Spring Harbor, N. Y.

So Davenport triumphantly went east instead of west; the enthusiasm for Cold Spring Harbor returned immediately and at full blast. In a whirlwind of activity he set up an office in New York, conferred with

architects, searched pet shops and animal shows for material, engaged workers, secured the cooperation of eight investigators as associates of the Station,
and developed the interest and good will of wealthy Cold Spring Harbor
neighbors. His appointment as Director of the new laboratory was voted
on January 19th, 1904. "Glory be! The Ex. com. of C. I. met and approved all my plans . . ." began his letter home, which was signed,
instead of an affectionate phrase and "Charlie," "Yours unworthily, Chas.
B. Davenport." In large letters in his journal he wrote "THIS IS A RED
LETTER DAY!"

Culmination (1904-1919)

The position was secure, but the program was as ill-defined as it had
always been, and as it was to remain. The issues of the campaign had
been geographic and personal; the specific experiments that were to solve
the problems of evolution had been subordinated as relatively unimportant
details. Varying lists of experiments had been proposed, but the differences
in the successive lists did not represent progressive critical thought. So a
laboratory was established with a staff and a building—but without a well-
planned program. In March, the director wrote, "I have little notion of
just what we shall do. We shall reconnoiter the first year." Six years
after his retirement, in his monograph on the post-natal development of
the head, he wrote, . . . "This paper will, it is hoped, be regarded as
a reconnaissance in this unexplored field." Returning to 1904, his report to the Executive Committee on two months' work showed much activity, which included the purchase of canaries, finches, long-tailed fowl
and a Manx cat, and the employment of an animal caretaker, formerly at
the University of Chicago. A Staff had been chosen:

"Mr. Frank E. Lutz, research assistant, will devote himself to making
reconnaissance of the variability offered by the animals of Long Island.
Mr. George H. Shull, research assistant, will reconnoiter the field of plant
variability and gather seed plants for heredity experiments in 1905. Miss
Lutz will reconnoiter the field of germ plasms for suitable material for
hybridization experiments. My own work will be largely a reconnaissance
of capacity for maintaining, breeding and crossing wild animals in captivity
and also the study of the behavior of unit characters in hybridization of
domestic races of birds.

"Of the Associates, Drs. Britton and MacDougal are to provide seeds
of 100 species of wild annual plants, which we shall try to breed at the
Station for ten years. They will cooperate in studying the results. Dr.
Castle will continue his study of the behavior of unit characters in breeding
races of small mammals. Prof. E. L. Mark will make cytological studies
in heredity. Dr. H. E. Crampton will breed large silk moths to study
effects of inbreeding and other conditions upon variability. Prof. Moenk-
haus will make hybridization experiments on fishes to get material for the
study of the behavior of the paternal germ nuclei when united in the fer-

tilized egg. Dr. Raymond Pearl will assist in the reconnaissance on animal variability. Prof. Wilson will make cytological studies."

Davenport was anxious to bring the workers on experimental evolution throughout the country into friendly relations with his Station. But when one of these associates wrote to him as an advisor, he would have none of it. "Extraordinary letter from Wilson, suggesting that I take up the matter of Cold Spring Harbor Laboratory with him as advisor. Saw Billings—not the place of advisors to interfere with settled matters." (Diary.) The protest was apparently so strong that thereafter Davenport went freely on his own way, unadvised, although the minutes of the Executive Committee of the Institution show that Prof. E. B. Wilson was not acting, as Davenport supposed, as a member of the Zoological Advisory Committee, but as a newly appointed special advisor of the Carnegie Institution on the organization and work of the Department of Experimental Biology.

The thirty years of Charles Davenport's directorship of the Station for Experimental Evolution were so full of such varied activities, involved so many people, organizations and different kinds of problems, that a whole volume on this period alone would be full of interest. However, for him personally, these years represented the development, in musical terms, of themes already stated, as different situations called into prominence different elements of his personality. There were climaxes, indeed many of them, and quiet intervals were few, but there was a changelessness about the man that seemed to exempt him from the influences of increasing age and accumulated experience. The charm of his boyish eagerness never faded. Novelty never ceased to attract him and, as a pioneer with many of the characteristics and methods of a frontiersman, he could not resist the temptation of breaking ground when he caught sight of a new field. In 1934, the year of his retirement, he was talking on the gene with two of his old biological friends at the annual meeting of the National Academy of Sciences. "Both were highly opposed to the gene—a clear case of 'old fogyism.' It is too bad that we reach a stage when new ideas are abhorrent to us."

While Davenport responded to general trends of thought, and "old fogyism" certainly did not appear, the price of the unquenchable enthusiasm was the failure of increasing experience to foster a more judicious or philosophical point of view. It seemed as though he was so busy responding to the immediate situation that there was no time to profit by the experience. At one moment, at least, he recognized the truth and full significance of this situation. His account of this rare moment of insight includes much of his familiar self. He was writing from England in 1909:

> "This is one of the most memorable days of my life. About 9 o'clock I started for Bexley in Kent to see a moth and butterfly breeder, a Mr. Newman. I got from him a lot of ideas on breeding Lepidoptera—understand why *we failed!* He rears hybrids and will send me some. After

about 70 minutes with him, the day being delightful, I decided (tho' I had a return ticket to London) to walk across country, 3½ miles to St. Paul's Crag where one of the Cook sons has a large poultry farm. The walk along the road . . . was delightful—under great trees most of the way. I had a pleasant and profitable visit at Cook's and got points on poultry raising. . . . The day continued delightful and I was tempted to go to Downe [home of Charles Darwin]—only 6 miles away. Woodward had written me that Geo. Darwin will take a party there on the 25th, i.e. the delegates—but I wanted to see it—*not* in a crowd. I found the place after a little trouble; and found it used as a girls' school. The school mistress directed me to the sand walk and here I spent over an hour. It is a wonderful place and seems to me to give the clue to Darwin's strength— solitary thinking out of doors in the midst of nature. I would give a good deal for such a walk and have planned such. But the first step is to get the factory corner from the Jones' and Hewletts! Then I would build a brick wall around it, as Darwin did around his garden, so as to shut off from the curious world. Fill it with trees and shrubs and have a walk there. . . . It could be accessible by a tunnel from our garden along the side of the ravine. . . . I know you will laugh at this, but it means success in my work as opposed to failure. I must have a *convenient,* isolated place for *continuous* reflection."

The factory corner was secured, but the wall and the walk were never built. There was no lack of rapid walking, but this was a cult of strenuousness, not opportunity for reflection. He did not permit himself to relax and enjoy life passively, or to meditate and give original ideas a chance to float up from his subconscious mind. Life was too full of action for pondering on the meanings of his results, or for critical evaluation, or even exactitude. His patience was extremely limited; methodical persistence, being unnatural and disciplined, appeared in temporary spurts.

In the early years of the new laboratory, Davenport personally undertook breeding experiments with snails, mice, house flies, moths, sow-bugs, trout and cats; but publishable results were not obtained, owing, in most cases, to difficulties with breeding techniques. Canaries and chickens, however, did breed and provided the basis for four beautifully illustrated publications. These, with a series of papers with E. G. Ritzman on sheep, constitute his major experimental contributions to genetics. There were, besides, a large number of brief notes, annual reports, reviews and addresses on animal genetics. The chicken papers represented a real advance over the quality of the previous work, although the meticulous oriental accuracy of the illustrations painted by Morita was missing from the tables. The canary paper gives shocking evidence of speed too great either for consistent tables or for sound logic. Such speed, with such effects, became habitual.

From 1900 to 1904 he had bred mice of all the basic colors without finding a genetic interpretation and accordingly concluded that there were unquestionably broader principles of heredity than those discovered by Mendel. He continued to hold this doubting attitude toward Mendelism

Bios   27

*1929.   At the celebration of the
25th anniversary of the organization of the Station for Experimental Evolution.*

1932.

until the supporting evidence began to appear from all sides and Bateson visited him. Unlike Bateson and the rediscoverers in 1900, Davenport had not been prepared to recognize the epochal significance of Mendel's work, although he was probably the first in this country to write about Mendelism. This was in a review of the de Vries and Correns first Mendelian papers (Biol. Bull., June '01), which, according to his diary, was written on November 27th, 1900; but in August, 1901, his vice-presidential address for Section F, AAAS, on "Zoology of the Twentieth Century," made no reference to de Vries, Correns or Mendel, although it did voice numerous prophesies including . . . "laws of mingling qualities in hybrids, reversion and prepotency will acquire a cytological explanation and this may seriously modify the theory of fertilization." But as soon as the general acceptance of Mendelism was apparent, he became a staunch supporter and proceeded to make human applications. As usual, the succession of projects was in response to suggestions picked up from outside, rather than the orderly development of thought based on his own previous conclusions.

The series of papers on human heredity—eye, skin and hair color, and hair form—with Gertrude C. Davenport as senior author, began to appear in 1907. At this time only one of the four publications on avian genetics had been published. "Although not strictly within the scope of experimental work, the necessity of applying the new knowledge to human affairs has been too evident to permit us to overlook it." (Annual Report, Carnegie Institution, 1909.) Thus began the active interest in Eugenics that grew rapidly and brought to an early close his participation in genetic experiments.

The fascination of new projects often put Davenport into the position of preparing a final paper after his enthusiasm had become focused on another subject. Indeed, publications on subjects several steps back continued to bob up; statistical studies on natural variation came out as late as 1909 and 1910, and Bryozoa reappeared in 1913 and 1918. Thus the range of simultaneous activities continued to increase. Whether the acceptance of manifold obligations be considered a virtue, or the untoward result of insatiable ambition, no human being could have carried out with high effectiveness his ever-growing program.

The year 1910 marked a major climax. Goose Island, off the Connecticut shore, had recently been purchased by Carnegie Institution for isolation experiments. At Cold Spring Harbor a house had been built for cat breeding. In a 21-acre tract, acquired early in the year, J. Arthur Harris's beans were growing; here too were G. H. Shull's *Lychnis* cultures, but Shull himself was spending a large part of his time at Santa Rosa, analyzing the Burbank records, as he had been doing since 1906. A special laboratory was being equipped for the biochemical work of Ross A. Gortner, and a cave excavated for the adaptation experiments of A. M. Banta. Anne Lutz was working in a gorgeous field of evening primroses, and the chicken runs were still well filled. The annual appropriation for the Station had grown from $12,000 in 1905, to $40,970. The Biological Laboratory was running along as usual, with Mrs. Davenport slaving from dawn till dark for the dining room, and Miss Goodrich, Davenport's private secretary, carrying many of the administrative details. But Davenport's dominant interest had become eugenics.

On this year (1910) he reported: "The work in human heredity has grown to such proportions and its outlook is so vast that it became evident that the Director of this Department could not cope with it alone. Much assistance was needed. Fortunately, at an opportune moment, assistance was forthcoming. . . . The outlook for the development of this very practical offshoot of our work is bright."

At the first of the year, as chairman of the Eugenics Section of the American Breeders Association, Davenport had been filled with plans for great developments, which would require much money. He proceeded to go after it. The first step was to list the wealthy Long Islanders from Who's Who. Mr. E. H. Harriman had recently died. His daughter Mary had been a student at the Biological Laboratory in 1906, and so Davenport attempted to interest her in his great project. He corresponded, and, on January 13th, called upon her.

"Feb. 1st. Spent evening on a scheme for Miss Harriman. Probably time lost.

"Feb. 3rd. Sent off letter to Miss Harriman.

"Feb. 12th. Letter from Miss Harriman on luncheon on eugenics.

"Feb. 16th. To Mrs. Harriman's to lunch. . . . All agreed on the desirability of larger scheme. *A Red Letter Day* for humanity!"

Mrs. E. H. Harriman had agreed to found the Eugenics Record Office; her donations finally totaled considerably more than half a million dollars.

After the delays of the formidable Carnegie Institution Committees and Board, this prompt and unexpected action was enormously exhilarating. Davenport was stirred more deeply than he had been for years, and his elation was sustained by the continued materialization of plans almost as soon as they were made. The jubilation found immediate expression in a burst of companionship and attention to his family. More than once he went to see one daughter play basket ball; he took his other daughter on walks and to the Sportsman's Show and bought her a canoe. There were walks with his wife to the Sand Spit and by their private right of way to Laurel Hollow; he went with her to a dinner of graduates of Kansas State University; he took her to the opera. In the following January, Charles Benedict Davenport, junior, was born.

In May, Mrs. Harriman bought a 75-acre estate with a large residence up the hill from the Station for Experimental Evolution, ". . . to be the campus of a great institution devoted to race and family stock improvement." A course for field workers in eugenics was announced by the Biological Laboratory, and the workers that summer ran their calculating machines in the Carnegie building; in the evening, exhausted, they flippantly sang:

> "We are Eu-ge-nists so gay
>   And we have no time for play,
> Serious we have to be
>   Working for posterity.

Chorus: "Ta-ra-ra-ra-boom-de-ay,
>   We're so happy, we're so gay,
> We've been working all the day,
>   That's the way Eu-gen-ists play.

> "Trips we have in plenty too,
>   Where no merriment is due.
> We inspect with might and main,
>   Habitats of the insane.

> "Statisticians too are we,
>   In the house of Carnegie.
> If to future good you list,
>   You must be a Eu-ge-nist."

The Eugenics Record Office was opened on October 1st, 1910, under the auspices of the American Breeders Association, with Charles Davenport as director and H. H. Laughlin as superintendent. In 1912 a Board of Scientific Directors was chosen, consisting of A. G. Bell, chairman, Wm. H. Welch, Irving Fisher, L. F. Barker, E. B. Southard, and C. B. Davenport, secretary.

The direct personal relation with the founder, which permitted such rapid progress, developed into warm friendship. Mrs. Harriman's interest was keen, especially in research that would have a lasting value. She listened appreciatively to all Davenport's accounts of progressing organization, plans for the building to be erected, and, later, the results of his studies. Her admiration fortified him and he came to depend upon her sympathy. Her death in 1932 was a serious blow. There was no one else to whom he could talk of his work with such satisfaction.

Man remained Davenport's subject for the rest of his life, excepting a brief return to experimental work in 1925 and 1927 (mouse endocrines). The publication of Heredity in Relation to Eugenics, the year following the opening of the Eugenics Record Office, established him as the leading American figure in eugenics. This book massed evidence covering a wide range of traits; while this was impressive, its continued usefulness was reduced by its hasty preparation and the lack of critical judgment in lumping together, indiscriminately, cases with ample and with insignificant evidence. The topics of Davenport's special studies covered a wide range.

Besides the more familiar subjects of stature, body build and longevity, he investigated goiter, otosclerosis, neurofibromatosis, pellagra, epilepsy, mental disorders, temperament, mental attributes of naval officers, mongoloid dwarfs, twinning, sex-linkage, and race crossing. The implications of eugenics for state, church, medicine and society in general were discussed wherever he could find a platform and an audience. His scientific background and associations gave him the prestige of an authority in the eyes of those inclined to accept his position on the social and political aspects of eugenics. But the opposition of many, instead of quickening the search for more accurate and convincing evidence, called forth a defensive attitude which led to exaggerated emphasis and dulled objective thinking.

Through the initiative of a Committee on Anthropology of the National Academy of Sciences, Davenport was commissioned (1918) as a major in the Sanitary Corps and assigned to the Surgeon General's Office to summarize the physical records of recruits, with Lt. Col. Albert G. Love, M.D. Working through the greater part of the year in Washington, with frequent visits to Cold Spring Harbor to keep things going, Davenport cooperated in the preparation of four volumes largely filled with tables and graphs showing the frequency of the different conditions and defects recorded and their geographical distribution. This work gave an anthropometrical trend to Davenport's interests that grew and occupied more and more of his time, especially after retirement. By measuring institutionalized children as they grew, notably those at Letchworth Village, N. Y., he collected data for extensive studies on growth curves and changes in proportions accompanying growth. With the hope of creating a renewed interest in child development, he published, in 1936, How We Came By Our Bodies, a popular book of 401 pages and 236 illustrations describing, from the standpoint of current sciences, the course of development and the role and origin of genes.

Overload (1919-1934)

At the time the work on army records began, Davenport had been directing simultaneously three institutions at Cold Spring Harbor for a period of eight years. During this time the growing concentration of his interest in eugenics did not prevent the material expansion of all three laboratories. The Eugenics Record Office had acquired a new building; so had the Station for Experimental Evolution, as well as an enlarged staff; the Biological Laboratory gained an endowment fund. The association of the three laboratories seemed to offer great strength; each had its own field and each had much to gain from the others. The one director for all seemed to guarantee effective cooperation. The vision and the goal were excellent, but the requirements for realization, as was so often the case, were not recognized. To secure support, put up buildings, and engage workers is not enough. There remains the need of fostering the integration of the work of highly individualistic investigators, in itself a

Personnel of the Department of Genetics, with John C. Merriam, president, and Mr. W. M. Gilbert, executive secretary, of Carnegie Institution of Washington, Cold Spring Harbor, May 12th, 1934—the year of Davenport's retirement.

FRONT ROW: Alice Laanes, Ruth Millar, Ethel Burtch, Margaret Martin, Ethyl Hunt, Martha Taylor, Margaret Hoover, Margaret Findley, Lillian Frink, Inger Andersen.

SECOND ROW (seated): Mary Holmes, Dr. A. Dorothy Bergner, Dr. J. Lincoln Cartledge, Dr. Robert Bates, Dr. Milislav Demerec, Dr. Oscar Riddle, Dr. Albert F. Blakeslee, Dr. John C. Merriam, Dr. Charles B. Davenport, Mr. Walter M. Gilbert, Dr. Harry H. Laughlin, Dr. Clarence Moran, Dr. James S. Potter, Dr. E. Carleton MacDowell, Dr. Morris Steggerda, Miriam North, Guinevere Smith, Elizabeth McKee.

THIRD ROW: J. Dixon McGlohon, Madeleine Wilkins, Gabriel A.

Lebedeff, Margaret Kuntz, Eunice White, Isabel Griffin, Harriet Smart, Jennie Schultz, Maria Cartledge, Amos Avery, Elizabeth Avery, Catherine Carley, Edith Harrigan, Alice Hellmer, Hilda Wullen.

FOURTH ROW: George Macarthur, M. J. Murray, Leslie Peckham, Dr. Benjamin Speicher, Theophil Laanes, Floyd Matson, James Banta, E. L. Lahr, Dr. J. P. Schooley, George Smith, William Drager, Louis Stillwell.

FIFTH ROW (back): William Schneider, Paul Holm, William Fagan, Peter Campbell, Edward Burns, Marie Bucuris, Jack Bucuris, Stanley Brooks, John Johnson, Clifford Valentine, Harry White, Domenico Sepe.

biological problem of the highest order of complexity. Further, the successful administration of three adjacent laboratories with differing financial status, and governing boards with different points of view, called for extraordinary scrupulousness, tact, and understanding to avoid pitfalls. Instead of integration within and between the three laboratories, difficulties continually arose that blocked effective cooperation. Davenport gave his time and energy to the limit, but the days were not long enough to satisfy all the claims on his attention and every move involved a choice, deliberate or not, between competing loyalties—loyalties to different institutions, to staffs, to family, to friends, to innumerable outside interests, to his early training, to his scientific ideals, his ethics, his objectives.

There was no time for, and, indeed, seemed to be little interest in, the deeper general significance of even his own studies. Groups of scientists attempting to build lasting foundations of knowledge require leadership, but leadership of a very special kind and on a philosophical plane. Effective operation of a group depends on mutual confidence, but this is not inspired by a leader whose own deep lack of confidence is covered by assuming a role of great independence and surpassing ability. This role served its purpose much of the time, but it broke down in the face of criticism. Davenport considered it was not hard for him to control his emotions; and his emotional life, in fact, was not highly developed in some fields, but he could not accept criticism objectively. High praise was eagerly received; adverse criticism tore down his defenses and allowed the specter of in-feriority to stare him in the face. It became an affront, an attack on his integrity. Emotions took charge; resentment would rise, not at the ideas presented so much as at the loss of his defense. In speech he might become confused and say the very opposite of his intention, or invent excuses. In writing, his greater fluency laid bare the wounded pride, which at times tried to restore itself by accusing the critic of personal animus, or calling names and belittling the charges. Thus valuable criticism was tossed aside. His intolerance of criticism was frankly enough admitted to himself, and in the presence of those who might criticise, he was increasingly silent, constrained and ill at ease. This applied to many colleagues, and especially to those of his own staff, in whom an insidious atmosphere of apprehension was developed by his practice of saying little and announcing plans by action.

A certain amount of administrative simplification was brought about by the Carnegie Institution in accepting from Mrs. Harriman the ownership of the Eugenics Record Office (1918) and a large endowment to support it. Later (1921) this Office and the Station for Experimental Evolution were combined coordinately as the Department of Genetics. The Institution also urged Davenport to resign the directorship of the Biological Laboratory. But these moves failed to concentrate his activities. The work in the Surgeon General's Office at this time greatly extended his program. Although he retired from the directorship of the Biological Laboratory

34   Bios

(1923) he accepted even broader responsibilities. At that time the Brook-
lyn Institute of Arts and Sciences desired to relinquish its control; and so
Davenport planned, solicited memberships and funds, and incorporated the
Long Island Biological Association to assume ownership of this laboratory.
Officially he was only its secretary, but he remained the leading spirit of
this Association.

The unrelenting urge kept driving him. The peace of satisfaction
remained out of reach. The wider the recognition and the higher the distinction, the more he had to live up to and defend. To convince him that
his brave picture of himself was true, ever more evidence was required, ever
more plans to obtain it.

Retirement and Retrospect (1934-1944)

On retirement (1934) a room in the Eugenics Record Office was assigned to Davenport, who became an Associate of the Carnegie Institution,
with a grant for the completion of his anthropometrical studies. During his
last ten years he wrote forty-seven papers, a book, and the fourth edition
of his STATISTICAL METHODS. Three of these papers were published post-
humously with the editorial supervision of Dr. Morris Steggerda. He took
an active part in civilian defense as an Air Raid Warden of the Nassau
County Defense Council, and gave many hours to plane spotting as a Recognized Observer of the Ground Observers Corps of the Army Air Forces.

Financial worries still followed him. He attempted, unsuccessfully,
to make money by breeding mice on a commercial scale; a special house
was built for this near his home and the equipment assembled, but the
mice did not thrive. The reduction of taxes seemed very important and he
took a prominent part in organizing tax payers' associations into larger
organizations, as well as in fighting against the tide of increasing acceptance of social responsibilities by the government. In one instance he led an
active campaign against the establishment of a county medical laboratory
which was strongly urged by physicians. The year of his retirement he published this statement: "The expenses of government have thus multiplied
due to the ever-increasing demands of the people, instigated by the welfare
workers, the city planners, the engineers, and the technologists, who,
through skillful advertising, create a public demand for their services and
products. . . . —and in times of reduced personal income without reduction in the cost of government we are, many of us, on the verge of starv-
ation."

In looking back over this dynamic life, which seemed ever to be dealing with new ideas and new projects, there dawns a realization that part
of this seeming newness was a reflection of Davenport's boyish eagerness
and enthusiasm. Indeed, rather than newness and originality, the persistence of early-established interests may more truly represent the man.

As a boy he was fascinated by organization, and he never stopped organizing. Besides the Long Island Biological Association, he was a founder of the Galton Society, the Aristogenic Association, The Eugenic Research Association, the Tax Payers League, the Cold Spring Harbor Whaling Museum. At Chicago he was involved in an attempt to form a Western Branch of the American Society of Naturalists . Membership in a society usually meant action; it might be committee work, collection of funds, or reorganization. He held many offices, including the presidency or vice-presidency of ten societies. A classification of his 64 memberships gives a rough measure of his interests: natural history (6), zoology (5), genetics (2), eugenics (7), anthropology (5), medicine (5), general science (11, nine being foreign), civic (10, several of these "Tax Payers"), social (5), and miscellaneous (8). Three societies awarded him honorary membership and one an honorary presidency; he was a Gold Medalist (1923) of the National Institute of Social Sciences; he was elected to the American Philosophical Society in 1907 and to the National Academy of Sciences in 1912.

The boy editor never stopped editing. He was on the editorial boards of Biometrika, Journal of Experimental Zoology, Zeitschrift fur Rassenkunde, Zeitschrift fur Menschliche Vererbung und Konstitutionslehre, Psyche, Journal of Physical Anthropology, Eugenical News (which he started in 1916), Growth and Human Biology. His projected Archives of Heredity and Variation, and Biochemical Journal, did not appear.

The real estate office boy never lost his delight in real estate transactions. He knew the procedures; he could write out a deed and survey a lot. The Harvard instructor bought for his family a house and lot that kept him in debt for many years. A lot was bought at Chicago early in 1902; six acres and a house on the shore at Cold Spring Harbor were bought personally in March 1904, to rent to laboratory personnel and to build apartment houses for them. ". . . . So we will have control from the stream . . . to the watering trough opposite Edmund Jones. Quite an empire for us, isn't it?" Next year a 19-acre farm was purchased in Mrs. Davenport's name, and later property was bought in Syosset. The last purchase that Davenport negotiated was a tract of 32 acres for the Biological Laboratory. He solicited the funds, arranged an exchange of Carnegie Institution property, steered the transaction, and had the property mapped out for a new community of biologists' homes.

The boy of the Saturday expeditions never lost his love of travel. Hard walking became a religion; wherever he was he walked, whether in the White Mountains, English lanes, or at home; he would walk half way to New York to attend a committee meeting, or rise at 4 a.m. to walk to Oyster Bay, *en route* to a Boston appointment. Many would try to give him a lift, especially in later years, as he charged, hatless, along local roads, but only occasionally—very apologetically—would he accept. To many

36   Bios

parts of the country he journeyed, and all the way across, by train or auto-
camping trip; across the Atlantic repeatedly and once to Australia; to Nassau, Bermuda, Jamaica, Yucatan, Canada, Gaspe; to meetings here, meetings there. Bateson once wrote him when in London, "Come as soon as
you like, only when you do come try to stop for more than half an hour
without rushing off to Vienna or Kamchatka."

The repressed boy in later years repressed himself. Personal satisfaction was somehow wrong, and yet, denied, it emerged on a deeper level
as prime mover. One of his deep regrets, he wrote, was that he ". . . .
was constantly prevented from doing agreeable things by the arising of
doubts that inhibited the agreeable action." And, in another place (1934),
"Effort and fatigue, are our greatest blessings, because they enable us to
appreciate the better their temporary absence." He was highly sensitive
to beauty, but could not accept the enjoyment of beauty as a value in life.
Repression accumulated pressure. This escaped explosively, forcing him into situations that clamped on new forms of repression and maintained the
intensity of the drive. This imperious drive was at once his handicap and
his special contribution. The picture is brought into high relief by contrast with that of his younger sister Frances G. Davenport. They had
much in common, both by nature and as investigators in the Carnegie Institution of Washington; but she, under her mother's guidance, grew up
with poise and confidence, happily adjusted to a single life devoted to the
search for knowledge and to the open enjoyment of beauty. In her in-
conspicuous and scholarly way, as a member of the Department of Historical Research, she found satisfaction in digging patiently and deeply,
and in publishing a few volumes of permanent value.

The boy who "played with baby after breakfast" never lost his love
of children. The late arrival of a son drew out from its hiding the tender
affection that was so keen and yet so sensitive; the affection that brought
weeping grief to the grown man when his mother died sixteen years before, now brought expanding joy and new confidence. Little Charlie
prospered and grew, as did his parents' delight in him. The warm tones
of the father's voice trumpeting over the neighborhood for son to come to
supper, told the story of pride and devotion. And then the poliomyelitis
epidemic of 1916 carried the lad off. Both parents, prostrated, retired to
a sanitarium. Davenport returned after some weeks, but commuted from
New York. Buried deeper than ever was the sensitive affection. At
times grown people would catch glimpses of this lovely side of him—
children always had the power to draw it forth, and he was charming
with them. All the measuring of babies and children was not without its
immediate rewards. During his last years a district school was built across
the road from his home . He would spend much time talking to the classes
on all manner of subjects. He would take the children to gather frogs'
eggs each spring; they would eat their lunch together by the pond and have
a happy time. With them he was relaxed; there was no defense and no

embarrassment, and he was content. Those children probably saw more clearly than anyone else the real Charles Davenport, and they loved him.

The eager boy who planned and planned, made one last fatal plan. Word came in January 1944 that a killer whale had been beached at the eastern end of Long Island, and Davenport determined to secure its skull for the Whaling Museum, of which he was Curator and Director. Many difficulties were overcome in moving the head to his home. Instead of using the slow but easy method of maceration in a pond, he undertook to boil it, and for a fortnight he labored far into the nights in the intense heat of a seething cauldron in an open shed, with the bitter winter cold pressing in from all sides. He became so permeated with the nauseating smell that heads were turned when he approached, and at the last Staff meeting he attended he sat off by himself. He caught a cold and still worked on. The job was far from finished, when a type of pneumonia developed that resisted all modern drugs. At the hospital, he thought the whale was to blame. But he had asked of himself one thing too many. He died on February 18th, 1944.

Only an extraordinarily resilient constitution could have sustained so long the continuous pressure Charles Davenport put upon himself in his struggle to overcome a deeply implanted sense of inferiority. Instead of destructively turning this in upon himself, he turned it outward. Instead of seeking immediate physical gratification and displaying the bombastic self-assurance typical of many ambitious men, he identified himself with great concepts—Science, Experimental Morphology, Advancement of Knowledge, Evolution, Improvement of Mankind. As a result he lived a full life and played a constructive part that gained him wide recognition. Such an interpretation gives meaning to many of the strange inconsistencies that appeared and opens the door to unqualified admiration of the valiant way he faced his great problem. It was a grand struggle.

Bibliography of Charles B. Davenport

Excluded from this list are: Reviews, preliminary abstracts, reports and addresses, reports of meetings, brief notes, notices, announcements, articles and correspondence in newspapers.

1890

Cristatella: The Origin and Development of the Individual in the Colony. Bull. Mus. Comp. Zool., 20:101-151.

1891

Preliminary Notice on Budding in Bryozoa. Proc. Amer. Acad. Arts and Sci., 25: 278-282.

Observations on Budding in Paludicella and Some Other Bryozoa. Bull. Mus. Comp. Zool., 22:1-114. 12 plates.

1892

The Germ Layers in Bryozoan Buds. Zool. Anz., No. 396, 1-3.

38   Bios

1893

On Urnatella Gracilis. Bull. Mus. Comp. Zool., 24:1-44. 6 plates.
Note on the Carotids and the Ductus Botalli of the Alligator. Bull. Mus. Comp.
    Zool., 24:45-50. 1 plate.
Studies in Morphogenesis: I. On the Development of the Cerata in Aeolis. Bull.
    Mus. Comp. Zool., 24:141-148. 2 plates.

1894

Studies in Morphogenesis: II. Regeneration in Obelia and Its Bearing On Differen-
    tiation in the Germ-Plasm. Anat. Anz., 9:283-294. 6 figures.

1895

Studies in Morphogenesis: III. On the Acclimatization of Organisms to High
    Temperatures. Arch. f. Ent. Mech., 2:227-249. By C. B. D. and W. E. Castle.
Studies in Morphogenesis: IV. A Preliminary Catalogue of the Processes Concerned
    in Ontogeny. Bull. Mus. Comp. Zool., 27:173-199. 31 figures.

1896

Studies in Morphogenesis: V. On the Acclimatization of Organisms to Poisonous
    Chemical Substances. Arch. f. Entwickelungsmechanik, 2:564-583. 3 figures.
    By C. B. D. and H. V. Neal.
Studies in Morphogenesis: VI. A Contribution to the Quantitative Study of Corre-
    lated Variation and the Comparative Variability of the Sexes. Proc. Amer.
    Acad. Arts and Sciences, 32:87-97. By C. B. D. and C. Bullard.
Harvard Admission Requirements. Harvard Grad. Mag., December.
Discussion. Botany as an Admission Requirement. Educational Rev., May.

1897

On the Determination of the Direction and Rate of Movement of Organisms by Light.
    Jour. Physiol., 21:22-32. 1 figure. By C. B. D. and W. B. Cannon.
Experimental Morphology. Part I. Effect of Chemical and Physical Agents Upon
    Protoplasm. The Macmillan Co., N. Y., 294 pp.
The Role of Water in Growth. Proc. Boston Soc. Nat. Hist., 28-73-84.
A Contribution to the Study of Geotaxis in the Higher Animals. Jour. Physiol., 22:
    99-110. By C. B. D. and H. Perkins.

1898

Morphogenesis. In "The Biological Problems of Today." Science, 7:158-161.
A Precise Criterion of Species. A. The General Method. Science, 7:685-690.
The Fauna and Flora About Cold Spring Harbor, L. I. Science, 8:685-689.

1899

The Importance of Establishing Specific Place Modes. Science, 9:415-416.
Statistical Methods With Special Reference to Biological Variation. John Wiley Sons,
    N. Y. (2nd ed. 1904; 3rd ed., 1914; 4th ed., 1936).
Experimental Morphology, Part II. The Macmillan Co., N. Y. 508 pp. 140 figures.
Phototaxis of Daphnia. Science, 9:368. By C. B. D. and F. T. Lewis.
The Aims of the Quantitative Study of Variation. Biological Lectures, Marine Biolog-
    ical Laboratory, 267-272.
Synopses of North American Invertebrates. I. Fresh Water Bryozoa. Amer. Nat.,
    33:593-596.
Articles on Zoology and General Biology in the New International Encyclopedia.
    Dodd, Mead & Co., New York.

1900

Introduction to Zoology. The Macmillan Co., N. Y. by C. B. D. with Gertrude C. Davenport. 2nd ed. (1911) title changed to Elements of Zoology. 508 pp.

On the Variation of the Statoblasts of Pectinatella Magnifica From Lake Michigan, at Chicago. Amer. Nat., 34:959-968. 9 figures.

A History of the Development of the Quantitative Study of Variations. Science, 12: 864-870.

On the Variation of the Shell of Pecten Irradians Lamarck From Long Island. Amer. Nat., 34:863-877. 2 figures.

1901

Mendel's Law of Dichotomy in Hybrids. Biol. Bull., 2:307-310.

Zoology of the Twentieth Century. Science, 14:315-324.

The Statistical Study of Evolution. Pop. Sci. Mo., 59:447-460.

1902

Variability, Symmetry and Fertility in an Abnormal Species. Biometrika, 1:255-256.

The Relation of the American Society of Naturalists to Other Scientific Societies. Science, 15:244-246.

Biological Experiment Station for Studying Evolution. Carnegie Inst. Wash. Year Book, 1:280-282.

1903

The Proposed Biological Station at the Tortugas. Science, 17:945-946.

The Animal Ecology of the Cold Spring Sand Spit, With Remarks on the Theory of Adaptation. The Decennial Publications of the Univ. of Chicago. The University of Chicago Press, Chicago, 10:157-176.

Quantitative Studies in the Evolution of Pecten. III. Comparison of Pecten Opercularis From Three Localities of the British Isles. Proc. Amer. Acad. Arts and Sciences, 39:123-159.

The Collembola of Cold Spring Beach, With Special Reference to the Movements of the Poduridae. Cold Spring Harbor Monographs. II, 32 pp. 1 plate.

A Comparison of the Variability of Some Pectens From the East and the West Coasts of the United States. Mark Anniversary Volume, 121-136. Plate IX. Henry Holt and Co., New York.

1904

Color Inheritance in Mice. Science, 19:110-114.

Wonder Horses and Mendelism. Science, 19:151-153.

Report on the Fresh-Water Bryozoa of the United States. Proc. U.S. Nat. Museum, 27:211-221. Plate VI.

Introductory Address at Opening of the Station For Experimental Evolution. Carnegie Inst. Wash. Year Book, 3:33-34.

Reports of Station For Experimental Evolution, Department of Genetics. Carnegie Inst. Wash. Year Books, 3-33, 1904-1934.

Animal Morphology in Its Relation to Other Sciences. Science 20:697-706.

Studies on the Evolution of Pecten. IV. Ray Variability in Pecten Varius. Jour. Exper. Zool., 1:607-616. By C. B. D. and Marian E. Hubbard.

1905

Evolution Without Mutation. Jour. Exper. Zool., 2:137-143.

The Origin of Black Sheep in the Flock. Science, 22:674-675.

Announcement of Station for Experimental Evolution, Cold Spring Harbor, Carnegie Inst. Washington, 4 pp. 3 plates.

40   Bios

Some Mendelian Results in Animal Breeding. Proc. Soc. Exper. Biol. and Med. 2:30.
Species and Varieties, Their Origin by Mutation. Science 22:369-372.

1906

On the Imperfection of Mendelian Dominance in Poultry Hybrids. Proc. Soc. Exper.
    Biol. and Med., 3:31-32.
Inheritance in Poultry. Carnegie Inst. Wash. Pub. 52, 136 pp. 17 plates.
The Mutation Theory in Animal Evolution. Science, 24:556-558.

1907

Cooperation in Science. Science, 25:361-366.
Heredity and Mendel's Law. Proc. Wash. Acad. Sci., 9:179-187.
Inheritance in Pedigree Breeding of Poultry. Report Amer. Breeders' Assn., 3:26-33;
    also, Daily National Live Stock Reporter, Jan. 19, 1907.
Dominance of Characteristics in Poultry. Report 3rd Internat. Conf. on Genetics,
    1906; Roy Hort. Soc. 138-139.
Recent Advances in the Theory of Breeding. Proc. Amer. Breeders' Assn. 3:132-135.
Heredity of Eye-Color in Man. Science, 26:589-592. By Gertrude C. Davenport and
    C. B. D.   ,

1908

Heredity of Hair-Form in Man. Amer. Nat. 42:341-349. By Gertrude C. Davenport
    and C. B. D.
The American Breeders' Association. Science, 27:413-417.
Degeneration, Albinism and Inbreeding. Science, 28:454-455.
Recessive Characters. Science, 28:729.
Inheritance in Canaries. Carnegie Inst. Wash. Pub. 95, 26 pp. 3 plates.
Heredity of Some Human Physical Characteristics. Proc. Soc. Exper. Biol. and Med.,
    5:101-102.
Recent Advances in the Theory of Breeding. Rep. Amer. Breeders' Assn. 3:132-135.
Determination of Dominance in Mendelian Inheritance. Proc. Amer. Philos. Soc.
    47:59-63.
Elimination of Self-Colored Birds. Nature, 78:101.
Aufgabe Und Einrichtung Einer Biologischen Versuchsanstalt Für Tierzucht. Jahrb.
    f. wissen. und praktische Tierzucht. III. Jahrg; 28-30.

1909

Heredity of Hair Color in Man. Amer. Nat., 43:193-211. By Gertrude C. Davenport
    and C. B. D.
Some Principles of Poultry Breeding. Report Amer. Breeders' Assn., 5:376-379.
    (Author given by error as James E. Rice).
A Suggestion As to the Organization of the Committee on Breeding Poultry. Report
    Amer. Breeders' Assn., 5:379-380.
The Factor Hypothesis in Its Relation to Plumage Color. Report Amer. Breeders'
    Assn., 5:382-385.
Influence of Heredity on Human Society. Annals of Amer. Acad. of Political and
    Social Science, 34:16-21.
Prepotency in Pigment Colors. Report Amer. Breeders' Assn., 5:221-222. By Ger-
    trude C. Davenport and C. B. D.
Inheritance of Characteristics in Domestic Fowl. Carnegie Inst. Wash. Pub. 121, 100
    pp., 12 plates, 2 figures.
Mutation. In "Fifty Years of Darwinism." 160-181. New York.

BIOS   41

Heredity in Man. Harvey Society Lectures, 1908-09, 280-290.
Fit and Unfit Matings. Bull. Amer. Acad. Med., 11:657-670. 4 figures.

1910

Eugenics—The Science of Human Improvement by Better Breeding. Henry Holt &
    Co., N. Y. 35 pp.
The Imperfection of Dominance and Some of Its Consequences. Amer. Nat. 44:129-
    135.
Imperfection of Dominance. Amer. Breeders' Mag., 1:39-42.
Inheritance of Plumage Color in Poultry. Proc. Soc. Exper. Biol. and Med., 7:168.
The New Views About Reversion. Proc. Amer. Phil. Soc. 49:291-296.
Variability of Land Snails (Cerion) in the Bahama Islands With Its Bearing on the
    Theory of Geographical Form Chains. Science, 31:600.
Heredity of Skin Pigment in Man. Amer. Nat., 44:641-731. By Gertrude C. Dav-
    enport and C. B. D.
Dr. Galloway's "Canary Breeding." Biometrika, 7:398-400.

1911

Euthenics and Eugenics. Pop. Sci. Mo., 78:16-20.
Heredity in Relation to Eugenics. Henry Holt & Co., N.Y. XI + 298 pp. 175 fig.,
    1 plate.
The Transplantation of Ovaries in Chickens. Jour. Morphology, 22:111-122.
Imperfection of Dominance. Report Amer. Breeders' Assn., 6:29-32.
Report of Committee on Eugenics. Report Amer. Breeders' Assn., 6:91-93.
Characters in Mongrel vs. Pure-Bred Individuals. Report Amer. Breeders' Assn., 6:
    339-341.
The Study of Human Heredity. Methods of Collecting, Charting and Analyzing Data.
    Eugenics Record Office Bull. No. 2, 30 pp. By C. B. D., H. H. Laughlin, D. F.
    Weeks, E. R. Johnstone and H. H. Goddard.
A First Study of Inheritance of Epilepsy. Jour. Nerv. and Mental Disease, 38:641-
    670. 31 figures and 11 tables. Reprinted as Bull. No. 4, Eugenics Record Office.
Eugenical Limitations to the Prevention of Infant Mortality. Amer. Assn. for Study
    and Prevention of Infant Mortality.
Another Case of Sex-Limited Heredity in Poultry. Proc. Soc. Exper. Biol. and Med.,
    9:19-20.
The Biological Laboratory at Cold Spring Harbor, New York, U. S. A. Internat.
    Revue d. ges. Hydrobiol. u. Hydrogr., 4:223-226.

1912

The Origin and Control of Mental Defectiveness. Pop. Sci. Mo., 80:87-90.
The Trait Book. Eugenics Record Office Bull. No. 6, 52 pp. 1 colored plate, 1 figure.
Light Thrown by the Experimental Study of Heredity Upon the Factors and Methods
    of Evolution. Amer. Nat., 46:129-138.
The Nams. The Feeble-Minded As Country Dwellers. The Survey, 27:1844-1845.
Eugenics and the Physician. N. Y. Med. Jour. for June 8. 13 pp.
Sex-Limited Inheritance in Poultry. Jour. Exper. Zool., 13:1-18. 8 plates.
The Family-History Book. Eugenics Record Office Bull. No. 7, 101 pp. 6 figures,
    5 plates.
The Inheritance of Physical and Mental Traits of Man and Their Application To
    Eugenics. Chapter VIII in "Heredity and Eugenics." The University of Chicago
    Press, 269-288.

42   Bios

The Geography of Man in Relation to Eugenics. Chapter IX in "Heredity and
    Eugenics." The University of Chicago Press, 289-310.
Horns in Sheep As a Typical Sex-Limited Character. Science 35:375-377. By T. R.
    Arkell and C. B. D.
The Nature of the Inheritance of Horns in Sheep. Science, 35:927. By T. R. Arkell
    and C. B. D.
The Hill Folk. Report on a Rural Community of Hereditary Defectives. Eugenics
    Record Office Mem. No. 1, 56 pp. 4 text figures, 3 charts. By Florence H. Dan-
    ielson and C. B. D.
The Nam Family. A Study in Cacogenics. Eugenics Record Office Mem. No. 2,
    85 pp. 4 text figures, 4 charts. By Arthur H. Estabrook and C. B. D.
Some Social Applications of Modern Principles of Heredity. Trans. 5th Internal.
    Cong. on Hygiene and Demography. 1-4.
Importance of Heredity to the State. Quart. Minnesota Educational, Philanthropic,
    Correctional and Penal Institutions, 12:23-51. Also, Indiana Bull. of Charities
    and Corrections, 94th Quarter, 1913. (Sept.)
Relation of Eugenics to Religion. Homiletic Rev., 63:8-12.
Let Church and State Cooperate. The Medical Times (June), 167.
How Did Feeble-Mindedness Originate in the First Instance? The Training School,
    9:87-90.
Heredity in Nervous Disease and Its Social Bearings. Jour. Amer. Med. Assoc., 59:
    2141-2142.
Eugenics in Its Relation to Social Problems. The N. Y. Assoc. for Improving the
    Condition of the Poor. Pub. No. 70, 7 pp.

                          1913
A Suggested Classification of Writings on Eugenics. Science, 37:370.
A Reply to Dr. Heron's Strictures. Science, 38:773-774.
Inheritance of Some Elements of Hysteria. Read at a meeting of alienists and
    neurologists held under the auspices of the Chicago Medical Society, June 23-25,
    2 pp.
Heredity, Culpability, Praiseworthiness, Punishment and Reward. Pop. Sci. Month-
    ly, 82:33-39.
State Laws Limiting Marriage Selection Examined in the Light of Eugenics. Eugenics
    Record Office Bull. No. 9, 66 pp. 2 charts, 3 tables.
Clonal Variation in Pectinatella. Amer. Nat., 47:361-371. By Annie P. Hench-
    man and C. B. D.
Heredity of Skin-Color in Negro-White Crosses. Carnegie Inst. Wash. Pub. 188,
    106 pp. 4 plates.

                          1914
The Importance to the State of Eugenic Investigation. 1st Nat. Conf. on Race Bet-
    terment, Battle Creek, Mich. 6 pp.
Reply to the Criticism of Recent American Work by Dr. Heron of the Galton
    Laboratory. A Discussion of the Methods and Results of Dr. Heron's Critique.
    Eugenics Record Office Bull. No. 11, 25 pp.
A Contribution, in "The Church, the People and the Age." Edited by Robert Scott
    and George W. Gilmore. Funk and Wagnalls Co., N. Y. 429-430.
The Origin of Domestic Fowl. Jour. Heredity, 5:313-315.
The Bare Necks. Jour. Heredity, 5:374.
Eugenics. In "Reference Handbook of Medical Sciences," 151-155.

BIOS   43

Medico-Legal Aspects of Eugenics. The Medical Times (Oct.), 3-15.
Skin Color of Mulattoes. Jour. Heredity, 5:556-558.
The Eugenics Programme and Progress in Its Achievement. In "Eugenics: Twelve
University Lectures." Dodd, Mead and Co., N. Y.

1915

Inheritance of Temperament. Proc. Soc. Exp. Biol. and Med., 12:182.
The Aboriginal Rock-Stencillings of New South Wales. Sci. Mo., 1:98-99. 2 figures.
The Feebly Inhibited. I. Violent Temper and Its Inheritance. Jour. Nervous and
Mental Disease. 42:593-628. Also, Eugenics Record Office Bull. No. 12.
The Feebly Inhibited: (A). Nomadism or the Wandering Impulse With Special
Reference to Heredity. (B) Inheritance of Temperament. Carnegie Inst. Wash.
Pub. 236, 158 pp. 89 figures.
The Value of Scientific Genealogy. Science, 41:337-342.
Field Work an Indispensable Aid to State Care of the Socially Inadequate. Read at
42nd Annual Session of the Nat. Conf. of Charities and Corrections, May 15,
16-19.
The Racial Element in National Vitality. Pop. Sci. Mo., 86:331-333.
A Dent in the Forehead. Jour. Heredity. 6:163-164.
Health and Heredity. In "Educational Hygiene," by R. W. Rapier, N. Y. 45-58.
Huntington's Chorea in Relation to Heredity and Eugenics. Proc. Nat. Acad. Sci.,
1:283-285.
The Heredity of Stature. Science, 42:495.
Hereditary Fragility of Bone (Fragilitas Osseus, Osteopsathyrosis). Eugenics Record
Office Bull. No. 14, 31 pp. By H. S. Conard and C. B. D.
How to Make a Eugenical Family Study. Eugenics Record Office Bull. No. 13, 35 pp.
4 charts and 2 tables. By C. B. D. and H. H. Laughlin.

1916

Huntington's Chorea in Relation to Heredity and Eugenics. Eugenics Record Office
Bull. No. 17, 195-222. (Reprinted from Amer. J. Insanity, 73:195-222).
Heredity of Albinism. Jour. Heredity, 7:221-223.
The Hereditary Factor in Pellagra. Arch. Internal Med., 18:1-28. 38 figures. (Re-
printed as Eugenics Record Office Bulletin No. 16).
The Form of Evolutionary Theory That Modern Genetical Research Seems to Favor.
Amer. Nat., 50:449-465.
Eugenics As a Religion. Paper read at Golden Jubilee Celebration of the Battle
Creek Sanitarium, 8 pp.
Introduction in "The Jukes in 1915" by Arthur H. Estabrook. Carnegie Inst. Wash.
Pub. 240.
Brief Studies in Heredity of 73 Well-Known Persons. (unsigned), Eugenical News,
1-3, 1916-1918.
Heredity of Stature. (Abstract) Proc. XIX Internat. Congress of Americanists.

1917

The Effect of Race Intermingling. Proc. Amer. Philos. Soc., 56:364-368.
The Personality, Heredity and Work of Charles Otis Whitman, 1843-1910. Amer.
Nat., 51:5-30.
Charles Otis Whitman. Proc. Amer Acad. Arts and Sci., 52:877-878.
Inheritance of Stature. Genetics, 2:313-389. Reprinted as Eugenics Record Office
Bull. No. 18. 19 text figures, 33 tables.

On Utilizing the Facts of Juvenile Promise and Family History in Awarding Naval Commissions to Untried Men. Proc. Nat. Acad. Sci., 3:404-409.

Introduction to "Survey of mental disorders in Nassau County, New York, July-October, 1916." Psychiatric Bull., 2:6-10.

Family Performance As a Basis For Selection in Sheep. Jour. Agric. Res., 10:93-97. By E. G. Ritzman and C. B. D.

1918

Hereditary Tendency To Form Nerve Tumors. Proc. Nat, Acad. Sci., 4:213-214.

Moss Animalcules (Bryozoa). Chapter XXVIII, First edition of Ward and Whipple's "Fresh-Water Biology," 947-956.

Multiple Neurofibromatosis (Von Recklinghausen's Disease) And Its Inheritance: With Description of a Case. Amer. Jour. Med. Sci., 156:507-541. (By S. A. Preiser and C. B. D.) (Reprinted as Eugenics Record Office Bull. No. 19. 34 pp. 36 text figures).

Report of the Surgeon General of the Army to the Secretary of War, 1918. 46-116; 164-226. Govt. Printing Office, Washington.

1919

The Genetical Factor in Dental Research. Jour. Dental Res., 1:9-11.

The Biological Laboratory at Cold Spring Harbor. Bull. Brooklyn Inst. Arts and Sci., April, 8 pp.

Eugenics in Relation to Medicine. Oxford Loose Leaf Med.

Standard Methods in Research Surveys. Proc. Nat. Conference of Social Work. 4 pp.

Naval Officers: Their Heredity and Development. Carnegie Inst. Wash. Pub. 259. (By C. B .D. and Mary T. Scudder). iv+236 pp. 60 charts.

A Comparison of White and Colored Troops in Respect to Incidence of Disease. Proc. Nat. Acad. Sci., 5:58-67. (By A. G. Love and C. B. D.).

Physical Examination of the First Million Draft Recruits: Methods and Results. (Compiled under direction of the Surgeon General, War Department). Office of the Surgeon General, Bull. No. 11, 1-521. By A. G. Love and C. B. D.).

Immunity of City-Bred Recruits. Arch. Internal Med., 24:129-153. (By A. G. Love and C. B. D.).

Defects Found in Drafted Men: Statistical information compiled from the draft records under the direction of the Surgeon General, M. W. Ireland. (Printed for the use of the Senate Committee on Military Affairs, 66th Cong. 1st Sess., Wash.). 359 pp. Govt. Printing Office. (By A. G. Love and C. B. D.).

A Strain of Multiple Births. Jour. Heredity, 10:382-384.

1920

Influence of the Male in the Production of Twins. The Med. Rec., 97:509-511.

Influence of the Male in the Production of Human Twins. Amer. Nat., 54:122-129.

Defects Found in Drafted Men. Scient. Mo., 10:5-25; 125-141. (By C. B. D. and A. G. Love).

Heredity of Twin Births. Proc. Soc. Exper. Biol. and Med., 17:75-77.

Defects Found in Drafted Men. Statistical information compiled from the draft records showing the physical condition of the men registered and examined in pursuance of the requirements of the Selective Service Act. 1663 pp. Govt. Printing Office, Washington.

A Comparison of Some Traits of Conformation of Southdown and Rambouillet Sheep and of Their F_1 Hybrids, With Preliminary Data and Remarks on

Variability in F_2. N. H. Agr. Exper. Sta. Tech. Bull. No. 15. 32 pp. 7 plates. (By E. G. Ritzman and C. B. Davenport).

The Best Index of Body Build. Quart. Pubs. Amer. Statis. Assoc., Sept. 341-344.

The Mean Stature of American Males. Quart. Pubs. Amer. Statis. Assoc., Dec., 484-487.

Height-Weight Index of Build. Amer. Jour. Phys. Anthrop., 3:467-475.

Heredity of Constitutional Mental Disorders. Psychol. Bull., 17:300-310. Reprinted as Eugenics Record Office Bull. No. 20. 11 pp.

1921

Comparative Social Traits of Various Races. School and Soc., 14:344-348.

Army Anthropology. Med. Dept. U. S. Army in the World War, Vol. 15. Statistics, Pt. 1, 635 pp. (By C. B. D. and A. G. Love). Govt. Printing Office, Washington.

Research in Eugenics. Science, 54:391-397.

1922

Address on Dr. Herbert S. Jennings' Research. Johns Hopkins Alumni Mag. 10:87-92.

Normal Changes in Body Build During Development. Eugenical News, 7:85-86.

Multiple Sclerosis From the Standpoint of Geographic Distribution and Race. Arch. Neurol. and Psychiatry, 8:51-58.

Edward Laurens Mark, Ph.D., LL.D. Harvard Alumni Bull., 24:822-826.

The Researches of Alfred Goldsborough Mayor. Science, 56:134-135.

1923

The Ecology of Epilepsy. II. Racial and Geographic Distribution of Epilepsy. Arch. Neurol. and Psychiatry, 9:554-566.

Hereditary Factors in Body Build. Proc. Soc. Exper. Biol. and Med., 20:388-390.

The Work of the Eugenics Record Office. Eugenics Review (April), 313-315.

Body Build and Its Inheritance. Proc. Nat. Acad. Sci., 9:226-230.

Heredity of Bone Defects. Eugenical News, 8:59-61

Comparative Social Traits of Various Races, Second Study. Jour. Applied Psychol., 7:127-134. (By C. B. D. and Laura T. Craytor).

The Deviation of Idiot Boys From Normal Boys in Bodily Proportions. Proc. 47th Ann. Session of the Amer. Assoc. for the Study of Feeble-minded: 8 pp. (C. B. D. and Bertha E. Martin).

Hereditary Influence of the Immigrant. Jour. Nat. Inst. Soc. Sci., 8:48-49.

1924

Influence des Glandes Endocrines sur l'Hérédité (translated by Dr. A. Thooris) Bull. de la Soc. d'Etude des Formes Humaines 2e Annee, No. 4, 379-387.

Radio-Ulnar Synostosis. Arch. Surgery, 8: 705-762. (By C. B. D. and Louise A. Nelson).

Influence of Endocrines on Heredity. Proc. 48th Ann. Session of the Amer. Assoc. for Study of Feeble-Minded, 132-144.

Body-Build and Its Inheritance. Carnegie Inst. Wash. Pub. 329. vi+176 pp., 9 plates, 53 figures.

1925

Body-Build: Its Development and Inheritance. Eugenics Record Office Bull. No. 24: 42 pp., 25 figures, 2 plates and 9 tables.

46   Bios

What Proportion of Feeblemindedness Is Hereditary? Investigation and Reports. Assoc. Res. in Nerv. and Mental Dis., 3:295-299.

Notes on Physical Anthropology of Australian Aborigines and Black-White Hybrids. Amer. Jour. Phys. Anthrop., 73-94.

Regeneration of Ovaries in Mice. Jour. Exper. Zool., 42:1-12.

Chromosomes, Endocrines and Heredity. Sci. Mo., 20:491-498.

The Normal Interval Between Human Births. Eugenical News, 10:86-88. (By Anne W. March and C. B. D.).

Family Studies on Mongoloid Dwarfs. Proc. of the 49th Annual Session of the Amer. Assoc. for the Study of the Feeble-Minded. 266-286. By C. B. D. and Grace Allen.

Evidence for Evolution. New Types of life. Sci. Mo., 21:135-136.

Heredity of Nasal Brow. Eugenical News, 10:126-127. (By Grace Allen and C. B. D.).

Heredity and Culture As Factors in Body-Build. Reprinted from Public Health Reports 40:2601-2605. (Reprint No. 1053). Govt. Printing Office, Washington. (By C. B. D. and Louise A. Nelson).

1926

Human Metamorphosis. Amer. Jour. Phys. Anthrop., 9:205-232.

Notes Sur l' Anthropologie des Aborigines Australiens et des Metis Blancs et Noirs. (Traduction de Mlle. M. Renaud). Bull. de la Soc. d'Etude des formes humaines. Année 4:3-22.

The Skin Colors of the Races of Mankind. Nat. Hist. 26:44-49.

A Remarkable Family of Albinos. Eugenical News, 11:50-52. (By Grace Allen and C. B. D.).

Rudolf Martin: A Biographical Sketch. Nat. Hist., 26:99-100.

Coordinates in Anthropometry. Anthropol. Anz., 3:30-32.

The Nature of Hereditary Mental Defect. Proc. 50th Ann. Session of the Amer. Assoc. for the Study of Feeble-Minded. 8 pp.

Human Growth Curve. Jour. Gen. Physiol., 10:205-216.

Heredity of Disease. Jour. Amer. Med. Assoc., 87:664-667.

Some Wool Characters and Their Inheritance. N. H. Agric. Exp. Sta. Tech. Bull. No. 31, 58 pp. (By C. B. D. and E. G. Ritzman).

1927

Measurement of Men. Amer. Jour. Phys. Anthrop., 10: 65-70.

Effects of Operations Upon the Thyroid Glands of Female Mice on the Growth of Their Offspring. Jour. Exp. Zool., 48:395-440. (By C. B. D. and W. W. Swingle).

Heredity of Human Eye Color. Bibliographica Genetica, 3:443-463.

Is Weight Hereditary? In "Your Weight and How to Control It," ed. by Morris Fishbein, 77-82.

Guide to Physical Anthropometry and Anthroposcopy. Eugenics Res. Assoc. (No. 1 Handbook Series), Cold Spring Harbor, 53 pp.

Alfred Goldsborough Mayor. 1868-1922. Mem. Nat. Acad. Sci., 21:1-14. 8th Memoir.

1928

Control of Universal Mongrelism. How a Eugenist Looks at the Matter of Marriage. Good Health, 10-11:31 (June).

BIOS   47

Is There Inheritance of Twinning Tendency From the Father's Side? Zeitschr. f. induk. Abstammungs- u. Vererbungslehre, Supplementband 1:595-602.

Crime, Heredity and Environment. Jour. Hered., 19:307-313.

Nasal Breadth in Negro x White Crossing. Eugenical News, 13:36-37. (By C. B. D. and Morris Steggerda).

Heredity and Longevity. Proc. 3rd Race Betterment Conference, Battle Creek, 15-18.

Aims and Methods in Anthropometry. Inst. International d'Anthropol. 3rd Session, Amsterdam, 22-29.

Race Crossing in Jamaica. Sci. Mo., 27:225-238.

Are There Genetically Based Mental Differences Between the Races? Science, 68: 628.

1929

Do Races Differ in Mental Capacity? Human Biol., 1:70-89.

Laws Against Cousin Marriages; Would Eugenicists Alter Them? Eugenics, 2: 22-23.

Race Crossing in Jamaica. Carnegie Inst. Wash. Pub. 395. IX + 516 pp. 29 plates. (By C. B. D. and Morris Steggerda).

1930

The Mingling of Races. Chap. XXIII in "Human Biology and Racial Welfare," ed. by E. V. Cowdry, 553-565.

Intermarriage Between Races; A Eugenic or Dysgenic Force? Eugenics, 3:58-61. (Discussion by C. B. D., Hrdlicka, Newman and Herskowitz).

As the Twig's Bent. Chapter in "The Alien in our Midst," ed. by Madison Grant and Charles Stewart Davison, 49-53.

Light Thrown by Genetics on Evolution and Development. Sci. Mo., 30:307-314.

J. Arthur Harris. Science, 71:474-475.

Interracial Tests of Mental Capacity. In Proc. and papers of 9th International Congress of Psychology.

Relation Between Physical and Mental Development. Eugenical News, 15:79-81.

The Mechanism of Organic Evolution. Jour. Wash. Acad. Sci., 20:317-331.

Some Criticisms of "Race Crossing in Jamaica". Science, 72:501-502.

Adolescent Spurt in Growth. Mem. vol. 60th birthday of Prof. V. Ruzicka, 35-44, Prague, Czechoslovakia (Abst. in Am. J. Phys. Anthrop., 14:85).

Litter Size and Latitude. Arch. f. Rass- und Gesellschaftsbiologie, 24:97-99.

Hearing in Children When Both Parents Have Otosclerosis. Trans. Amer. Laryngological, Rhinological and Otological Soc., 7 pp.

Intelligence Quotient and the Physical Quotient: Their Fluctuation and Intercorrelation. Human Biol., 2:473-507. (By C. B. D. and Blanche M. Minogue).

Sex Linkage in Man. Genetics, 15:401-444.

1931

Heredity and Disease. Eugenical News, 16:189-191.

Individual vs. Mass Studies in Child Growth. Proc. Amer. Philos. Soc., 70:381-389.

Some Results of Inbreeding on Fecundity and on Growth in Sheep. N. H. Agr. Exper. Sta. Tech. Bull. No. 47, 1-27. (By E. G. Ritzman and C. B. D.).

1932

Heredity and Disease. Sci. Mo., 34:167-169.

The Growth of the Human Foot. Amer. Jour. Phys. Anthrop., 17:167-211.

48   Bios

The Development of Eugenics. In "A decade of progress in Eugenics." Third Inter-
    nat. Congress of Eugenics, 17-22.
On the Need of Checking in Anthropometry. In "A decade of progress in Eugenics."
    Third Internat. Congress of Eugenics, 45:46.
Relation Between Pathology and Heredity. Eugenical News, 17:105-109.
Mendelism in Man. Proc. Sixth Internat. Congress of Genetics, 1:135-140.
The Genetical Factor in Endemic Goiter. Carnegie Inst. Wash. Pub. 428, 56 pp.,
    9 charts, 6 text figures.

1933

Body-Build and Its Inheritance. Proc. of the Assoc. for Res. in Nervous and Mental
    Disease, 14:21-27.
A Scientist's Viewpoint. Jour. Calendar Reform, 3:109.
The Inheritance of Disease. To What Extent and in What Circumstances Are
    Cancer, Tuberculosis, Goiter and Syphilis Inheritable. Sci. Amer., 149:162-164.
An Alleged Case of Inheritance of Acquired Characters. Amer. Nat., 67:549-558.
Evidences of Man's Ancestral History in the Later Development of the Child. Proc.
    Nat. Acad. Sci., 19:783-787.
The Crural Index. Amer. Jour. Phys. Anthrop, 17:333-353.
The Genetic Factor in Otosclerosis. Arch. Otolaryngology, 17:135-170, 340-383,
    503-548. (By C. B. D., Bess L. Milles and Lillian B. Frink).
Heredity and Its Interaction With Environment. In White House Conference on
    Child Health and Protection; Sec. 1: Growth and Development of the Child.
    Part I: General Considerations, 13-43.

1934

The Thoracic Index. Human Biol., 6:1-23.
Critical Examination of Physical Anthropometry on the Living. Proc. Amer. Acad.
    Arts and Sci., 69:265-284. (By C. B. D., M. Steggerda and W. Drager).
Ontogeny and Phylogeny of Man's Appendages. Proc. Nat. Acad. Sci., 20:359-364.
The Value of Genealogical Investigation to the Promotion of the Welfare of Our
    Families and Our Nation. Amer. Pioneer Records, 2:143-144.
Heredity and Eugenics in Relation to Medicine. Oxford Dictionary of Medicine,
    (Amer. Branch, Oxford University Press), 1:501-520.
Critique of Curves of Growth and of Relative Growth. Collecting Net, 9:95-98;
    Cold Spring Harbor Symposia on Quantitative Biology, 2:203-206.
Child Development From the Standpoint of Genetics. Sci. Mo., 39:97-116.
Reply to Sec. Wallace's Article, "The Scientist in An Unscientific Society." Sci.
    Amer., Aug. 77-78.
Better Human Strains. 25th Annual Report Board of Visitors, Letchworth Village,
    49-51.
How Early in Ontogeny Do Human Racial Characters Show Themselves? Eugen
    Fischer-Festband, Zeitschr. f. Morph. u. Anthrop., 34:76-78.

1935

Influence of Economic Conditions on the Mixture of Races. Zeitschr. f. Rassenkunde,
    1:17-19.
The Development of Trunk Width and the Trunk Width Index. Human Biol., 7:
    151-195.
Variations in Proportions Among Mammals, With Special Reference to Man. Jour.
    Mam., 16:291-296.

A Provisional Hypothesis of Child Development. Proc. Amer. Philos. Soc., 75:537-548.

Child Development From the Standpoint of Genetics. Carnegie Inst. Wash. Supplementary Publs. No. 11, 20 pp 22 text figures.

1936

Reginald Gordon Harris. Science, 83:47-48.

Developmental Changes in Facial Features. (Abstract), Amer. Jour. Phys. Anthrop., 21:9.

Growth Curve of Infants. Proc. Nat. Acad. Sci., 22:639-645. (By C. B. D. and William Drager).

Dr. Little the Man, and What He Did. Village Views, spec. ed. July.

Eugenische Forschung und Ihre Praktische Anwendung in den Vereinigten Staaten. Der Erbarzt, 29:97-98.

Causes of Retarded and Incomplete Development. Amer. Assoc. Mental Deficiency, 41:208-214.

How We Came by Our Bodies. Henry Holt & Co., N. Y. 401 pp.

Discussion of Mortality of Tall Men. Annual Meeting Proc. Assoc. Life Ins. Med. Directors of Amer., 23:182-183.

1937

An Improved Technique for Measuring Head Features. Growth, 1:3-5.

Some Principles of Anthropometry. Amer. Jour. Phys. Anthrop., 23:91-99.

Postnatal Growth of the External Nose. Proc. Amer. Philos. Soc., 78:61-77.

Home of the Ancon Sheep. Science, 86:422.

Interpretation of Certain Infantile Growth Curves. Growth, 1:279-283.

Hereditary Strength. Chapter I in "Implications of Social-Economic goals for Education." Nat. Educ. Assoc. of U. S. 126 pp (By C. B. D., Chloe Owings, Ernest R. Groves, Leta S. Hollingsworth and Warren S. Thompson).

George Davidson. Nat. Acad. Sci. Biographical Memoirs, 18:189-217.

1938

Genetics of Human Inter-Racial Hybrids. Current Science. Special Number on "Genetics" (March) 34-36.

Eugenics. In "How to Live" by Irving Fisher and Haven Emerson. Funk and Wagnalls Co., New York, 389-395.

Bodily Growth of Babies During the First Postnatal Year. Contrib. to Embryol. No. 169, 271-305. Reprinted from Carnegie Inst. of Wash. Pub. 496. 1 text figure, 4 tables and 13 graphs.

Growth Lines in Fossil Pectens as Indicators of Past Climates. Jour. Paleontology, 12:514-515.

1939

Post-Natal Development of the Human Outer Nose. Proc. Amer. Philos. Soc., 80:175-355.

Growth. Ann. Rev. Physiol., 1:81-108. (By C. B. D., O. Rahn, M. E. Mayer, H. W. Chalkley and D. M. Pace)

Edward Laurens Mark. Bios., 10:69-82.

Frederick Augustus Porter Barnhard. Nat. Acad. Sci. Biographical Memoirs, 20:259-272.

The Genetical Basis of Resemblance in the Form of the Nose. Kultur and Rasse,

50   Bios

Festschrift zum 60. Geburtstag Otto Reches, pp. 60-64. J. F. Lehmann's Ver-
lag München/Berlin.
The Progress of Eugenics. Medical Blue Book Journal, 10:39-41.
The Relation Between Change in Basal Metabolism and Growth During Adolescence.
Child Development, 10:181-202. (By C. B. D., Olive Renfrew and W. D. Hal-
lock).

1940

Analysis of Variance Applied to Human Genetics. Proc. Nat. Acad. Sci., 26:1-3.
Developmental Curve of Head Height/Head Length Ratio and Its Inheritance.
Amer. Jour. Phys. Anthrop., 26:187-190.
Post-Natal Development of the Head. Proc. Amer. Philos. Soc., 83:215 pp.
Three Lectures: Human Variability and Mate Selection. Some Social Applications of
Eugenics. Heredity in Relation to Eugenics. In Medical Genetics and Eugenics,
Women's Medical College of Pennsylvania, Phila.   .
Adolescent Development of the Sella Turcica and the Frontal Sinus Based on Con-
secutive Roentgenograms. Amer. J. Roent. Rad. Therapy, 44:665-679. (By
C. B. D. and Olive Renfrew).

1941

Post-Natal Development of the Head. Scient. Mo., 52:197-202.
The Early History of Research With Drosophila. Science, 93:305-306.
Account of My Anthropometric Work in Institutions. Amer. Jour. Mental Defic-
iency, 45:343-345.
Responsive Bone. Proc. Amer. Philos. Soc., 84:65-70.
Auxology. Amer. Jour. Mental Deficiency, 46:181-182.

1943

Harry Hamilton Laughlin. Science, 97:194-195.
The Development of the Head. Amer. J. Orthodontics and Oral Surgery, 29:541-547.

1944

Post-Natal Development of the Human Extremities. Proc. Amer. Philos. Soc. 88:
375-455.
Dr. Storr's Facial Type of the Feeble-Minded. Amer. Jour. Mental Deficiency, 48:
339-344.

1945

The Dietaries of Primitive Peoples. Amer. Anthropologist, 47:60-82.

Appendix 2

(Reprinted from the Cold Spring Harbor booklet, *The Reginald Harris Building Dedication Ceremony, May 27, 1982.*)

Reginald Harris circa 1930.

REGINALD HARRIS AND THE BIOLOGICAL LABORATORY

Biological research at the Laboratory ... should be free from undue emphasis on any specialized group. The cyclic periods of interest in this or that department of original investigation should find the Biological Laboratory ready to lend its facilities, unhampered by previous commitments, to any kind of worthy research.

Reginald Harris
1925 Annual Report of the
Biological Laboratory

The Biological Laboratory at Cold Spring Harbor was established in 1890 as a department of the Brooklyn Institute of Arts and Sciences. The founding purposes of the Laboratory were to offer an exceptional location for general biological instruction, and research opportunities for advanced students and investigators. The Laboratory, however, was actually used only during the summer.

Neighboring the Biological Laboratory was the Department of Genetics of the Carnegie Institution. Since 1904 the Institution had operated a year-round research program, the Station for Experimental Evolution. The Eugenics Records Office, founded in 1910 also at Cold Spring Harbor, came under the control of the Carnegie Institution in 1918. The Station for Experimental Evolution merged with the Eugenics Records Office in 1921 forming the Department of Genetics. Research at the Department of Genetics flourished while the teaching program at the Biological Laboratory did not. The rapid growth of Carnegie was eclipsing the identity of the Laboratory.

By 1921, largely due to a paucity of funds, the Biological Laboratory was in serious trouble. The emphasis on research had rapidly declined as had the caliber of students. In addition to not being able to buy needed equipment and supplies, the Laboratory could not afford to purchase land necessary for its development and expansion. "The Biological Laboratory," summarized a memorandum concerning the crisis, "has become quite outdistanced by that at Woods Hole." The Marine Biological Laboratory at Woods Hole was established in 1888 on much the same principles as Cold Spring Harbor. The climax to the crisis came when Woods Hole rather than the Biological Laboratory received the funds necessary to establish the much wanted and needed Biophysics laboratory. Dr. Charles B. Davenport, Director of all the biological programs at Cold Spring Harbor, had hoped that the physiology research to be performed in the Biophysics laboratory, would revitalize the suffering institute. Drastic action had to be taken in order for the Biological Laboratory at Cold Spring Harbor to survive.

23 WALL STREET
NEW YORK   July 21, 1927.

Mr. Reginald G. Harris,
            Director,
    L. I. Biological Association,
        Cold Spring Harbor,
            New York.

Dear Sir:

    I received your letter of
July 11th, and am willing to contribute $500. annually, for three
years, to the Long Island Biological
Association. I enclose cheque for
$500. in payment of this year's con-
tribution.

            Yours very truly,

W. K. VANDERBILT
GRAND CENTRAL TERMINAL
NEW YORK

                            February 3rd, 1926.

Mr. Chas. B. Davenport,
    Secretary, Long Island Biological Asso., Inc.,
        Cold Spring Harbor, L. I.

Dear Sir:-

        Your letter of January 27th has just reached me, in
which you state that Mr. Schiff has offered to increase his subscription
to $10,000 provided the other three principal donors, including
Mr. Matheson, Mr. Field and myself, would do likewise.

        I should be pleased to increase my subscription to
$10,000 and will forward my check to Mr. A. W. Page, c/o Doubleday,
Page & Company, Garden City, who I understand is the Treasurer of
the Association, whenever you are ready for closing of the title,
and in the meantime, as I am going south for a cruise on ARA I have
notified my attorneys, Anderson & Anderson, to be good enough to
communicate with you so that in case there is no hitches as far as titles
are concerned, my check will be forthcoming in due time through their
firm at your request.

        Trusting to have the pleasure of seeing you next Summer,
I remain

            Yours very truly,

In 1924 control of the Biological Laboratory was transferred to the Long Island Biological Association (LIBA), a membership of biologists and neighbors whose "raison d'être" was the maintenance and improvement of the Laboratory. Reginald Gordon Harris, at the age of twenty six, became the Laboratory's first year-round in-residence director. The Biological Laboratory, under Dr. Harris' direction, was to go through a period of enormous growth and reconstruction.

Dr. Harris cited three immediate goals in the 1924 Annual Report: "The acquisition of additional land for immediate use and future extension.... a much needed Research Laboratory.... [and] a boat suitable for taking students to distant collecting grounds and for dredging." By the end of 1925 LIBA's Board of Directors had approved the purchase of forty acres of Townsend Jones' land lying to the west and north of the original grounds; the Laboratory had acquired a boat from Dr. Walter B. James which would be available for use by the summer of 1926; and one of the existing buildings, now known as Wawapex, had been refitted as a research laboratory. The title to thrity-two and a half acres (Carnegie purchased the remaining acreage) was transferred to the Laboratory in April 1926, and soon after a laboratory for the study of Botanical and

Dredging Boat

Physiological Science, now known as Delbrück Laboratory, was erected. The George Lane Nichols Memorial Laboratory was completed in 1928, and the Doctor Walter B. James Memorial Laboratory in 1929.

While under the tutelage of the Brooklyn Institute of Arts and Sciences, the Biological Laboratory was primarily known as a summer biology institute concentrating on instruction with research taking on a secondary role. Dr. Harris succeeded in altering this image. Firstly, he increased the summer research staff and activities, and was committed to developing a year-round research staff. In 1928 Dr. Hugo Fricke, a biophysicist, was employed as the Laboratory's first permanent resident research staff member, and already by 1929 the Laboratory was engaged in four year-round research programs. One of these was the Physiology of Reproduction, in which Dr. Harris was an investigator. Secondly, requirements for admission to the summer courses became more stringent with concomitant reduction in the number of students per class. "The students," explained Dr. Harris, "no longer determine the scientific atmosphere of the Laboratory."

James Laboratory, 1929

For the Laboratory and its new Association the 1920's were marked by an atmosphere of great optimism. Harris reported in 1927 that "Only four or five years ago the Biological Laboratory was a source of serious concern.... The question ... was no less vital than whether or not the Laboratory could continue to exist.... The question today has nothing to do with death. It is concerned with growth." Only a few years were to pass, however, until the country was struck by the depression. During this period, while many organizations, businesses, and individuals were going bankrupt, the Laboratory continued to survive and tried to grow, instituting the Cold Spring Harbor Symposia on Quantitative Biology in 1933. Reporting on the future of the Laboratory in the midst of the economic hardships of the day, Harris in 1932 wrote:

George Lane Nichols Lab, 1928

After looking at the past, present and the future, there are three courses open to a man.... One is to commit suicide. Another is to change one's mode of life and one's aims.... The third is to build with broader foundations for the future, since, knowing the exigencies and upsets of life, one is only the more determined to make one's life of the greatest possible satisfaction and value to himself and to his fellows.

All three of the possibilities have been, and are being, adopted in these times by men, and by organizations and institutions of men.

The Biological Laboratory is no exception. A choice must be made.

An old saying is, that something worth doing at all is worth doing well. Applied to institutions, I should say that an institution which is worth keeping alive is worth developing to its greatest useful destiny.

We formally stated a few years ago that ... the Laboratory's greatest useful destiny was to advance the frontiers of knowledge of biology (1) directly through its own research, and (2) indirectly through providing a common meeting and working place for scientists from many institutions: a clearing house for discoveries in biology, lending its influence to the useful adoption of new methods of biological research.

Harris' great fund raising talents were largely responsible for carrying the Laboratory through each difficult year without a deficit. In order to meet the demands of the budget Harris extended his appeal to outside organizations. In 1931 the Carnegie Corporation responded with a grant of $15,000 and the Rockefeller Foundation with a grant of $20,000. He barraged financiers with appeal letters often refusing to take no for an answer. Responding to a LIBA member's letter that she could only commit herself to "urgent relief measures" Harris in 1931 wrote "... is it more important to have clothing on the back or food in the stomach of any given person than it is to have adrenal cortical hormone in the blood?"

```
                                        January 28, 1925.

Mr. Victor C. Twitty
  411 East 50th St.
    Indianapolis,,Ind.

Dear Mr. Twitty:

        In the absence of Dr. Harris, I wish to thank you on behalf
of the Long Island Biological Association for your contribution of
$1.00 as a sustaining member for 1925.


                        Sincerely yours,



                        Secretary to Dr. Harris.
```

Perhaps the most important impact Dr. Harris had on the history of the Biological Laboratory at Cold Spring Harbor was his fostering of a closer relationship between biology, which was changing from a descriptive to a quantitative science, and the other basic sciences. In the 1932 Annual Report Harris wrote that "...quantitative biology, an exact science, must more and more approach and pass into historical domains of the older exact sciences of mathematics, physics, and chemistry."

The first outcome of this merging was the establishment of a biophysics laboratory in 1928, headed by Dr. Fricke and later located in James Laboratory. The second was the initiation in 1933 of the Cold Spring Harbor Symposia on Quantitative Biology. "There is, of course, nothing new in conferences or in symposia," explained Harris in the 1933 Annual Report, "but the union of the two during a month and the inclusion of mathematicians, physicists, and chemists in the conferring group will, we hope, partially realize for biology an opportunity which it has had for some time." The first conference, *Surface Phenomena,* and the resulting Symposium publication were such a success that even though no outside support had been granted to the first Symposium, the Rockefeller Foundation funded $5,000 toward the second and $7,000 toward the third.

Unfortunately Dr. Harris' term as Director was brought to an abrupt halt. He died of pneumonia on January 7, 1936 after working very hard on bringing Symposia Volume III to completion in only three and one-half months. Cold Spring Harbor Laboratory is indebted to Reginald Gordon Harris for enabling it to mature into the renowned research institute that it is today.

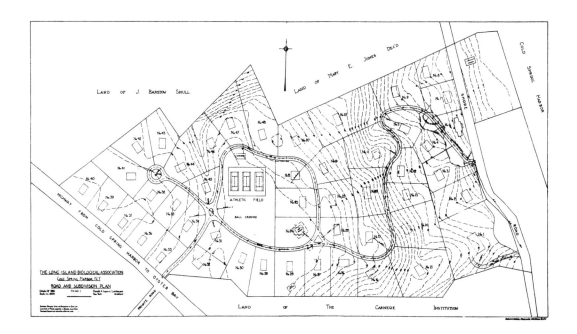

1926 Road and Subdivision Plan

In 1925 LIBA authorized the purchase of the Townsend Jones property adjacent to the Laboratory. Dr. Harris explained that this land "would provide building sites for immediate expansion and for scientists who might desire to establish their summer homes at Cold Spring Harbor in close proximity to the Biological Laboratory."

Appendix 3

(Reprinted from *Genetics* **67**: 1-3)

MILISLAV DEMEREC

1895–1966

MILISLAV DEMEREC was an outstanding geneticist. He was Director of the world-famous laboratories at Cold Spring Harbor (The Biological Laboratory and the Department of Genetics of the Carnegie Institution of Washington) during what might easily prove to be their Golden Era. Furthermore, he was a kind and good person. He died on April 12, 1966.

Dr. DEMEREC was a quiet—nay, silent—man. Thus, despite eleven years association with him, I know remarkably little about his early life; the few facts that are known are to be found in his own *curriculum vitae* or must be learned from his friend of early school days, ALOJZ TAVČAR of Zagreb. In brief, he was born at Kostajnica, Croatia (Jugoslavia) on January 11, 1895. He attended high school and college in or near Zagreb. Following World War I he attended the College of Agriculture at Grignon, France and then became a student of R. A. EMERSON at the College of Agriculture, Cornell University. After obtaining his degree in 1923 for work on the genetics of maize, he went to Cold Spring Harbor as a staff member at the Department of Genetics, Carnegie Institution of Washington; it was this laboratory of which he was Director from 1943 until 1960. Somewhat earlier, in 1941, he had acquired the directorship of the physically adjacent laboratory, The Biological Laboratory—a position he also held until 1960. Following his retirement from the laboratories at Cold Spring Harbor, he was appointed Senior Geneticist at the Brookhaven National Laboratory for a period of five years. At the age of 71, while organizing a new laboratory at C. W. Post College, he died.

To know these details is not to know Dr. DEMEREC. He was, as I said, a quiet man. In part, he was shy; in part, preoccupied with his own thoughts. During evening parties at his home, he wanted no more than to have the conversation and activity involve everyone else while he, silent but contented, sat to one side in his favorite chair. What experiments were planned, what sources of funds were identified, what symposia topics were selected, or what committee business was organized during these social evenings only Dr. DEMEREC knew. Promptly at ten o'clock, if no senior guest had already done so, Dr. DEMEREC would arise, look at the clock with a startled exclamation, and announce that the morrow was coming. And so the evening would end.

Dr. DEMEREC was physically a strong man. His day began before six each morning working at his house—gardening in the spring and summer, otherwise going over data or papers in his study. As Director of two laboratories, he arrived on the grounds at eight with the first workers, supervised the work schedules, discussed financial matters, listened to and commented on new experimental data described by various colleagues and staff members he happened to encounter, went over carefully the data from his own laboratory staff, and, in addition, carried out the multitude of other tasks that fell on his shoulders either through

Genetics **67**: 1–3 January, 1971.

committee work, editorial duties, or the affairs of various scientific societies and congresses. He was an exceptionally well-organized man but that in itself was not enough, he needed, and possessed, tremendous physical stamina to withstand this routine work load.

One of his remarkable abilities was that which enabled DEMEREC to evaluate a problem, reach a decision, and then put the matter out of his mind. This ability reduced the strain of supervising the day-by-day operations on the laboratory grounds, of supervising the housing and dining of hundreds of summer guests, and of assuming responsibility for the research efforts of numerous postdoctoral fellows and research associates. In each of these matters, he serenely did the best he could under the circumstances and then refused to fret over what might otherwise have been.

The Cold Spring Harbor Symposia on Quantitative Biology were Dr. DEMEREC's special joy. With the exception of the periodically recurrent topic, *The Gene*, suggestions for appropriate subjects came from his many friends and colleagues; he invariably favored those proposals that were just beyond what at the moment was fashionable research. His only instruction to the members of each year's Program Committee was for them to prepare the ideal program as if unlimited funds were at their disposal; raising funds for these meetings was his responsibility and, should he fail to find enough, it was then that the program would be modified. He had a complete and well-justified faith that funds are easier to find for excellent symposia than for mediocre ones. The only foreign scientist who could not be brought to the meetings at Cold Spring Harbor despite DEMEREC's repeated efforts was J. B. S. HALDANE; federal regulations concerning visas effectively barred HALDANE's entry into the country. Ironically, HALDANE finally came to the United States to participate in a symposium sponsored by the space agency.

The summer courses, including the nature study course for children of the scientific and neighboring communities, were also a source of great pride to Dr. DEMEREC. The Phage Course, first under the direction of MAX DELBRÜCK and then for many years under that of the late MARK ADAMS, was in many respects the birthplace of molecular genetics. The talent that was involved in P. U. (Phage University) at one time or another during DEMEREC's years at Cold Spring Harbor is nearly unbelievable; four Nobel Laureates come to mind immediately but the list of truly superb research men who passed through this course would be at least twenty times as great.

The physical nature of the gene remained throughout Dr. DEMEREC's life his main research problem. During the course of his work he passed with deceptive ease from maize to Delphinium to flies and to microorganisms leaving behind a body of excellent data in the literature on each. The climax of his work was undoubtedly the finding that the linear arrangement of genes (DEMEREC was reluctant to adopt new terminology such as "cistron") controlling tryptophan synthesis in *Salmonella typhimurium* corresponds to the order of biochemical reactions that these genes mediate. "The assembly line has finally been found," was HALDANE's comment upon hearing of this work. Subsequent studies on *S. typhi-*

murium involved analyses of the fine structure of its chromosome (including its large "silent" regions), a comparison of this structure with the corresponding one of *Escherichia coli*, and a study of the consequences of transferring genes from the chromosome of one of these species to that of the other. In short, DEMEREC had started a systematic attack on evolutionary genetics at the molecular level.

Dr. DEMEREC succumbed to a heart attack. His death reminded me of a morning many years before when we walked together from Jones Laboratory (on the waterfront at Cold Spring Harbor) up the steps toward Bungtown Road. About midway up the rather long slope Dr. DEMEREC paused, turned toward the harbor and its surrounding hills, and gazed at the familiar but always beautiful view. Then, quietly as if he were discussing the sea gulls in the mud flats of the inner harbor, he said that his doctor had told him to take it easy and so he would like to rest for a moment. "Anything serious?" I asked as we stood there. "Doughhhh, just my heart, nothing really serious." Many times during my subsequent years at Cold Spring Harbor I saw Dr. DEMEREC pause halfway up those steps and turn to gaze out over the harbor; to the casual observer it would appear that he never tired of his laboratory and its surroundings. And, of course, he never did.

BRUCE WALLACE

Name Index

Subject Index